Materials Horizons: From Nature to Nanomaterials

Materials are an indispensable part of human civilization since the inception of life on earth. With the passage of time, innumerable new materials have been explored as well as developed and the search for new innovative materials continues briskly. Keeping in mind the immense perspectives of various classes of materials, this series aims at providing a comprehensive collection of works across the breadth of materials research at cutting-edge interface of materials science with physics, chemistry, biology and engineering.

This series covers a galaxy of materials ranging from natural materials to nanomaterials. Some of the topics include but not limited to: biological materials, biomimetic materials, ceramics, composites, coatings, functional materials, glasses, inorganic materials, inorganic-organic hybrids, metals, membranes, magnetic materials, manufacturing of materials, nanomaterials, organic materials and pigments to name a few. The series provides most timely and comprehensive information on advanced synthesis, processing, characterization, manufacturing and applications in a broad range of interdisciplinary fields in science, engineering and technology.

This series accepts both authored and edited works, including textbooks, monographs, reference works, and professional books. The books in this series will provide a deep insight into the state-of-art of Materials Horizons and serve students, academic, government and industrial scientists involved in all aspects of materials research.

Review Process

The proposal for each volume is reviewed by the following:

1. Responsible (in-house) editor
2. One external subject expert
3. One of the editorial board members.

The chapters in each volume are individually reviewed single blind by expert reviewers and the volume editor.

N. Manjubaashini · T. Daniel Thangadurai ·
D. Nataraj · Sabu Thomas

Graphene Quantum Dots

The Emerging Luminescent Nanolights

Springer

N. Manjubaashini
National Centre for Nanoscience and Nanotechnology
University of Madras
Chennai, Tamil Nadu, India

T. Daniel Thangadurai
Department of Chemistry
KPR Institute of Engineering and Technology
Coimbatore, Tamil Nadu, India

D. Nataraj
Department of Physics
Bharathiar University
Coimbatore, Tamil Nadu, India

Sabu Thomas
IIUCNN
Mahatma Gandhi University
Kottayam, Kerala, India

ISSN 2524-5384 ISSN 2524-5392 (electronic)
Materials Horizons: From Nature to Nanomaterials
ISBN 978-981-97-5721-3 ISBN 978-981-97-5722-0 (eBook)
https://doi.org/10.1007/978-981-97-5722-0

This Springer imprint is published by the registered company Springer Nature Singapore Pte Ltd.
The registered company address is: 152 Beach Road, #21-01/04 Gateway East, Singapore 189721, Singapore

Preface

Outline

This cutting-edge transdisciplinary textbook comprises nineteen chapters, which cover the history and development of carbon nanostructures, synthesis and characterization techniques, and unique physicochemical properties of GQDs. It focuses on various applications such as agriculture, healthcare, environmental, anti-microbial and antibacterial, and other miscellaneous and multi-functional applications of GQDs. Furthermore, the density functional theory (DFT) is employed to calculate the tunable energy gap and dipole moment to explain the fluorescence phenomenon of GQDs. It also discusses the mathematical modeling and simulation in GQD and their nanocomposites concerning the biomedical and environmental domains. Since almost every analytical approach is covered in this book, it is an invaluable resource to graduates and postgraduates, engineers, research scholars, material scientists, engineers, and technologists from industry, who work in related fields (material science, chemistry, physics, nanoscience and nanotechnology, and biomedical and environmental) are expected to read and utilize it. Furthermore, this book helps readers solve basic and practical difficulties in graphene-based materials and devices for modern innovations.

The authors sincerely thank the contributions of leading graphene researchers from industry, academia, government, and private research institutes all over the world.

Usefulness of the Book to the Readers

A great opportunity lies in the future development of novel GQD-based nanomaterial products which are now being employed in a variety of fields, including electronics, biomedicine, environmental remediation, and agriculture. As the title indicates, the objective of this book is to provide a fairly comprehensive overview of the recent achievements in the field of GQDs. Materials with enticing properties at very low amounts of chemicals are required by modern technologies. We can expect beneficial qualities at extremely low amounts of GQD-based nanomaterials due to the high aspect ratio. Furthermore, many experts have consented to share their extensive experience and knowledge in the field of GQD-based nanomaterials and applications. This book is divided into different sections based on the findings of professionals in specific disciplines. This technical book's nature may make it a valuable reference or textbook for a wide range of scientists, industrial practitioners, graduate and undergraduate students, and other professionals working in the fields of nanoscience and engineering, materials science, surface science, bioengineering, and chemical engineering. We proposed a single volume with nineteen chapters to provide a clear understanding of GQD's fundamentals and fascinating attributes to the readers.

Important Features of the Book

It is a handbook that explores the many unique characteristics of GQDs, as well as their potential and advanced applications. This molecular-scale material is versatile, with outstanding electronic, mechanical, electrical, cytotoxicity, biocompatibility, and thermal properties with applications in a variety of disciplines. Interestingly, chemists, physicists, and engineers are all interested in developing GQD functional materials. Though many excellent books have already been written on various aspects of GQDs, however, to our knowledge, there is no single book that consolidates information across all aspects of GQDs, engineering, and biomedical applications in a very detailed manner, spanning the entire gamut of up-to-date literature citations, current market, and patents. Therefore, we narrated all the considerable progress made in the experimental and theoretical research of several properties of GQDs and their composites in recent years. In summary, this book aims to provide an in-depth investigation of the present state of the art in GQDs, composites, hybrid structures, and other related topics.

Chennai, India N. Manjubaashini
Coimbatore, India T. Daniel Thangadurai
Coimbatore, India D. Nataraj
Kottayam, India Sabu Thomas

Contents

About the Authors

Dr. N. Manjubaashini is currently working as a research associate under Department of Biotechnology at National Centre for Nanoscience and Nanotechnology, University of Madras, Chennai, India. She obtained her Bachelor of Engineering (EEE) and Master of Technology (M.Tech.) degrees from Anna University, India, in 2012 and 2014, respectively. She was a senior research fellow in the Council of Scientific and Industrial Research programme when she earned her Ph.D. in Technology from Anna University, India, in 2019. Her major research focuses on nanostructured materials for chemical and biological analyte sensing, and Bionanocomposites for healthcare and environmental applications. She has published more than 20 research articles and 5 review articles in peer-reviewed international journals, 2 authored textbooks, and 1 book chapter. She is a life member of Society of Biomaterials and Artificial Organs India (SBAOI).

Dr. T. Daniel Thangadurai is a professor of Chemistry and the head for the Centre for Nanoscience and Technology at KPR Institute of Engineering and Technology, India. He received his Ph.D. in Chemistry from Bharathiar University, India, in 2001. He has 6 postdoctoral experiences at KAIST, POSTECH, and CNU of South Korea, CNRS-UMR 8601 of France, and USM of Malaysia, during the period 2001–2010. Subsequently, he worked as an assistant professor at Kook-Min University, South Korea (2010–2013). He has published more than 110 research articles in high impact journals, holds copyright of one Indian patent, and authored 2 textbooks, and 3 book chapters. Carbon-based nanomaterials, biological applications of nanostructured materials, and chromofluorogenic sensors are the areas of his research interest.

Dr. D. Nataraj is currently a professor at Department of Physics, Bharathiar University, India. He obtained his Ph.D. in Physics in 2001 from Bharathiar University, India. He did his postdoctoral research work at NIAIST, Hokkaido, and RCIQE, Hokkaido University, Japan, as an STA and COE fellow during 2001–2003 and 2003–2005, respectively. His research areas include understanding the photophysics of quantum nanomaterials and understanding the dynamics of excitons. He has published more than 75 research articles and holds rights for two Indian patents. He is a life member at Luminescent Society of India and Magnetic Society of India.

Prof. Sabu Thomas is the director of the International and Interuniversity Centre for Nanoscience and Nanotechnology, Mahatma Gandhi University, Kottayam, Kerala, India. He was the former vice chancellor of Mahatma Gandhi University, Kottayam, Kerala, India. Prof. Thomas is an outstanding leader with sustained international acclaims for his work in nanoscience, polymer science and engineering, polymer nanocomposites, elastomers, polymer blends, interpenetrating polymer networks, polymer membranes, green composites and nanocomposites, nanomedicine, and green nanotechnology. In collaboration with India's premier tyre company, Apollo Tyres, Prof. Thomas' group invented new high-performance barrier rubber nanocomposite membranes for inner tubes and inner liners for tyres.

Chapter 1
History and Development of Carbon Materials

1.1 Overview

From primeval times, people have been aware of carbon, and one of the first forms of carbon that they used was charcoal. Charcoal was first used by ancient societies for drawing, but they also later learned that it could be used in metallurgy to remove metals from ores. C60 is the name given to the molecule composed of 60 carbon atoms arranged in a sphere, or buckyball, by Richard Smalley, Robert Curl, and Harold Kroto. This finding created new opportunities for materials made of carbon with special qualities [1]. Graphite started to be used more frequently in the sixteenth century. Graphite, originally believed to be a type of lead, was used in writing instruments (pencils) because of its lubricating qualities. Since ancient times, people have valued diamonds as another type of carbon for their durability and brilliance. Over time, the understanding of the structure of diamonds and how they relate to carbon changed. An important turning point was reached in 1985 with the discovery of fullerenes [2].

Graphene is a revolutionary material with applications in electronics, energy storage, and other fields due to its exceptional mechanical strength, thermal properties, and electrical conductivity. During the second half of the twentieth century, carbon fiber composites became more and more popular in the aerospace industry. These materials are used in high-performance sports equipment, spacecraft, and aircraft because of their excellent strength-to-weight ratios, which come from carbon fibers embedded in a matrix [3]. The use of activated carbon in purification procedures is not new. It is useful for various industrial applications, air purification, and water treatment due to its large surface area and adsorption capacity. Energy storage technologies have advanced due to the widespread use of carbon materials, such as graphite and different types of carbon nanomaterials, in batteries and supercapacitors. Applications for carbon materials in medicine include drug delivery and imaging systems, as well as carbon nanoparticles [4].

N. Manjubaashini et al., *Graphene Quantum Dots*, Materials Horizons: From Nature to Nanomaterials, https://doi.org/10.1007/978-981-97-5722-0_1

In the late twentieth century, carbon allotropes became a popular research topic and eventually took center stage in nanoscience across nearly all domains. The carbon era and nanotechnology/nanoscience have benefited from the enrichment of carbon allotropes from well-known diamond and graphite to carbon nanomaterials in recent decades [5]. Carbon nanomaterials' light weight, excellent thermal and electrical conductivity, high strength, and high stability led to the development of nanoscience in the fields of electronics, optoelectronics, medicine, mechanical engineering, catalysts, and energy. The challenges posed by silicon-based electronic devices were overcome by the use of carbon nanomaterials in electronics to comply with Moor's Law [6]. Carbon nanomaterials are perfect for conducting skeletons for electrochemical materials because of their high conductivity and stability. Carbon nanotubes (CNTs) are cylindrical structures with extraordinary electrical, thermal, and mechanical properties that were discovered in 1991 by Sumio Iijima. Since then, CNTs have been used in a variety of industries, such as electronics, materials science, and nanotechnology [7]. A single layer of carbon atoms arranged in a hexagonal lattice, known as graphene, was isolated in 2004 by Andre Geim and Konstantin Novoselov.

1.2 Introduction

With an atomic number of six, carbon is found in Group IV of the periodic table. The three allotropic forms of carbon are graphite, diamond, and fullerene, in that order. Carbon-based materials have many uses because they can be chemically combined with other carbon-based materials and form strong covalent bonds with a wide range of elements [8]. As a result, they have extraordinary properties like high strength, high density, and high hardness. Because of their exceptional qualities, carbon materials are a great option for many applications involving advanced technology [9].

It is challenging to gain or lose electrons to form fully occupied or unoccupied orbitals in chemical reactions due to its peculiar electronic structure. Modifying carbon hybrids with different sp^3, sp^2, and sp^1 hybridizations or adding additional foreign atoms can change the desired characteristics [10]. The diverse orbital configurations of C hybridization, such as sp^1, sp^2, and sp^3, may offer an ideal environment for the discovery and development of novel C materials. For instance, the sp^3 configuration, in which each C atom is tetrahedrally surrounded by four neighboring C atoms with a 109.5° directed bond, confers a strong σ bond to an adjacent atom [11, 12]. Regarding the three-fold coordinated sp^2 configuration, three of the four valence electrons form σ bonds in plane-style sp^2 orbitals that are directed trigonally [13]. The sp^2 atom's fourth electron is located in a $p\pi$ orbital, which is perpendicular to the σ bonding plane. With a π orbital on one or more neighboring atoms, this π orbital forms a weaker π bond [14].

1.3 Evolution of Carbon

Almost all organic materials are made of carbon networks; carbon is one of the elements that is most abundant on earth. We use carbon daily in many products, such as newspapers, pencils, refrigerators, and activated carbon. Among the non-metal elements that are most prevalent on Earth, carbon is essential to our day-to-day existence [15]. With 0D (fullerenes), 1D (carbon nanotubes), 2D (graphene), and 3D (graphite/diamond) structures, carbon is a very diverse family of materials [16] (Figs. 1.1 and 1.2). Carbon is derived from biomass, heavy oils and residues, thermosetting and thermoplastic polymers, coal tar, and polymer [17]. The historical evolution of carbon materials is listed below.

- Charcoal may have been the first carbon material to be used in a practical sense.
- Carbon was discovered in burned ashes dating back to the prehistoric era.
- During the Neolithic era, carbon was also used to adorn ceramic paintings.
- Graphite, the first carbon allotrope, was discovered and used for mold marking and casting in the sixteenth century [16].
- Diamond was originally found in India approximately 3,000 years ago. When the diamond was burned by Nobel laureate Antoine Lavoisier in 1772, it was discovered to be composed of carbon.
- The discovery of fullerenes was made in 1985 by Kroto et al., who were awarded the 1996 Nobel Prize in Chemistry [18].
- The scientific era of graphene is the twenty-first century.
- 2011 saw the synthesis of aluminum, steel, and iron from carbon.
- Carbon electrodes are used for silicon production.
- Lithium-ion batteries are made of graphite, carbon fibers, and carbon nanotubes.
- The highly oriented pyrolytic graphite (HOPG) was synthesized in 2004 by Geim and Novoselov, and graphene has garnered significant attention.

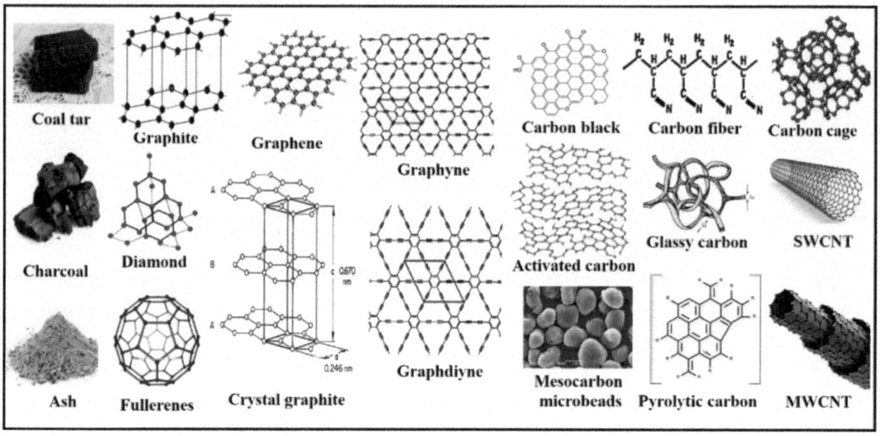

Fig. 1.1 Various structures of carbon allotropes

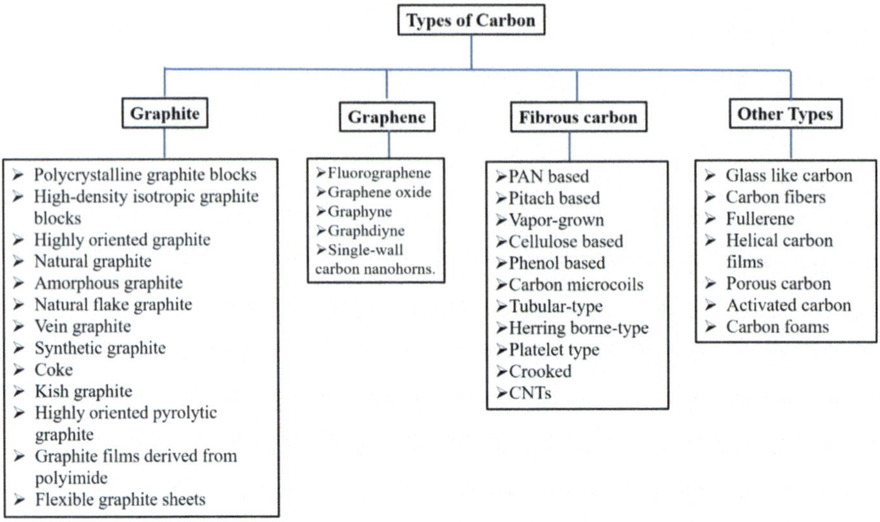

Fig. 1.2 Different forms of carbon

- In the 1970s and 1980s, carbon nanotubes were also produced through chemical vapor deposition (CVD) with structural characterization.
- At approximately 1800, the introduction of carbon materials as battery electrodes was performed [19].
- Large-sized carbon rods, known as graphite electrodes due to their crystalline graphite structure, were produced industrially by heat treatment at temperatures as high as 3000 °C and have been used as electrodes for iron refining since 1878.
- Before 1960, the basic science and technology of classic carbons, artificial graphite blocks, carbon blacks, and activated carbons were developed.
- In 1964, the formation of optically anisotropic spheres in pitches, mesophase spheres, and their coalescence sparked several fundamental studies on the structure of the spheres, their growth and coalescence mechanism, and the formation of bulk mesophase. These studies also produced new carbon products, such as needle-like cokes, which were crucial raw materials for high-power graphite electrodes, high-performing carbon fibers based on mesophase pitches, and mesocarbon microbeads for various applications.
- The invention of carbon fibers from poly(acrylonitrile), pyrolytic carbons by CVD process, and glass-like carbons from thermosetting resins, all of which were entirely different from the carbon materials used before 1960, can be considered the beginning of the era of new carbons.
- Before 1960, four distinct carbon materials were identified as having practical uses in a range of industries. These materials included natural diamonds, carbon blacks for ink and rubber reinforcement, artificial graphite blocks for steel refining, and activated carbons for water purification.

- Three carbon materials were developed in 1960: glass-like carbons, pyrolytic carbons, and carbon fibers. These materials differed greatly from classic carbons in terms of both their properties and methods of production.
- The discovery of carbon materials' good biocompatibility circa 1970 paved the way for the creation of several prostheses, including tooth roots and heart valves.
- It was discovered that adding a small amount of carbon fibers to cement paste produced noticeable reinforcement in concrete around 1985.
- Another era for carbon materials began in 1985 with the discovery of buckminsterfullerene C60, a carbon cage with 60 carbon atoms. This was followed by the discovery of other carbon cages, including C70, C86, and so on.
- The first report of a cage (cluster) made of 60 carbon atoms, or C60 (buckminsterfullerene), was made in 1985. The soot was produced by laser irradiation on a graphite block, and the structure of the block was made up of 20 hexagons with 12 pentagons of carbon atoms.
- Carbon nanotubes were first reported in 1991, and, subsequently, single-wall carbon nanotubes were discovered.
- Following the discovery of single-wall carbon nanotubes in 1991, reports of multi-walled carbon nanotubes were made.
- In 2004, there was a report of a single hexagonal carbon layer. The discovery of these new carbons, or nanocarbons, spurred interest in nanoscale science and technology and advanced the field's scientific advancement.
- The preparation of a single, 2D sheet of carbon atoms, called graphene, was first reported in 2004.
- The discovery stage of carbon dots occurred between 2004 and 2006; the initial developing stage occurred between 2007 and 2011; the explosively developing stage occurred between 2011 and 2021; and the well-developed stage of carbon dots lasted until the present day [20].

1.4 Structure of Carbon Materials

A hexagonal carbon layer is the basic building block of the structure of carbon-based materials in the graphite family. Graphite crystal, hexagonal graphite with ABAB stacking regularity, and rhombohedral graphite with ABCABC stacking regularity are all produced by regular stacking of these layers [21]. Because it is simple to compare rhombohedral and hexagonal graphite, rhombohedral graphite is frequently expressed in a hexagonal system. As a result, two unit cells in each system are connected with equivalent points,thick lines denote rhombohedral unit cells, while double lines denote hexagonal unit cells (Fig. 1.3). The hexagonal structure of graphite shows P63/mmc space group with their interplanar spacing has 0 0 0, 2/3 1/3 0, 0 0 1/2 and 1/3 2/3 1/2 planes with the lattice parameters as $a_0 = 0.2462$ and $c_0 = 0.6708$ nm. Likewise, the rhombohedral graphite structure shows an R3m space group, with the lattice parameters of $a_0 = 0.3635$ nm, $\alpha = 39.49°$ [22].

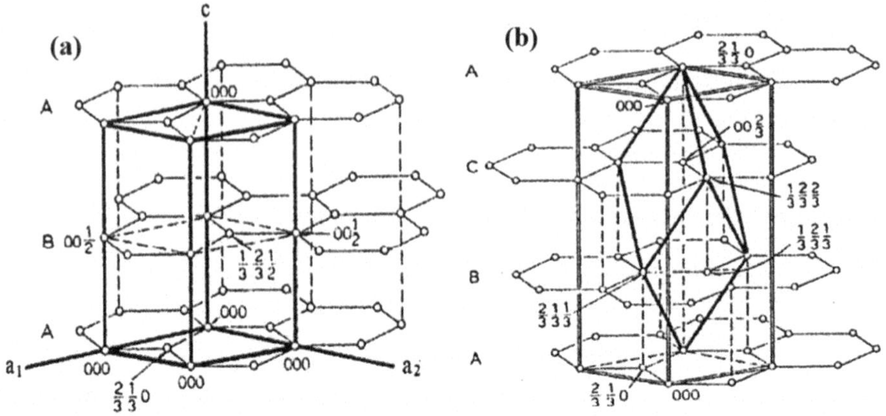

Fig. 1.3 Structures of graphite derivatives: **a** hexagonal structure and **b** rhombohedral structure. Reprinted with permission from © Handbook of Advanced Ceramics (2013)

1.4.1 Structures of One-Dimension (1D) Carbon

Based on sp^2 hybridization, carbon nanotubes are 1D allotropes of carbon with a cylindrical nanostructure. There are three common types of carbon nanotubes: single-wall (SWCNT), double-wall (DWCNT), and multi-wall (MWCNT) [23]. Two or more concentric SWCNTs make up the DWCNTs and MWCNTs. SWCNT, DWCNT, and MWCNT typically have diameters of 0.6–2.5, 1–3, and 2–100 nm, respectively. There are two types of carbon nanotubes: armchair and zigzag [24].

1.4.2 Structures of Two-Dimension (2D) Carbon

It was discovered that the fundamental component of many carbon-based materials, such as graphite, carbon nanotubes, and fullerene, is graphene, a common 2D carbon material. Graphene, with a lattice constant of approximately 2.46 Å and a space group of P63/mmc, is a hexagonal planar structure created by sp^2 hybridization [25]. The inner graphene also contained lattice defects with five or seven rings, which caused the graphene sheets to curl. A zero bandgap and tapered symmetry of the conducting and valent bands at six Dirac points (K point) point to a semimetal character [26]. Since the effective rest mass is zero around the Dirac points and the energy of electrons is linear with the wave vector, graphene's Fermi velocity is getting closer to that of light. Pure graphene's average electron-free path is submicrometric, comparable to trajectory transport, indicating a great deal of promise for fast devices [27].

1.4.3 *Structure of Three-Dimension (3D) Carbon*

Among the most prevalent 3D carbon allotropes are diamond and graphite. They have been studied for a very long time [28]. New 3D carbon structures have emerged recently, such as T-carbon, glassy carbon, cubic carbon, and amorphous diamond. By inserting a carbon tetrahedron into the locations of the diamond atoms, a cubic crystalline carbon with the space group of Fd3m, or T-carbon, is produced [29]. Large specific surface area, a substantial pore structure, and a comparatively high adsorption capacity characterize active carbon (AC) [30]. AC contains oxygen, hydrogen, and other elements in addition to carbon. The cost of AC is less than that of expanded graphite, graphene aerogel tubes, and other 3D carbon materials. AC is frequently used in the separation of mixtures and purification of liquids [31].

1.4.4 *Crystal Structure of Graphite*

There are two crystalline forms of carbon: graphite and diamond. It was demonstrated that each carbon atom in a diamond is covalently bonded to four other carbon atoms arranged tetrahedrally in an early application of X-ray diffraction. The bond between carbon atoms is 1.54 Å. Because it would take the breaking of numerous strong bonds to rupture this isotropy structure, a diamond's extreme hardness is a direct result of its configuration. Because diamond turns into graphite so quickly at atmospheric pressure and temperatures above 1500 °C, it is useless. Naturally, diamond has a wide range of significant applications in mild to low temperatures [32].

A large variety of low-cost precursors can be used to prepare a wide range of materials at relatively low cost; porous carbons have indeed garnered a lot of attention. Because of their diverse porous structure, resistance to basic and acidic environments, low cost, easy accessibility, good recycling characteristics, low density, and most importantly the ability to introduce different functionalities through a wide range of activation, functionalization, and carbonization methods, porous high-surface-area carbons are typically good catalyst supports [33]. Materials made of porous carbon are categorized as microporous (with pore sizes less than 2 nm), mesoporous (with pore sizes from 2 to 50 nm), and macroporous (with pore sizes greater than 50 nm) based on their pore diameters [34].

1.5 Fullerenes

As the pentagons are introduced into the layer, as in the case of corannulene, the bonding nature of fullerenes differs slightly from that of graphite based on the sp^2 hybridization of carbon atoms [35]. From the smallest closed shell of C_{60} to giant fullerenes, the repetition of these curved layers can produce a variety of closed shells with different sizes. By maintaining closed-shell morphology, the addition of extra hexagons among pentagons in C_{60} results in giant fullerenes [36]. Because all other carbon materials are not dissolved into any organic solvent, the majority of fullerenes behave like molecules and can be dissolved into an organic solvent, giving rise to a distinctive color. Thanks to developments in surface science, clusters have been studied, starting with chains of carbon atoms and moving on to rings of carbon atoms up to 15 atoms in size [37].

1.6 Diamond

Diamond is the metastable cubic form of carbon crystal. Three-dimensional extensions of purely covalent chemical bonds make up the sp^3-hybridized carbons that make up diamonds. Because a diamond contains no p electrons, it is extremely hard and electrically insulating. A long-range periodic and regular repetition of the CeC bond is necessary to form a diamond crystal [38]. The purest form of crystalline carbon is diamond. Tetrahedral bonds hold a number of its carbons together. The carbon atoms that make up each tetrahedral unit are bonded to four other carbon atoms, which are subsequently joined to other carbons. As a result, the carbon atoms are arranged in three dimensions in an allotrope of carbon [39].

The glitter and shine of diamonds are a result of intense heat and pressure that occurs far below the surface of the Earth. The crystal structure of a diamond is known as an FCC lattice or face-centered cubic. Every carbon atom is connected to four other carbon atoms in a typical tetrahedron, or triangle prism [40]. Because of their cubic shape and extremely symmetrical atom arrangement, diamond crystals can take on a wide range of shapes that are known as crystal nature. The most common crystal form is the diamond form or eight-sided octahedron. Additionally, cubes, dodecahedra, and combinations of these shapes can be seen in diamond crystals [41]. Instead of having perfectly smooth faces, real diamond crystals have raised or indented triangular growths called "trigons" on them. Diamonds will split precisely along these lines rather than breaking in a jagged way because they have flawless cleavage in four directions. The diamond crystal has fewer chemical bonds in the plane of its octahedral face than in other orientations, which is why the cleavage lines are visible [42].

1.7 Structure of Activated Carbon

A raw form of graphite with a highly porous structure and different-sized pores is called activated carbon. Its surface area is significantly increased by the numerous molecular cracks and crevices that it possesses. The internal surface area of activated carbon can reach up to 1500 m^2/g. This makes it possible for activated carbon to perform the process of adsorption, which is the process by which molecules of a gas or liquid are held in place by the interior or exterior surface of a solid [43]. Activated carbon's structure is essentially similar to that of pure graphite. Carbon–carbon bonds bind the hexagonal layers of the activated carbon molecules together.

- Activated charcoal has found application in medicine and adsorption—in 1773.
- Using the adsorption of gases on the surface of charcoal, it was possible to recognize the adsorptive power of charcoal—in 1773.
- Applying charcoal to different aqueous solutions for decolorization. This is the first time charcoal has been used in a liquid phase—in 1785.
- One can use charcoal made from animal tissue to remove color from an aqueous solution—in 1793.
- Sugar companies employed charcoal derived from wood to remove the color from sugar syrup—in 1794.
- Proven application of charcoal for decolorization of sugarbeet liquor—in 1805.
- First activated carbon was made by heating blood with potash and combining physical and chemical activation. Compared to bone char, it is 50 times more effective—in 1822.
- Made char by heating a mixture of flour, tar, and magnesium carbonate—in 1856.
- Made carbon ready to filter portable water—in 1862.
- Made activated carbon out of coconut shells and researched gas adsorption—in 1865.
- Activated carbon was prepared using phosphate heating and waste from paper mills—in 1868.
- Two processes were used to develop commercial activated carbon: (i) using metal chlorides before carbonization, and (ii) selectively oxidizing charcoal at a high temperature using carbon dioxide—in 1900.
- Using the Ostrejko method, activated carbon was first manufactured industrially and sold under the brand name—in 1911.
- A study found that heating Epoint with zinc chloride increased its decolorizing capacity—in 1913.
- Investigating the sorption phenomenon on activated carbon through the application of the capillary condensation method—in 1931 [44].

1.8 Applications of Carbon-Based Materials

Carbon-based materials have drawn interest due to their important contribution to the field of materials science (Fig. 1.4). These materials include carbon nanotubes (CNTs) with one or more walls, carbon nanofibres (CNF), fullerenes, carbon dots, and graphene and its derivatives (graphene oxide, reduced graphene oxide, and functionalized graphene nanosheets), in addition to the more recent generation of carbon materials such as graphene and its derivatives. High surface area, low cost, chemical stability, ultrahigh optical, thermal, and mechanical properties, remarkable electrical conductivity, and highly developed and tunable porosity are just a few of their many impressive qualities [45]. Water purification, catalysis, oxygen reduction reactions,

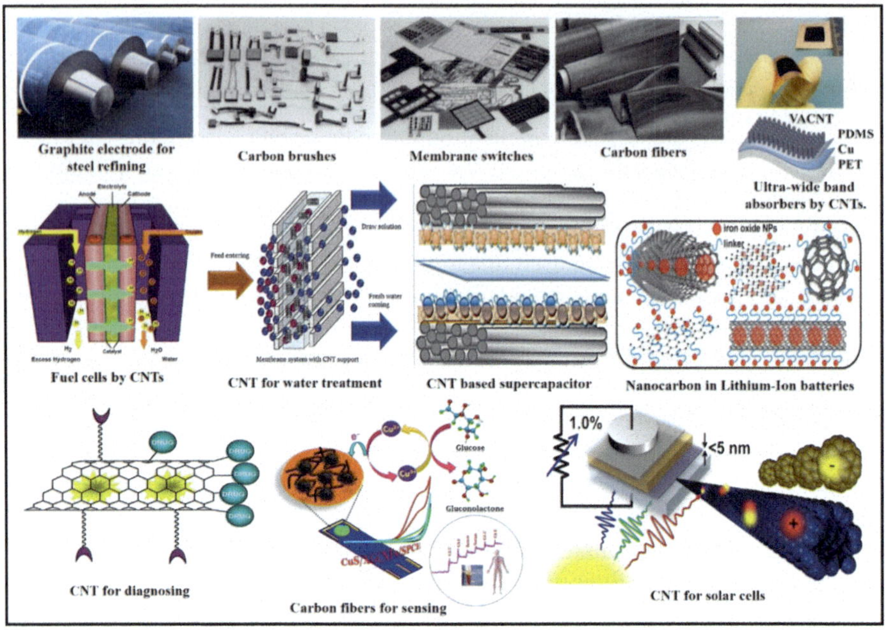

Fig. 1.4 Discrete applications of carbon-based materials

clean energy conversion and storage, polymer reinforcement, solar steam generation, biomedical engineering, advanced information encryption, and CO_2 capture are just a few of the processes that have found them appealing due to these qualities [46].

- Metal processing—artificial graphite blocks
- Electrical and electronic devices—polycrystalline graphite blocks
- Nuclear applications—Polycrystalline graphite [47]
- Molecular sieving carbons [48]
- Mesoporous carbons for car canisters [49]
- Porous carbons for electric double-layer capacitors and hybrid capacitors
- Carbon materials for energy storage—Rechargeable batteries [50]
- Electrochemical capacitors and pseudocapacitance [51]
- Storage of hydrogen and methane gas [52]
- Carbon materials for environment remediation [53]
- Carbon materials for sorption of viscous fluids [54]
- Sorption of biomedical fluids [55].

References

1. Mokoloko LL, Forbes RP, Coville NJ (2022) The Transformation of 0-D carbon dots into 1-, 2- and 3-D carbon allotropes: a minireview. Nanomaterials 12:2515
2. Liang J, Xiang C, Zou Y, Hu X, Chu H, Qiu S, Xu F, Sun L (2020) Spacing graphene and Ni-Co layered double hydroxides with polypyrrole for high-performance supercapacitors. J Mater Sci Technol 55:190–197
3. Maslov MM, Grishakov KS, Gimaldinova MA, Katin KP (2020) Carbon vs silicon polyprismanes: a comparative study of metallic sp^3-hybridized allotropes. Fuller Nanotub Carbon Nanostruct 28:97–103
4. Liu J, Ma Q, Huang Z, Liu G, Zhang H (2019) Recent progress in graphene-based noble-metal nanocomposites for electrocatalytic applications. Adv Mater 31:1800696
5. Jariwala DH, Patel D, Wairkar S (2020) Surface functionalization of nanodiamonds for biomedical applications. Mater Sci Eng C 113:110996
6. Roberto CD (2016) Handbook of Friction Materials and their Applications. https://doi.org/10.1016/B978-0-08-100619-1.00007-9, Chapter 1, Elsevier
7. Kim SJ, Cho MH, Cho KH, Jang H (2007) Complementary effects of solid lubricants in the automotive brake lining. Tribol Int 40:15–20
8. Setton R, Bernier P, Lefrant S (2014) Carbon molecules and materials, 1st edn. CRC Press, pp 512, https://doi.org/10.1201/b16832
9. Inagaki M, Feiyu K (2014) Materials science and engineering of carbon: fundamentals, 2nd edn. Butterworth-Heinemann, https://doi.org/10.1016/C2013-0-13699-9
10. Shi ZJ, Ma MG (2019) Synthesis, structure, and applications of lignin-based carbon materials: a review. Sci Adv Mater 11:18–32
11. Robertson J (1986) Amorphous carbon. Adv Phys 35:317–374
12. Robertson J (2002) Diamond-like amorphous carbon. Mater Sci Eng R 37:129–281
13. Mirderikvand F, Shamlouei HR, Samiey B (2022) Design and computational evaluation of structural, electronic and optical properties of the oligomers of nC20 fullerene (n=1-6). Diamond Related Mater 129:109381
14. Li J, Yin D, Qin Y (2023) Carbon materials: structures, properties, synthesis and applications. Manufacturing Rev 10:13
15. Inagaki M, Kang F (2014) Chap 1: Introduction, in materials science and engineering of carbon: fundamentals. https://doi.org/10.1016/B978-0-12-800858-4.00001-2
16. Sharif S, Arbab S, Saeed A, Shahzad K, Iqbal MA, Durrani AK, Maqsood A, Ikram M (2021) Introductory chapter: introduction to advanced carbon materials and innovative engineering.https://doi.org/10.5772/intechopen.95969
17. Jäger H, Frohs W, Hauke T (2021) Introduction: the future of carbon materials – the industrial perspective, industrial carbon and graphite materials: raw materials, production and applications, 1st edn. Edited by Hubert Jager, Wilhelm Frohs, Wiley
18. Kolluri DK, Boidin X, Desplanques Y, Degallaix G, Ghosh AK, Kumar M, Bijwe J (2010) Effect of natural graphite particle size in friction materials on thermal localisation phenomenon during stop-braking. Wear 268:1472–1482
19. Wang X, Tabakman SM, Dai H (2008) Atomic layer deposition of metal oxides on pristine and functionalized graphene. J Am Chem Soc 130:8152–8153
20. Li S, Li L, Tu H, Zhang H, Silvester DS, Banks CE, Zou G, Hou H, Ji X (2021) The development of carbon dots: From the perspective of materials chemistry. Mater Today 51:188–207
21. Inagaki M (2013) Chapter 2.1: Advanced carbon materials. In: Somiya S (ed) Handbook of advanced ceramics, 2nd edn. Academic Press, pp 25–60
22. Inagaki M, Kang F (2014) Chap 2: Fundamental science of carbon materials. In: Materials science and engineering of carbon: fundamentals. https://doi.org/10.1016/B978-0-12-800858-4.00001-2, Elsevier
23. Kausar A (2020) A review of fundamental principles and applications of polymer nanocomposites filled with both nanoclay and nano-sized carbon allotropes–graphene and carbon nanotubes. J Plast Film Sheeting 36:209–228

24. Jiang S, Li Y, Qian Y, Zhou J, Li T, Lin N, Qian Y (2019) Constructing a buffering and conducting carbon nanotubes-interweaved layer on graphite flakes for high-rate and long-term K-storage properties. J Power Sources 436:226847
25. Neumann C, Kaiser D, Mohn MJ, Fueser M, Weber NE, Reimer O, Goelzhaeuser A, Weimann T, Terfort A, Kaiser U, Turchanin A (2019) Bottom-up synthesis of graphene monolayers with tunable crystallinity and Porosity. ACS Nano 13:7310–7322
26. Guo H, Zhang R, Li H, Wang X, Lu H, Qian K, Li G, Huang L, Lin X, Zhang YY, Ding H, Du S, Pantelides ST, Gao HJ (2020) Sizable band gap in epitaxial bilayer graphene induced by silicene intercalation. Nano Lett 20:2674–2680
27. Zhang Y, Tang TT, Girit C, Hao Z, Martin MC, Zettl A, Crommie MF, Shen YR, Wang F (2009) Direct observation of a widely tunable bandgap in bilayer graphene. Nature 459:820–823
28. Sheng XL, Yan QB, Ye F, Zheng QR, Su G (2011) T-Carbon: a novel carbon allotrope. Phys Rev Lett 106:155703
29. Zhang J, Wang R, Zhu X, Pan A, Han C, Li X, Dan Z, Ma C, Wang W, Su H, Niu C (2017) Pseudo-topotactic conversion of carbon nanotubes to T-carbon nanowires under picosecond laser irradiation in methanol. Nat Commun 8:683
30. Ye X, Qi M, Chen M, Zhang L, Zhang J (2023) Zero to Three Dimension Structure Evolution from Carbon Allotropes to Phosphorus Allotropes. Adv Mater Interfaces 10:2201941
31. Li M, Mu B (2019) Effect of different dimensional carbon materials on the properties and application of phase change materials: A review. Appl Energy 242:695–715
32. Roberto CD (2016) Handbook of friction materials and their applications. WoodheadPublishing, p 174, https://doi.org/10.1016/B978-0-08-100619-1.00007-9
33. Yurum Y, Taralp A, Veziroglu TN (2009) Storage of hydrogen in nanostructured carbon materials. Int J Hydrogen Energy 34:3784–3798
34. De S, Balu AM, van der Waal JC, Luque R (2015) Biomass-derived porous carbon materials: synthesis and catalytic applications. ChemCatChem 7:1608–1629
35. Kroto HW (1992) C_{60}: Buckminsterfullerene, the celestial sphere that fell to earth. Angew Chem Int Edn 31:111–129
36. Dresselhaus MS, Dresselhaus G (1995) Fullerenes and fullerene-derived solids as electronic materials. Annual Rev Mater Sci 25:487–523
37. Dresselhaus MS, Terrones M (2013) Carbon-based nanomaterials from a historical perspective. IEEE 101:1522–1535
38. Ribeiro FJ, Tangney P, Louie SG, Cohen ML (2005) Structural and electronic properties of carbon in hybrid diamond-graphite structures. Phys Rev B 72:214109
39. Pierson HO (1994) Handbook of carbon, graphite, diamond and fullerenes properties. Elsevier, Processing and Applications
40. More RB, Bokros JC (2006) Biomaterials: Carbon. In: Webster JG (ed) Encyclopedia of medical devices and instrumentation. https://doi.org/10.1002/0471732877.emd023
41. Tappert R, Tappert MC (2011) The surface textures of diamonds. In: Diamonds in nature. Springer, Berlin, Heidelberg. https://doi.org/10.1007/978-3-642-12572-0-4
42. Pavlushin A, Zedgenizov D, Vasilev E, Kuper K (2021) Morphology and genesis of ballas and ballas-like diamonds. Crystals 11:17
43. Hajira T, Muhammad S (2021) Chapter 3: Using dyes to evaluate the photocatalytic activity. Interf Sci Tech 32:125–224
44. Sadashiv B, Shivashankar M (2017) History, method of production, structure and applications of activated carbon. Int J Eng. Res. & Tech. 6:495–498
45. Zhang F, Liu T, Hou G et al (2016) Hierarchically porous carbon foams for electric double layer capacitors. Nano Res 9:2875–2888
46. Opeyemi EA, Paul BO, Ugochukwu E, Lima EC (2022) Emerging trends in the application of carbon-based materials: a review. J Env Chem Eng 10:107260
47. Yazyev OV, Chen YP (2014) Polycrystalline graphene and other two-dimensional materials. Nat Nanotechnol 9:755–767
48. Lei L, Pan F, Lindbrathen A et al (2021) Carbon hollow fiber membranes for a molecular sieve with precise-cutoff ultramicro pores for superior hydrogen separation. Nat Commun 12:268

49. Jeong-Rae A, Hye-Min L, Byung-Joo K (2023) Highly mesoporous activated carbon fibers using high density polyethylene precursor with a multi-step stabilization technique for automobile carbon canister. J Ind Eng Chem 128:578–585
50. Gang W, Minghao Y, Xinliang F (2021) Carbon materials for ion-intercalation involved rechargeable battery technologies. Chem Soc Rev 50:2388–2443
51. Shohreh M, Ali A, Zongping S, Arami-Niya A (2022) Rational design of carbon-based materials for purification and storage of energy carrier gases of methane and hydrogen. J Energy Storage B 56:105967
52. Gill SS, Goyal T, Goswami M, Patel P, Das Gupta G, Verma SK (2023) Remediation of environmental toxicants using carbonaceous materials: opportunity and challenges. Environ Sci Poll Res Int 30:69727–69750
53. Ifeoluwa A, Tony B, van Koningsbruggen PJ, Yuan Q (2023) A viscosity study of charcoal-based nanofluids towards enhanced oil recovery. J Mole Liq 387:122615
54. Ye X, Fan Q, Shang L, Ye F (2022) Adsorptive carbon-based materials for biomedical applications. Engineered Regeneration 3:352–364
55. Tucek J, Kemp KC, Kim KS, Zboril R (2014) Iron oxide supported nanocarbon in lithium-ion batteries, medical, catalytic, and environmental applications. ACS Nano 8:7571–7612
56. Egorov NV, Sheshin EP (2020). Carbon-based field emitters: properties and applications. In: Gaertner G, Knapp W, Forbes RG (eds) Modern developments in vacuum electron sources. Topics in Applied Physics, 135. Springer
57. Jie W, Shengyang D, Bing D, Ya W, Xiaodong H, Hui D, Yongyao X, Xiaogang Z (2017) Pseudocapacitive materials for electrochemical capacitors: from rational synthesis to capacitance optimization. National Sci Rev 4:71–90

Chapter 2
Roadmap: From Carbon to Graphene Quantum Dots

2.1 Achievements of Carbon-Based Materials

Two Nobel Prizes for fullerenes and graphene and two Kavli Prizes in Nanoscience to M Dresselhaus and S Iijima for work on carbon nanotubes have resulted from recent research on carbon materials [1]. The majority of materials that have piqued the interest of carbon science researchers thus far are constructed using the benzene ring, a hexagon that is sp^2 bonded, as the main building block [2]. Graphene is produced by combining an infinite number of benzene rings in a two-dimensional (2D) fashion [3].

2.2 Graphite

There are only three forms of carbon that occur naturally: diamond, amorphous carbon, and graphite. The structure and atom bonding within each of the three naturally occurring allotropes distinguish them from one another: graphite has a honeycomb lattice structure, amorphous carbon lacks a crystalline structure, and diamond has a diamond lattice crystalline structure. Although carbon can take on many different forms, graphite is the most stable form under normal circumstances and is of the highest grade [4]. As a result, it is frequently used in thermochemistry as the reference state to define the heat formation of carbon-based compounds. It occurs naturally in three different forms: lump or vein, amorphous, and crystalline flake graphite. Depending on the form, it can be used for a variety of purposes. As was mentioned earlier, the structure of graphite is planar and layered, with carbon atoms arranged in a hexagonal lattice to form each layer [5]. The carbon atoms in these bonds, or covalent bonds as they are more formally called, are only 0.142 nm apart, and they are incredibly strong. Strong sp^2 hybridized bonds hold the carbon atoms together in a single, 2D layer of atoms. In graphite, there is a 0.335 nm gap

N. Manjubaashini et al., *Graphene Quantum Dots*, Materials Horizons: From Nature to Nanomaterials, https://doi.org/10.1007/978-981-97-5722-0_2

between each, 2D, one-atom-thick layer of sp^2-bonded carbon atoms. As previously mentioned, graphite is essentially made up of hundreds of thousands of separate layers of linked carbon atoms stacked one on top of the other in the crystalline flake form [6].

Carbon atoms form the hexagonal crystal structure of graphite, which has a dark gray metallic opaque. It is a carbonaceous crystalline mineral with an entire lamellar structure. To improve thermal conductivity, graphite was easily obtained [7]. The majority of focus has recently been on graphene's remarkable mechanical qualities, which make it the strongest 2D material, following ground-breaking experiments on the material and Geim and Novoselov's 2010 Nobel Prize [8]. Furthermore, because graphene planes are extremely flexible, it makes no difference if graphene layers are oriented along the matrix's sliding direction when there is friction; ultimately, regardless of the shape and orientation of the particles within the friction material, the graphene layers will be oriented parallel to the opposite surface [9]. With a thickness of just one atom, it is the thinnest compound ever discovered by humans and possesses an exceptional blend of qualities. It is the lightest, most flexible, and strongest compound ever found. Moreover, it is transparent, highly electrically and thermally conductive, and impermeable to molecules [10]. Owing to its special set of characteristics, graphene is anticipated to aid in the creation of new products and devices in a wide range of application areas, including biomedical technologies, sensors, photonics and optoelectronics, energy, electronics, composites and coatings, and electronics [11].

2.3 Graphene

Graphene, the lightest material known to science (0.77 mg), is a single layer of carbon atoms tightly arranged in a hexagonal honeycomb lattice (Fig. 2.1). It is an allotrope of carbon with a molecular bond length of 0.142 nm, atoms bonded in a plane [12]. Graphite is made up of layers of graphene stacked on top of one another, with an interplanar spacing of 0.335 nm. Van der Waals forces, which can be overcome during graphene exfoliation from graphite, hold the separate layers of graphene in graphite together [13]. The strongest known compound, with a tensile strength of 130 GPa and Young's modulus of 1 TPa—150,000,000 psi, is also the best heat conductor at room temperature, with values ranging from $4.84 \pm 0.44) \times 10^3$ to $5.30 \pm 0.48) \times 10^3\ W \cdot m^{-1} \cdot K^{-1}$. Research has also revealed that it is the best-known electrical conductor, with electron mobility at values of more than 200,000 $cm^2 \cdot V^{-1} \cdot s^{-1}$. Graphene's consistent light absorption throughout the visible and near-infrared regions of the spectrum ($\pi\alpha \approx 2.3\%$), as well as its possible application in spin transport, are other noteworthy characteristics [14].

Depending on the level of oxidation, graphene oxide can function as an insulator or a semiconductor, and its electrical and optical characteristics are highly adjustable. There are numerous applications for graphene oxide due to its tunable optical and electronic characteristics [15]. The primary areas of concern regarding

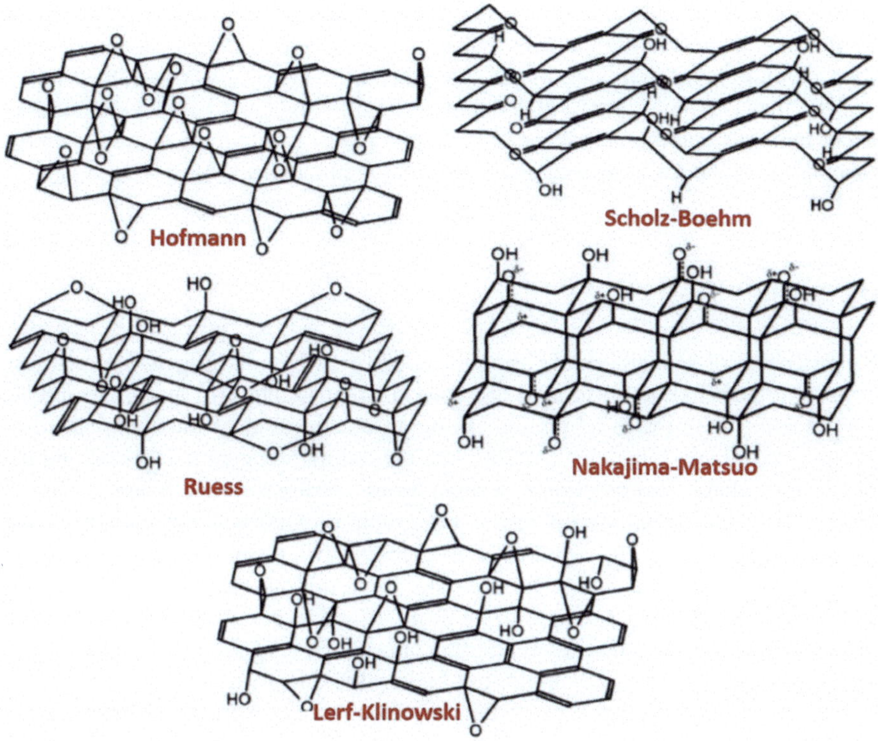

Fig. 2.1 Different forms of graphene structures. Reprinted with permission from @ Butterworth-Heinemann 219–525 (2014)

graphene oxide are its electronic characteristics, reduction reaction, chemical functionalization, and chemical structure. In the graphene layer, various oxygen species and bonding types, including epoxy, hydroxyl, carbonyl, and carboxylic groups, are present [16].

Graphene shows very high electrical conductivity due to its zero-overlap semimetal property. Six electrons make up a carbon atom: two are found in the inner shell and four are found in the outer shell [17]. In a single carbon atom, four outer shell electrons are available for chemical bonding; however, in graphene, each atom is joined to three other carbon atoms in a 2D plane, leaving one electron openly available for electronic conduction in the third dimension [18]. These extremely mobile electrons, which are found both above and below the graphene sheet, are known as π electrons. In graphene, these pi orbitals overlap and strengthen the carbon-to-carbon bonds. The valance and conduction bands, or bonding and anti-bonding of these pi orbitals, essentially control graphene's electronic characteristics [19].

These days, chemical vapor deposition (CVD), epitaxial growth, and mechanical or thermal exfoliation are typically used to produce graphene [20]. The chemical reduction of graphene oxide may be one of the most efficient methods for producing graphene on a large scale [21]. Interest in graphite oxide, which is created by oxidizing graphite, has increased dramatically since the first report on the mechanical exfoliation of monolayer graphene in 2004. People are looking for a way to produce graphene that is simpler, more affordable, more effective, and better yielding than current methods, can be massively scaled up compared to current methods, and is financially feasible for industrial or commercial applications [22].

2.4 Graphite Oxide (GO)

The molecules of graphite oxide are composed of carbon, hydrogen, and oxygen. Strong oxidizers like sulfuric acid are used to treat graphite to artificially create it. By reacting with the graphite and taking an electron out of the chemical reaction, these oxidizers function [23]. Because the oxidizing agent is reduced and the reactant is oxidized, this reaction is referred to as a redox reaction. Historically, the most widely used technique for producing graphite oxide was the Hummers and Offeman method, which involved treating graphite with a solution of sulfuric acid, potassium permanganate, and sodium nitrate. Nevertheless, more recent techniques have been developed that are said to be more effective, reaching 70% oxidization levels [24]. These techniques involve adding phosphoric acid along with sulfuric acid instead of sodium nitrate and employing higher doses of potassium permanganate. Since the oxidizing agents react with the graphite to increase the interplanar spacing between the layers, graphene oxide is essentially a by-product of this oxidization [25]. Graphene oxide can then be created by dispersing the fully oxidized compound in a base solution, like water. Although graphene oxide and graphite oxide differ greatly in structure, they are chemically very similar [26]. The interplanar spacing, which is brought about by water intercalation, between the individual atomic layers of the compounds is the primary distinction between graphene oxide and graphite oxide. Because of the oxidation process's increased spacing, which also interferes with the sp^2 bonding network, graphite oxide and graphene oxide are frequently referred to as electrical insulators [27].

2.4.1 Graphite Oxide to Graphene Oxide

Graphite oxide differs slightly from graphite, a three-dimensional (3D) carbon-based material composed of millions of graphene layers. By employing potent oxidizing agents to oxidize graphite, oxygenated functionalities are added to the graphite structure, expanding the layer separation and rendering the material hydrophilic [28]. Because of this characteristic, graphite oxide can be sonicated in water to exfoliate it and produce graphene oxide (GO), which is either single- or few-layered graphene. Thus, the number of layers is the primary distinction between GO and graphite oxide

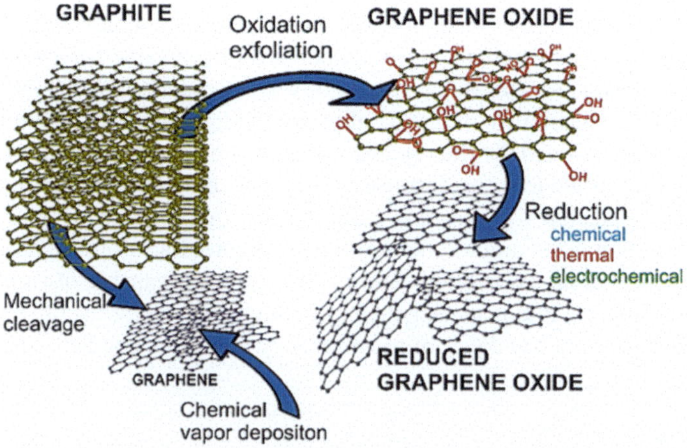

Fig. 2.2 Structural transformation of graphite, graphene, and graphene oxide to reduced graphene oxide

(Fig. 2.2). There are some layers of flakes and monolayer flakes in a Go dispersion even though graphite oxide is a multilayer system [29].

The individual graphene layers may ultimately suffer significant harm during the conversion of graphite oxide to GO, which will have additional repercussions when the compound is reduced even further. Additional damage is undesirable because individual graphene platelets are already harmed by the oxidation process from graphite-to-graphite oxide, which lowers their mean size [30]. Water is scattered throughout the monolayer and a few layers of graphene flakes make up GO. There are several ways to convert graphite oxide to GO. The most popular methods include stirring, sonication, and a mix of the two. While sonication is a very effective and time-efficient method of exfoliating graphite oxide, it can also severely damage graphene flakes, causing their surface size to decrease from microns to nanometers and producing a wide range of graphene platelet sizes [31].

2.4.2 *Properties of Graphene Oxide (GO)*

Due to the oxygen functionalities present, GO has the advantage of being easily soluble in a variety of matrixes and organic solvents, including water. This characteristic is still crucial when combining the material with polymer or ceramic matrixes to enhance its mechanical and electrical capabilities [32]. However, because its sp^2 bonding networks are disrupted, GO is frequently referred to as an electrical insulator in terms of electrical conductivity. It is necessary to reduce the GO to restore the honeycomb hexagonal lattice and electrical conductivity. It must be considered that the reduced graphene oxide obtained after the majority of the oxygen groups are eliminated tends to form aggregates, making it more challenging to disperse [33].

The properties of GO can be significantly altered by functionalization. The resultant chemically altered graphene may then be far more versatile for a wide range of uses. GO can be functionalized in a variety of ways, depending on the intended use. To improve the dispersibility of chemically modified graphene in organic solvents, amines can be substituted for the organic covalent functionalization of graphene in optoelectronics, biodevices, or a drug-delivery material [34]. Furthermore, it has been demonstrated that GO platelets could be functionalized with secondary and porphyrin-functionalized amines, thereby enhancing nonlinear optical performance. The development of an oxidization and reduction process capable of isolating individual carbon layers without altering their structure is crucial for the use of graphene oxide as a bridge in the synthesis of monolayer or few-layer graphene sheets [35].

2.5 Graphene Oxide to Reduced Graphene Oxide (rGO)

The process of reducing GO to reduced graphene oxide (rGO) is crucial because it greatly affects the quality of the rGO that is produced and, consequently, determines how close the structure of the rGO will be to that of pristine graphene [36]. Due to the relative ease of producing sufficient quantities of graphene to desired quality levels, rGO is the most obvious solution in large-scale operations where scientific engineers need to utilize large quantities of graphene for industrial applications such as energy storage [37]. Reduction can be accomplished in a variety of ways, as one might anticipate, but they are all dependent on chemical, thermal, or electrochemical processes (Fig. 2.3). While some of these methods can yield rGO of extremely high quality, comparable to pristine graphene, their execution can be laborious or time-consuming [38]. GO has previously been treated with hydrazine hydrate and the solution kept at 100 ºC for 24 h; GO has also been exposed to a hydrogen plasma for a brief period; GO has also been exposed to other strong pulse light sources, such as those generated by xenon flash tubes; and GO has been heated in distilled water to varying degrees for variable periods [39].

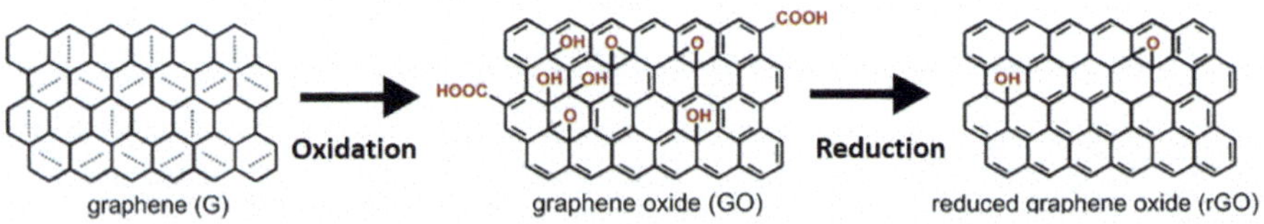

Fig. 2.3 Schematic of graphene to rGO preparation

2.6 Properties and Application of GO and rGO

GO and rGO exhibit distinct mechanical, electrical, and chemical properties as a result of their structural and chemical differences. The electrical conductivity of GO and rGO is one of the key distinctions between them [40]. GO exhibits semi-conducting or insulating behavior, whereas rGO has an exceptional electrical conductivity that is nearly on par with that of pure graphene. When oxygen-containing compounds are removed, rGO materials have a larger specific surface area than GO. The ability to be hydrophilic or hydrophobic is another unique quality [41]. While rGO exhibits hydrophobic behavior, GO demonstrates strong hydrophilic behavior. Furthermore, rGO's mechanical qualities are noticeably better than those of GO. However, when compared to rGO, GO exhibits far superior colloidal and dispersible qualities [42].

GO and rGO's intriguing qualities are applied in a wide range of applications. GO and rGO nanocomposites are widely utilized in solar cells, supercapacitors, biosensors, sensors, energy storage applications, and solar cells [43]. For each of these application areas, the large surface area of both materials is a significant benefit. The anode material for Li-ion batteries is made more capable and cyclically stable by using rGO's high conductivity [44]. However, the cathode of Li-ion batteries makes use of GO's chemical activity. GO and rGO are also used in lithium–sulfur and lithium–oxygen batteries for high-capacity applications, in addition to Li-ion batteries. Similarly, supercapacitors' capacitance can be raised by using GO and rGO [45]. It has been discovered that GO's DNA binding and biocompatibility are advantageous in biomedical applications, including drug delivery, diagnostics, and therapeutics [46]. Alzheimer's disease diagnosis and cancer treatment are two uses for GO-based nanocomposites. Overall, graphene derivatives GO and rGO show great promise as materials for graphene-based bio-based, electronics, sensors, and energy storage applications [47].

2.7 Graphene Quantum Dots (GQDs)

GQDs are minuscule nanomaterials that originate from graphene, a 2D crystalline structure with a thickness of one atom. GQDs and graphene share the same crystalline structure structurally. This semiconductor nanomaterial is made up of one or more layers of less than 10 mm wide graphene sheets [48]. Mostly made up of functional groups like oxygen and hydrogen, GQDs can bind electrostatically or through the π–π stacking interaction with a variety of aromatic biomolecules. Because of their edge effect and zero dimensions (0D), GQDs exhibit a strong quantum confinement effect that, according to theoretical predictions and experimental studies, induces photoluminescence (PL) [49]. Consequently, because of their special chemical and physical characteristics, including their non-toxicity, solubility, surface grafting, biocompatibility, inertness, and stable fluorescence activity, GQDs have attracted a great deal of attention, particularly in the domains of nanoscience and nanotechnology [50]. GQDs have applications in photovoltaics, bioimaging, optoelectronics, and other fields. Carbon-based materials such as graphite or GO, carbon nanotubes (CNTs), and carbon fibers (CFs) have been treated with high-resolution electron beams, microwave irradiation, electrochemical scissoring, chemical cutting, or hydrothermal cutting to create GQDs [51, 52].

A quantum dot (QD) is, broadly speaking, a semiconductor crystal with a size range of 1–10 nm. QD's unique size range allows for the occurrence of quantum phenomena that significantly improve optical properties. It is commonly known that when a QD is excited, its size will decrease and its energy and intensity will increase. Metals, semiconductors, and carbon-based materials can all be used to create QDs [53]. Because of their unique characteristics, semiconductor quantum dots are used as photodiodes and have many uses in analytical chemistry; however, their possible toxicity makes them unsuitable for use in biology and medicine [54]. This limitation of QDs was the prime driving source for finding new alternatives, and with the advancement in nanotechnology, quantum dots fabricated from graphene evolved as a more biocompatible source for biomedical applications [55]. Graphene is a two-dimensional (2D) carbon-allotrope sheet of single-layer sp^2 hybridized carbon with zero bandgap and exceptional thermoelectric properties.

2.7.1 Formation of GQDs

Graphene oxide is typically treated oxidatively using Hummer's method or graphite nanoparticles are exfoliated to create GQDs (Fig. 2.4). Partial pyrolysis of organic materials, such as sugar, polythiophene, and citric acid, has also occasionally been used to create GQDs [56]. Not only are the resulting GQDs heterogeneous within a batch, but when prepared using various synthesis techniques, their properties also differ significantly. Because of this, the genesis of GQD properties is largely unknown and the subject of conjecture [57]. It is clear that GQDs' defects as opposed to their HOMO/LUMO energy gap or other clearly defined features dominate their optoelectronic properties. From a structural perspective, GQDs are positioned in between molecules and macro-molecular dimensions, resembling big, diverse carbon-based molecules [58].

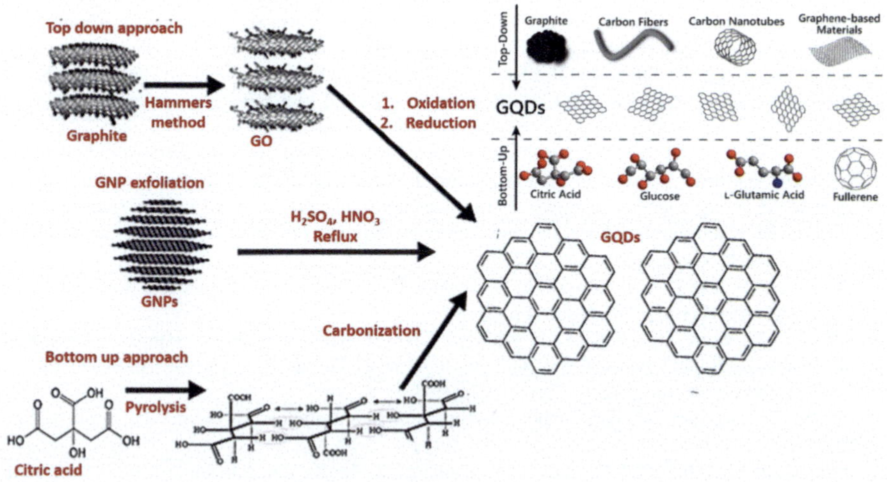

Fig. 2.4 Synthesis methods to produce GQDs via top-down approach, GNP exfoliation method, and bottom-up approach; inset image shows variant approaches of GQD formation

References

1. Zhang J, Terrones M, Park CR, Mukherjee R, Monthioux M, Koratkar N, Kim YS, Hurt R, Frackowiak E, Enoki T, Chen Y, Chen Y, Bianco A (2016) Carbon science in 2016: Status, challenges and perspectives. Carbon 98:708–732
2. Jain S, Shimpi NG (2023) Chapter 1: Carbon-based smart nanomaterials. In: Micro and nano technologies, carbon-based nanomaterials and nanocomposites for gas sensing. Elsevier, pp 3–24
3. Ferrari AC, Bonaccorso F et al (2015) Science and technology roadmap for graphene, related two-dimensional crystals, and hybrid systems. Nanoscale 7:4598–4810
4. Baig S, Ahmed M, Batool A, Bashir A, Mumtaz S, Ikram M, Saeed M, Shahzad K, Farooq MU, Maqsood A, Ikram M (2023) Introductory chapter: brief scientific description to carbon allotropes – technological perspective, graphene – a wonder material for scientists and engineers. IntechOpen
5. Georgakilas V, Perman JA, Tucek J, Zboril R (2015) Broad family of carbon nanoallotropes: classification, chemistry, and applications of fullerenes, carbon dots, nanotubes, graphene, nanodiamonds, and combined superstructures. Chem Rev 115:4744–4822
6. Hess P (2021) Bonding, structure, and mechanical stability of 2D materials: the predictive power of the periodic table. Nanoscale Horiz 6:856–892
7. Li M, Mu B (2019) Effect of different dimensional carbon materials on the properties and application of phase change materials: a review. Appl Energy 242:695–715
8. Gerstner E (2010) Nobel Prize 2010: Andre Geim & Konstantin Novoselov. Nature Phys 6:836
9. Cho MH, Ju J, Kim SJ, Jang H (2006) Tribological properties of solid lubricants (graphite, Sb_2S_3, MoS_2) for automotive brake friction materials. Wear 260:855–860
10. Abdel-Rahim YM, Darwish SM (2010) Generalized braking characteristics of friction pad synthetic graphite composites. Tribol Int 43:838–843
11. Tiwari SK, Sahoo S, Wang N, Huczko A (2020) Graphene research and their outputs: status and prospect. Journal of Science: Advanced Materials and Devices. 5:10–29
12. Ray SC (2015) Applications of Graphene and Graphene-oxide based Nanomaterials. Elsevier, https://doi.org/10.1016/C2014-0-02615-9

13. Chopra S, Pande K, Shantha Kumar V, Sharma JA (2023) Novel applications of carbon-based nanomaterials. CRC, Taylor & Francis.https://doi.org/10.1201/9781003183549
14. Singh SK, Nayak B, Singh TJ, Halder S (2023) Investigating the role of synthesized reduced graphene oxide and graphite micro-fillers on mechanical and fretting wear performance of glass fiber epoxy-based composite. High Perfor Poly 35:946–962
15. Brisebois PP, Siaj M (2020) Harvesting graphene oxide – years 1859 to 2019: a review of its structure, synthesis, properties and exfoliation. J Mater Chem C 8:1517–1547
16. Inagaki M, Kang F (2014) Chapter 3: engineering and applications of carbon materials. In: Inagaki, M, Kang F (eds) Materials science and engineering of carbon: fundamentals, 2nd edn. Butterworth-Heinemann, pp 219–525
17. Devrim A, Potasz P, Korkusinski M, Hawrylask P (2014) Graphene quantum dots. Springer. https://doi.org/10.1007/978-3-662-44611-9
18. Ghosh D, Sarkar K, Devi P, Kim KH, Kumar P (2021) Current and future perspectives of carbon and graphene quantum dots: From synthesis to strategy for building optoelectronic and energy devices. Renew Sustain Energy Rev 135:110391
19. Yun K, Govindaraju S (2013) Carbon and graphene quantum dots for biomedical applications. Elsevier
20. Ahmad SI, Sheikh Mohd G, Fatimah I, Natasya ZZ, Nadia ZN, Nurain A (2023) Graphene quantum dots: a comprehensive overview. Open Chem 21:20220285
21. Ghaffarkhah A, Hosseini E, Kamkar M, Sehat AA, Dordanihaghighi S, Allahbakhsh A et al (2022) Synthesis, applications, and prospects of graphene quantum dots: a comprehensive review. Small 18:2102683
22. Rodriguez-Fortea A, Moreno-Vicente A, Schardt S, Dunk P, Poblet JM (2019) Electronic structure and properties of boron-doped endohedral metalloheterofullerenes. In: ECS Meeting. IOP Publishing, vol 11. pp 792
23. Chung DDL (2001) Graphite intercalation compounds. In: Jürgen Buschow KH, Cahn RW, Flemings MC, Ilschner B, Kramer EJ, Mahajan S, Veyssière P (eds) Encyclopedia of materials: science and technology. Elsevier, pp 3641–3645
24. Kang JH, Kim T, Choi J, Park J, Kim YS, Chang MS, Jung H, Park KT, Yang SJ, Park CR (2016) Hidden second oxidation step of hummers method. Chem Mater 28:756–764
25. Chen X, Qu Z, Liu Z, Ren G (2022) Mechanism of oxidization of graphite to graphene oxide by the hummers method. ACS Omega 27:23503–23510
26. Ibarra-Hernández A, Vega-Rios A, Osuna V (2018) Synthesis of graphite oxide with different surface oxygen contents assisted microwave radiation. Nanomaterials 8:106
27. Smith AT, LaChance AM, Zeng S, Liu B, Sun L (2019) Synthesis, properties, and applications of graphene oxide/reduced graphene oxide and their nanocomposites. Nano Materials Science 1:31–47
28. Bacon M, Bradley SJ, Nann T (2014) Graphene Quantum Dots. Part Part Syst Charact 31:415–428
29. Chung S, Revia RA, Zhang M (2021) Graphene quantum dots and their applications in bioimaging, biosensing, and therapy. Adv Mater 33:1904362
30. Pandey A, Priya C (2023) Chapter 1: Functionalized graphene nanomaterials: Next-generation nanomedicine. In: Mallakpour S, Mustansar Hussain C (eds) Micro and nano technologies, functionalized carbon nanomaterials for theranostic applications. Elsevier, pp 3–18
31. Rowley-Neale SJ, Randviir EP, Abo Dena AS, Banks CE (2018) An overview of recent applications of reduced graphene oxide as a basis of electroanalytical sensing platforms. Appl Mater Today 10:218–226
32. Alev Devrim G, Pawel P, Marek K, Pawel H (2014) Graphene quantum dots, vol 172, 1st edn, Springer Berlin, Heidelberg, https://doi.org/10.1007/978-3-662-44611-9
33. Ansari SA (2022) Graphene quantum dots: novel properties and their applications for energy storage devices. Nanomaterials 12:3814
34. Novoselov KS, Geim AK, Morozov SV, Jiang DE, Zhang Y, Dubonos SV et al (2004) Electric field effect in atomically thin carbon films. Science 306:666–669

35. Yang K, Feng L, Liu Z (2016) Stimuli-responsive drug delivery systems based on nanographene for cancer therapy. Adv Drug Delivery Rev 105:228–241
36. Raslan A, del Burgo LS, Ciriza J, Pedraz JL (2020) Graphene oxide and reduced graphene oxide-based scaffolds in regenerative medicine. Int J Pharm 119226.https://doi.org/10.1016/j.ijpharm.2020.119226
37. Suk JW, Piner RD, An J, Ruoff RS (2010) Mechanical properties of monolayer graphene oxide. ACS Nano 4:6557–6564
38. Deepa C, Rajeshkumar L, Ramesh M (2022) Preparation, synthesis, properties and characterization of graphene-based 2D nano-materials for biosensors and bioelectronics. J Mater Res Tech 19:2657–2694
39. Rajkumari NP, Dolakashoria S, Goswami P (2021) Plant-based natural dye-stimulated visible-light reduction of go and physicochemical factors influencing the production of oxidizing species by a synthesized (rGO)/TiO_2 nanocomposite for environmental remediation. ACS Omega 6:2686–2698
40. Ling S (2019) Structure and synthesis of graphene oxide. Chin J Chem Eng 27:2251–2260
41. McCoy TM, Turpin G, Teo BM, Tabor RF (2019) Graphene oxide: a surfactant or particle? Current Opinion Collo Interf Sci 39:98–109
42. Lawal AT (2019) Graphene-based nanocomposites and their applications A review. Biosens Bioelectro 141:111384
43. Singh RK, Kumar R, Singh DP (2016) Graphene oxide: strategies for synthesis, reduction and frontier applications. RSC Adv 6:64993–65011
44. Viprya P, Kumar D, Kowshik S (2023) Study of different properties of graphene oxide (GO) and reduced graphene oxide (rGO). Eng Proc 59:84
45. Qazi UY, Javaid R (2023) Graphene utilization for efficient energy storage and potential applications: challenges and future implementations. Energies 16:2927
46. Bellier N, Baipaywad P, Ryu N, Lee JY, Park H (2022) Recent biomedical advancements in graphene oxide- and reduced graphene oxide-based nanocomposite nanocarriers. Biomater Res 26:65
47. Dideikin AT, Vul AY (2019) Graphene oxide and derivatives: the place in graphene family. Front Phys 6:149
48. Tajik S, DourandishZ ZK, Beitollahi H, Van Le Q, Jang HW, Shokouhimehr M (2020) Carbon and graphene quantum dots: a review on syntheses, characterization, biological and sensing applications for neurotransmitter determination. RSC Adv 10:15406–15429
49. Cui Y, Liu L, Shi M, Wang Y, Meng X, Chen Y, Huang Q, Liu C (2024) A review of advances in graphene quantum dots: from preparation and modification methods to application. J Carbon Res 10:7
50. Tian P, Tang L, Ten KS, Lau SP (2018) Graphene quantum dots from chemistry to applications. Mater Today Chem 10:221–258
51. Barati F, Avatefi M, Moghadam NB, Asghari S, Ekrami E, Mahmoudifard M (2023) A review of graphene quantum dots and their potential biomedical applications. J Biomater Appl 37:137–1158
52. Shin Y, Park J, Hyun D, Yang J, Lee H (2015) Generation of graphene quantum dots by the oxidative cleavage of graphene oxide using the oxone oxidant. New J Chem 39:2425–2428
53. Joshi PN, Kundu S, Sanghi SK, Sarkar D (2016) Graphene quantum dots - from emergence to nanotheranostic applications. In: Smart drug delivery system. InTech Open. https://doi.org/10.5772/61932
54. Novoselov KS, Geim AK, Morozov SV, Jiang D, Zhang Y, Dubonos SV, Grigorieva V, Firsov A (2004) Electric field effect in atomically thin carbon films. Science 306:666–669
55. Novoselov KS, Geim AK, Morozov SV, Jiang D, Katsnelson MI, Grigorieva IV, Dubonos SV, Firsov AA (2005) Two-dimensional gas of massless Dirac fermions in graphene. Nature 438:197–200
56. Schroeder KL, Goreham RV, Nann T (2016) Graphene Quantum Dots for Theranostics and Bioimaging. Pharm Res 33:2337–2357

57. Chung S, Revia RA, Zhang M (2021) Graphene quantum dots and their applications in bioimaging, biosensing, and therapy. Adv Mater 33:1904362
58. Shen J, Zhu Y, Yang X, Li C (2012) Graphene quantum dots: emergent nanolights for bioimaging, sensors, catalysis and photovoltaic devices. Chem Commun 48:3686–3699

Chapter 3
Introduction to Graphene Quantum Dots

3.1 Overview of Carbon and Its Related Materials

Carbon, an extraordinary element that is the main ingredient of coal, is one of the reasons life exists on the earth. It has just surprised us once more, this time in the form of graphene. The microscopic needle-like carbon nanotube (CNT), another allotrope of carbon, was responsible for the discovery of the football-shaped fullerene in 1985. CNTs were properly defined for the first time in 1991 [1]. These remarkable discoveries of carbon allotropes have attracted and piqued the interest of scientists across various disciplines. Graphitic carbon exists in four different forms: zero-dimension (0D), one-dimension (1D), two-dimension (2D), and three-dimension (3D) (Fig. 3.1). To create a single-layered nanomaterial, 2D graphene can be split from multi-layered graphite using the van der Waals force between the layers [2].

The strongest and thinnest material ever measured in the universe is a single-atom-thick carbon. The graphene and graphene-related series includes several materials with comparable nanostructures but various names that indicate the existence of one or more monolayers of graphene in the carbon material [3]. Examples of existing allotropes of carbon-based materials include amorphous carbon, graphite, diamond, CNT, graphene oxide (GO), graphene quantum dot (GQD), carbon dot (CD), carbon quantum dot (CQD), and fullerenes. Recently, these materials have drawn huge attention because of their numerous applications in the energy, healthcare, and environmental sectors as well as other industries due to their exceptional optoelectronic, electrical, and catalytic properties [4]. Since the nineteenth century, carbon, one of the most common elements, has been thoroughly explored, and research is still growing today [5]. The entire history of the evolution of carbon materials is shown in Fig. 3.2. Graphene, a flat graphitic sheet and one of the several carbon allotropes exhibits unique physicochemical properties. It is composed of an sp^2 hybridized carbon network with an interlayer spacing of 3.4 nm and a C–C bond distance of 1.42 nm [6]. The thermal conductivity of graphene's 2D crystal structure is in the

N. Manjubaashini et al., *Graphene Quantum Dots*, Materials Horizons: From Nature to Nanomaterials, https://doi.org/10.1007/978-981-97-5722-0_3

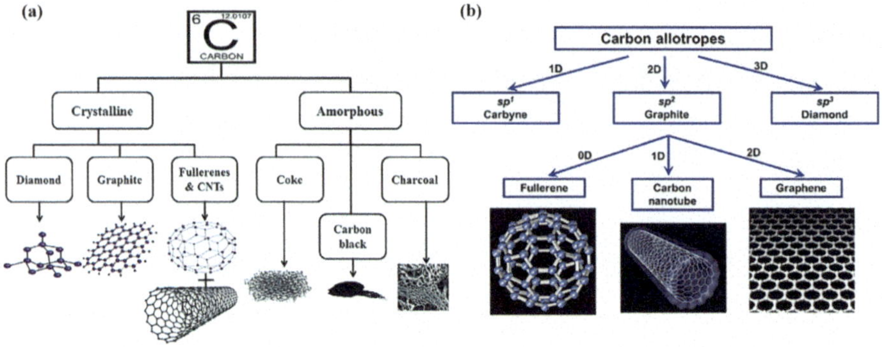

Fig. 3.1 Derivatives (a), and allotropes (b) of carbon

range of 5000 W/mK, electron mobility is 2.5×10^5 cm^2V^{-1} s^{-1} at room temperature, intrinsic strength is 130 GPa, Young's modulus is 1 TPa, the surface area is 2630 m^2g^{-1}, optical absorptions are 2.3% in the infrared, and it may be chemically functionalized [7].

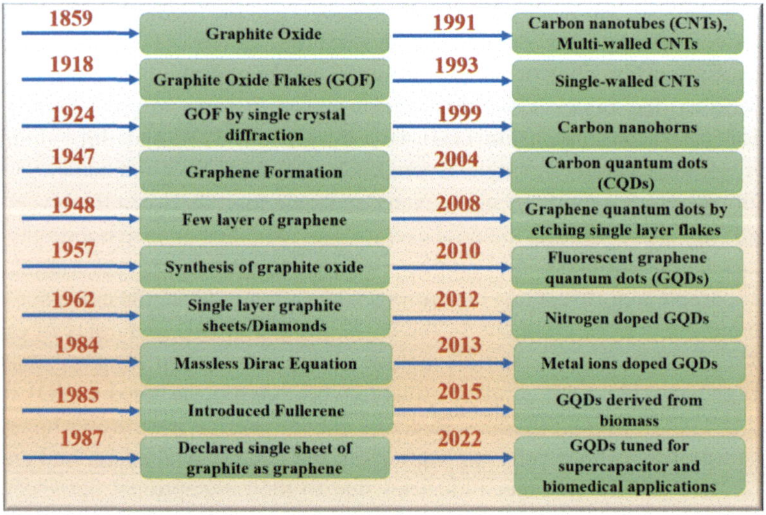

Fig. 3.2 History of the development of carbon materials

3.2 Outline of Graphene

In 2004, the "Scotch-tape approach" was utilized to isolate a single layer of material from highly oriented pyrolytic graphite (HOPG) to produce graphene [8, 9]. Graphene possesses an infinite exciton Bohr radius and a zero band gap, which makes it non-luminous and enables quantum confinement of infinite-sized specimens. At the K point, also known as the Dirac point of graphene, the conduction band (CB) and valence band (VB) combine to form two Dirac cones. Graphene electrons are known as massless Dirac fermions, having a Fermi velocity of up to 106 m/s. The straightforward Dirac Hamiltonian, which describes the mechanics of graphene's carrier, rather than Schrodinger's equation, adds to its allure [10].

Chiral nanostructures composed of metals and semiconductors have attracted a lot of attention as components for polarization-enabled optoelectronic devices. Like other nanotechnology domains, chiral nanostructures can be used in biology as well as other disciplines because their optical and electrical properties can be greatly improved by physical and chemical processes in graphene-based materials [11]. Epitaxial graphene islands generated by hydrocarbon breakdown on densely packed metal surfaces have a wide variety of shapes and sizes (Fig. 3.3). By using chemical vapor deposition (CVD), graphene nanoislands were made from the breakdown of propene on the Ni (111) surface with a regulated yield and shape. Carefully choosing the reaction/annealing conditions, may result in a transition from amorphous clusters to triangles and hexagons larger than 100 nm^2 and with sharp edges [12].

3.2.1 Graphene Derivatives and Their Properties

Graphene and its derivatives possess some unique physicochemical properties such as a large surface area of 2630 $m^2\ g^{-1}$, easy solubility, high drug-loading capacity, high purity, and easy cell membrane penetration capability. As mentioned earlier, it shows excellent room-temperature electron mobility (~200,000 $cm^2V^{-1}\ s^{-1}$), high thermal conduction (5300 $Wm^{-1}\ K^{-1}$), an outstanding tensile strength (130 Gpa), a young modulus of 1.0 TPa with 20% third-order elastic stiffness, and 97.4% transmittance with electric sheet resistance as low as 125 $\Omega\ m^{-1}$ [13]. It also undergoes several

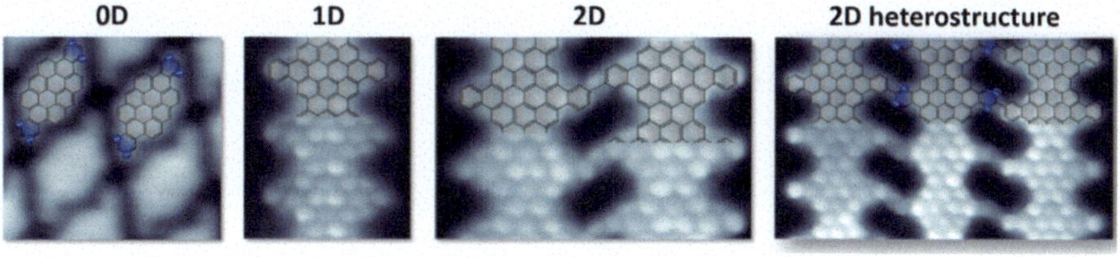

Fig. 3.3 Graphene islands in different dimensions

chemical reactions resulting in the formation of covalent linkage with different functional groups. These properties made this nanomaterial an attractive candidate for a broad range of researchers from chemists, to physicians, and biologists, to medical researchers, and it has become a promising candidate for cargo delivery, bioimaging, and therapy [14]. Based on the number of layers or alterations introduced, several kinds of graphene are available such as GO, reduced graphene oxide (rGO), GQD, monolayer graphene, few-layer graphene, etc. (Fig. 3.4) [15, 16].

After graphene was discovered, it was revealed that 0D structures were the only ones that could contain 1D graphene nanoribbons (GNRs). These 0D structures often referred to as GQDs, have very strong optoelectronic properties. These GQDs have properties that are superior to those of conventional semiconductor QDs [17]. Additionally, GQDs have many advantageous characteristics, such as outstanding optical and electrical properties, minimal environmental toxicity, ease of synthesis, economical effectiveness, and exceptional sustainability [18]. Due to these intrinsic characteristics and their numerous beneficial material properties, the viability of using GQDs has been established through countless applications in the material sciences, chemical sciences, biological sciences, and physical sciences [19]. Graphene has garnered a lot of interest due to its distinct physical characteristics and a wide range of exciting nanotechnology applications. However, graphene's usage in electrical

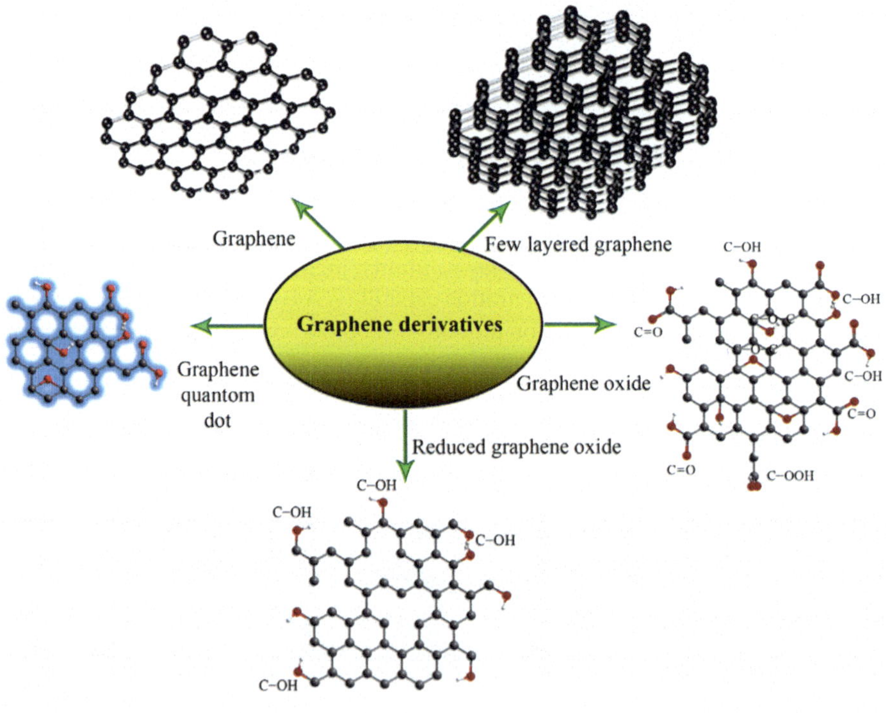

Fig. 3.4 Structure of various graphene derivatives

and optoelectronic devices is constrained by its zero-bandgap semiconductor nature [20]. Because there is no bandgap in pure graphene, optical light cannot be detected. However, a bandgap can be generated into GNRs and GQDs because of quantum confinement and edge effects [21].

3.3 Fundamentals of Quantum Dots (QDs)

A QD is a semiconductor crystal with a size between 1 and 10 nm. Because of their specific size range, QDs exhibit quantum phenomena, which significantly affect their optical properties. It is well known that the size of the QD affects the amount of energy and intensity of light emitted upon activation [22]. Materials based on carbon, semiconductors like selenium and cadmium, or metals like gold can all be used to prepare QDs. Semiconductor QDs are employed as photodiodes and have a range of functions in analytical chemistry because of their special properties, but their potential toxicity prevents them from being used in biology and medicine [8, 23]. This limitation served as the primary driving force behind the hunt for innovative QD substitutes. Graphene-based QDs become a more biocompatible source for biomedical applications with the advancement of nanotechnology.

It is common to use a class of different carbon compounds known as CDs, including CQDs, GQDs, carbon nanodots (CNDs), and carbonized polymer dots (CPDs). These CD family types can be divided into groups based on particular traits, surface groups, and carbon core structures [24]. Even though they are classified as CNDs, GQDs are distinct from CNDs in some ways. CNDs are nanoparticles (NPs) with photoluminescence (PL) properties that are quasi-spherical and have a diameter of less than 10 nm [25]. GQDs, on the other hand, are graphene nanosheets (GNSs) that can be divided into one, two, or more layers, each of which is less than 10 nm thick and 100 nm broad. They usually have functional groups (such as carboxyl, hydroxyl, carbonyl, and epoxide) at their edges as well, which can act as reaction sites and affect PL emission from the dots by altering their electron densities [26]. Depending on the specific synthesis method and whether the surface has been passivated, reduced, or otherwise modified, quantum yield (QY), a critical component for fluorescence (FL) materials, ranges in GQDs from 2 to 46% [27].

3.4 Uniqueness of Graphene Quantum Dots (GQDs)

Understanding the various expressions of the chirality of carbon nanostructures in symmetrically paired forms is essential for many scientific areas, including quasi-particle physics, optics, electronics, drug delivery, and biomedical imaging. GQDs are the most recent and basic type of nanocarbon useful for all of these disciplines [28]. Additionally, due to exciton confinement in GQDs and plasmonic effects in graphene sheets, this nanoscale material is similar to both noble metal and semiconductor nanostructures [29, 30]. Owing to the importance of chiral GQD research for comprehending chiroptical phenomena in delocalized states and their utilization in biology and medicine, the synthesis of symmetrically paired GQDs in dispersions is being done [31].

GQDs are anisotropic materials that mimic graphene. GQDs resemble CDs and graphene in terms of their morphological properties. In comparison to their huge edge dimensions, GQDs have very small vertex dimensions. This structure offers numerous opportunities for chemical functionalization since it enables one or more graphene sheets to have functional groups on their exterior. GQDs can interact with other nanomaterials to functionalize them, producing hybrid nanostructures or van der Waals heterostructures [32]. Additionally, GQDs can be functionalized using proteins, antibodies, and nucleic acids. The transducer surface can also be effectively improved by GQDs to interact with the targeted analyte. GQDs, an OD member of the carbon family, are sometimes compared to sliced graphene sheets. GQDs display unique properties as a result of the combined structural features of graphene that are endowed with edge effects and quantum confinement properties. As a result, GQDs are frequently utilized in many different types of technology, such as supercapacitors, solar cells, photodetectors, batteries, light-emitting diodes, and biosensors [33].

3.4.1 Synthesis of GQDs

GQDs are graphene nanostructures with significant quantum characteristics and edge effects, which provide potential photoluminescence properties. The size of GQDs' structures greatly depends on their optical properties such as quantum confinement effect, types, and formation of sp^2 sites into the structure. The GQDs bandgap may be improved by their size and surface properties. The GQD synthesis approach includes top-down (cutting of graphene sheets) and bottom-up synthesis (small precursors of aromatic compounds) [34]. The GQDs synthesized by the top-down approach often contain negatively charged, hydrophilic oxygen groups. Therefore, they are soluble in water, ethanol, and N, N-dimethylformamide (DMF) [35]. Generally, GQDs show 3 to 20 nm with one or more layers of graphene sheets with less than 5 nm thickness. GQDs have roughly circular or elliptical forms, along with hexagonal and quadrilateral structures. GQD has intriguing applications in the biomedical field, specifically for diagnostics, photothermal treatment using near-infrared (NIR) light, and bioimaging [36]. Ding et al. reported the development of a GQD-labeled anticancer nanocarrier for monitoring the transport, release, and response of doxorubicin by generating a large number of fluorescence signals [37]. The GQDs' superior qualities were revealed by the ease of fabrication, stable photoluminescence, high aqueous stability, long-term resistance to photobleaching, low cytotoxicity, excellent biocompatibility, chemical inertness, and cell compatibility [38]. Further, GQDs are used in batteries, supercapacitors, electronics, and agriculture sectors [39, 40].

The GQDs are synthesized mainly by graphene-based precursors such as CNTs, polycyclic aromatic hydrocarbons, biomass, etc. [41, 42]. The main flaw preventing a wide range of graphene-based materials application is their absence of band gap. For instance, despite their high switching rates, graphene-based transistors have poor on–off ratios owing to their zero band gap [43]. The performance of graphene devices operating in and beyond the visible regime is further limited by graphene's flat interband absorption rate, which is 2.3%. However, to alter the physical properties of graphene, a bandgap can be introduced by developing it into graphene quantum structures [44]. GQDs structural and morphological studies deliver the interlayer spacing of graphite as 0.33 nm while the in-plane lattice spacing ranges from 0.18 to 0.24 nm. Further, various defects, heteroatoms, and functional groups are added to GQDs by various synthesis techniques [45]. In optical spectra, the $C = C$ double bond structure of GQDs reveals $\pi \rightarrow \pi^*$ transition, to absorb photons over a broad spectrum of wavelengths. In general, it exhibits a strong UV absorption peak between 260 and 320 nm with a tail that extends into the visible spectrum. Due to the influence of the $n \rightarrow \pi^*$ transition, GQDs may also exhibit shoulder peaks from 270 to 390 nm region [46].

Interestingly, improved solubility, steady photoluminescence, photobleach resistance, and controllable luminescence are characteristics of GQDs, which successfully replace colloidal inorganic semi-conductive QDs [47]. According to Bentez-Martnez et al., small fluorescent NPs, also known as GQDs, offer unique properties that make them ideal research tools in a variety of disciplines [48]. In comparison to spherical CQDs, GQDs with a lamellar structure have a more regular crystalline structure with a high quantum yield. Oxidizing graphite to GO, which produces epoxy groups that rupture C–C bonds, was a crucial step in the synthesis process [49]. Figure 3.5 illustrates how these epoxy groups are further oxidized to produce stable carbonyl groups, which are responsible for GQDs water dispersibility. Later, it was proposed that green fluorescent fine crystalline GQDs may be synthesized using a modified high-temperature synthesis technique [50].

For making GQDs, coal is the most affordable resource with 20% of yield (Fig. 3.6). Therefore, the market for QDs could be considerably impacted by this revelation. Rice researchers revealed GQDs from relatively cheap carbon fiber and coal for high yield of GQDs with an inexpensive single-step chemical process [51]. The method utilized by Tour offers some control over the coal's size, which normally ranges from 2 to 20 nm, depending on the coal source. GQDs were synthesized by direct sonication to activate graphene into few-layer graphene sheets (FLGs) (Fig. 3.7). The FLGs were mixed in1:1 v/v mixture of ethanol and water, activated with KOH at a high temperature, and then sonicated for 2 h using a tip sonicator. The suspension is filtered with a 0.22 mm microporous membrane filter, and then further separated by dialysis [52]. The resulting activated GQDs (a-GQDs) have an average diameter between 1 to 8 nm with bluish fluorescence. Another technique involved exfoliating the small carbon fibers (CFs) with a resin-rich surface [21]. Concentrated H_2SO_4 and HNO_3 were applied to CFs over 24 h at various temperatures, resulting in GQDs with tunable diameters of 1–4 nm (at 80 °C), 4–8 nm (at 100 °C), and 7–11 nm (at 120 °C) (Fig. 3.8).

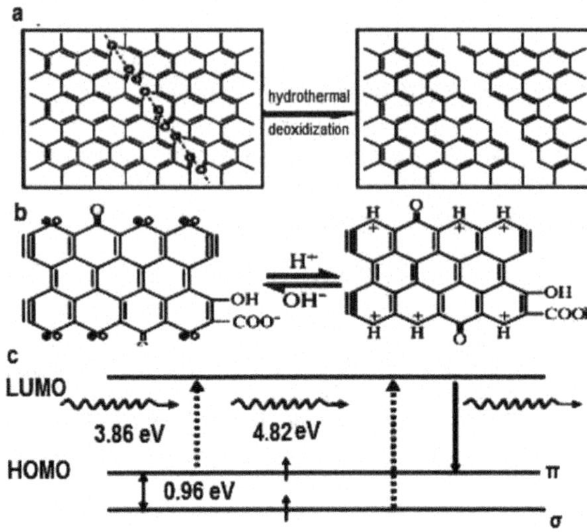

Fig. 3.5 Synthesis mechanism of GQDs via cutting of graphite sheets. GQDs were synthesized by reducing epoxy groups that were generated during the oxidation and cutting steps. Reprinted with permission from @ Materials Today 18 447–458 (2015)

Fig. 3.6 Schematic representation of the formation of GQD from coal

3.4.2 Properties of GQDs

GQDs have a graphene lattice, a honeycomb crystal structure composed of carbon atoms organized in six-atom rings. Each carbon atom forms the sp^2 hybridized feature and delocalized electrons in the orbitals perpendicular to the plane of the sheet as a result of its covalent connection to three more carbon atoms. These characteristics give them exceptional optical qualities. In GQDs, there are two distinct edge classes: armchair and zigzag. The armchair edges (Fig. 3.9A) show a structure akin to a carbyne and triple carbon bonds. Two valence electrons are not shared on any of the zigzag edges (Fig. 3.9B). These edge types have an impact on GQD shape that consequently affects electrical and optical properties. When the edges are identical, the corners are 120°; however, when an armchair edge intersects a zigzag edge,

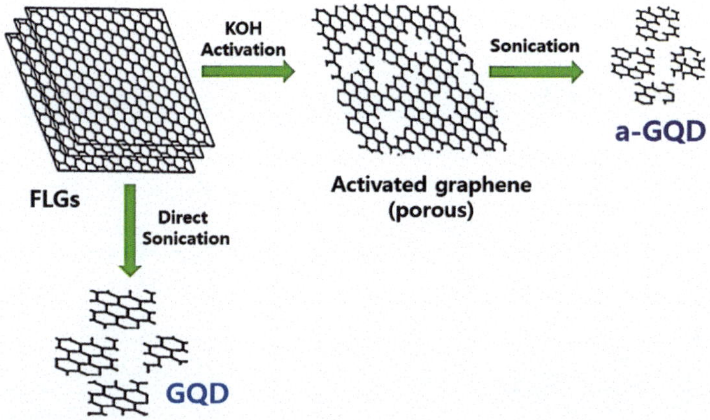

Fig. 3.7 Synthesis of GQDs by direct sonication of graphene sheets. Reprinted with permission from @ Elsevier 45–81 (2020)

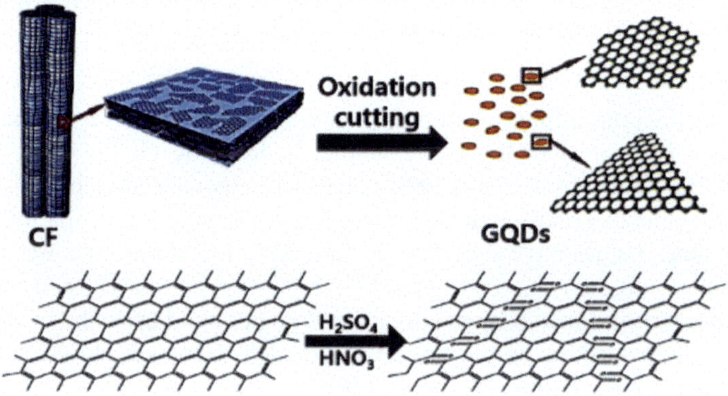

Fig. 3.8 Formation of GQDs from CFs by the cutting process. Reprinted with permission from @ TrAC Trends in Analytical Chemistry 54, 83–102 (2014)

the corners are 90°. A similar effect appeared when the GQD shape is hexagonal, which is achieved by generating six 120° angles, as seen in Fig. 3.9A and B. The rectangular GQDs have both types of edges, resulting in four 90° angles (Fig. 3.9C) [53]. However, the circular or oval shapes can also be determined when the corners cannot be fully developed. When 2D graphene is turned into 0D GQDs, the size becomes smaller than the exciton Bohr radius and their electron distribution is drastically changed [54]. As a result, size-dependent characteristics and quantum effects including quantum confinement and edge effects are induced [55, 56].

GQD electrostatic characteristics were related to their negatively charged carboxyl groups which have been conjugated as π-π linkages [57]. The surface of GQDs shows carboxyl and hydroxyl groups, which enhance their water-soluble nature. The functional groups that contain oxygen aid in the creation of a stable aqueous solution.

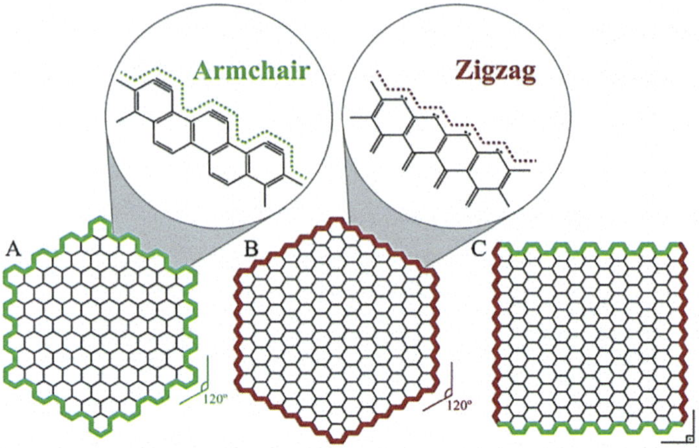

Fig. 3.9 GQD shapes include **A** armchair, **B** zigzag, and **C** hybrid armchair-zigzag GQDs, depending on the type of edge. The magnified images in (**A**) and (**B**) demonstrate the zigzag and armchair edge structures in detail. Reprinted with permission from @ Journal of Water Process Engineering 43, 102,249 (2021)

Nevertheless, the total affinity of the nanomaterial can be altered by adding certain groups to the surface of GQD. For example, hydrophilic GQDs were converted to hydrophobic GQDs by reacting with epoxy ring-opening after being combined with *N*,*N*′-dicyclohexyl carbon diimide [58]. GQD has good biocompatibility and is chemically inert. GQDs have shown to be incredibly low hazardous materials with in vitro studies; this may be owing to their high oxygen content and small size. GQDs are quickly eliminated by the organs and do not build up in the major tissues, according to in vivo studies employing animal models. Not even a high GQD dosage produced a discernible harmful response [59].

3.4.3 Applications of GQDs

GQDs are graphene NPs that combine the characteristics of graphene and QDs. GQDs are regarded as excellent prospects for nano-theranostics applications because of their exceptional biocompatibility and tunable physicochemical characteristics. This has inspired a tremendous amount of research interest in the disciplines of biology and medicine [60]. GO, fullerene, glucose, CNTs, and CFs have been used to prepare GQDs [61]. Physical techniques such as lithography, which can etch graphene to a width of roughly 20 nm, are expensive and inappropriate for the mass production of material [56]. Hydrothermal and electrochemical processes make it simple to make GQDs, but the cost of precursor materials including fullerene, graphite, CNTs, and CFs is normally high [62]. When the size of a semiconductor crystal is reduced to match the exciton Bohr radius of the bulk material, the boundary drastically alters the electron distribution, resulting in size-dependent properties like the band gap and energy relaxation dynamics. Numerous semiconductor materials have demonstrated quantum confinement, which has led to practical applications in photovoltaics, lasers, light-emitting diodes, and bioimaging [20].

GQDs are distinguished by their small size, modifiable surface edges, quantum confinement, and edge effects. They also have outstanding biocompatibility, chemical insensitivity, and minimal cytotoxic reactivity. GQDs' attributes can be altered via surface/edge functionalization or doping to improve their optoelectronic and physicochemical characteristics [63]. It is noteworthy that surface-functionalized GQDs with an organic fluorophore and heteroatom components are being investigated as prospective bases for the energy, healthcare, and environmental fields, among other industrial applications. GQDs are biocompatible, photo-stable, and have superior surface grafting. They also inherit improved thermal, electrical, and mechanical properties from graphene. There are many cutting-edge uses for technology, including taggants for security, anti-counterfeiting, and brand protection applications, bioimaging markers, antibacterial, anti-biofouling, disinfection systems, humidity and pressure sensors, batteries, flash memory devices, photovoltaic devices, and light-emitting diodes [64, 65] (Fig. 3.10).

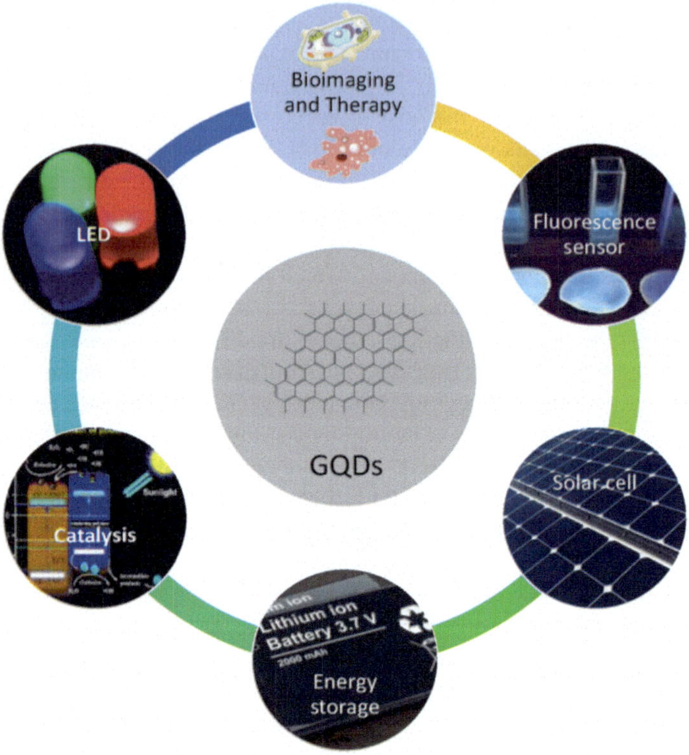

Fig. 3.10 Overview of applications of GQDs

References

1. Wick P, Louw-Gaume AE, Kucki M, Krug HF, Kostarelos K, Fadeel B, Dawson KA, Salvati A, Vazquez E, Ballerini L, Tretiach M, Flahaut E, Gauthier L, Prato M, Bianco A (2014) Classification framework for graphene-based materials. Angew Chem Int Edn 53:7714–7718
2. Deshmukh K, Joshi GM (2015) Embedded capacitor applications of graphene oxide reinforced poly(3,4-ethylene dioxythiophene)-tetramethacrylate (PEDOT-TMA) composites. J Mater Sci Mater Electron 26:5896–5909
3. Thangamani GJ, Deshmukh K, Sadasivuni KK, Chidambaram K, Ahamed MB, Ponnamma D, AlMaadeed MAA, Pasha SKK (2017) Recent advances in electrochemical biosensor and gas sensors based on graphene and carbon nanotubes (CNT) – a review. Adv Mater Lett 8:196–205
4. Kakaei K, Khodadoost S, Gholipour M, Shouraei N (2021) Core-shell polyaniline functionalized carbon quantum dots for supercapacitor. J Phys Chem Solids 148:109753
5. Tian P, Tang L, Teng KS, Lau SP (2018) Graphene quantum dots from chemistry to applications. Mater Today Chem 10:221–258
6. Erickson K, Erni R, Lee Z, Alem N, Gannett W, Zettl A (2010) Determination of the local chemical structure of graphene oxide and reduced graphene oxide. Adv Mater 22:4467–4472
7. Bai H, Li C, Shi G (2011) Functional composite materials based on chemically converted graphene. Adv Mater 23:1089–1115
8. Novoselov KS, Geim AK, Morozov SV, Jiang D, Zhang Y, Dubonos SV, Grigorieva V, Firsov A (2004) Electric field effect in atomically thin carbon films. Science 306:666–669

9. Rao CNR, Sood AK, Subrahmanyam KS, Govindaraj A (2009) Graphene: the new two-dimensional nanomaterial. Angew Chem Int Edn 48:7752–7777
10. Orlita M, Potemski M (2010) Dirac electronic states in graphene systems: optical spectroscopy studies. Semicond Sci Technol 25:063001
11. Suzuki N, Wang Y, Elvati P, Qu ZB, Kim K, Jiang S, Baumeister E, Lee J, Yeom B, Bahng JH, Lee J, Violi A, Kotov NA (2016) Chiral graphene quantum dots. ACS Nano 10:1744–1755
12. Olle M, Ceballos G, Serrate D, Gambardella P (2012) Yield and shape selection of graphene nanoislands grown on Ni(111). Nano Lett 12:4431–4436
13. Sturala J, Luxa J, Pumera M, Sofer Z (2018) Chemistry of graphene derivatives: synthesis, applications, and perspectives. Chemistry 24:5992–6006
14. Ramezani M, Alibolandi M, Nejabat M, Charbgoo F, Taghdisi SM, Abnous K (2019) Graphene-based hybrid nanomaterials for biomedical applications. In: Nurunnabi Md, McCarthy J (eds) Biomedical applications of graphene and 2D nanomaterials. https://doi.org/10.1016/B978-0-12-815889-0.00006-4, Elsevier
15. Wu ZS, Ren W, Gao L, Liu B, Jiang C, Cheng HM (2009) Synthesis of high-quality graphene with a pre-determined number of layers. Carbon 47:493–499
16. Hass J, Heer WAD, Conrad EH (2008) The growth and morphology of epitaxial multilayer graphene. J Phys Condens Matter 20:323202
17. Wang Y, Li X, Zhao X, Xiao L, Zeng H, Sun H (2016) Nonlinear absorption and low threshold multiphoton pumped stimulated emission from all-inorganic perovskite nanocrystals. Nano Lett 16:448–453
18. Yang ST, Wang X, Wang H, Lu F, Luo PG, Cao L, Meziani MJ, Liu JH, Liu Y, Chen M, Huang Y, Sun YP (2010) Carbon dots as nontoxic and high-performance fluorescence imaging agents. J Phys Chem C 113:18110–18114
19. Ghosh D, Sarkar K, Devi P, Kim KH, Kumar P (2021) Current and future perspectives of carbon and graphene quantum dots: From synthesis to strategy for building optoelectronic and energy devices. Renew Sustainable Energy Rev 135:110391
20. Li LS, Yan X (2010) Colloidal graphene quantum dots. J Phys Chem Lett 1:2572–2576
21. Peng J, Gao W, Gupta BK, Liu Z, Romero-Aburto R, Ge L, Song L, Alemany LB, Zhan X, Gao G, Vithayathil SA, Kaipparettu BA, Marti AA, Hayashi T, Zhu JJ, Ajayan PM (2012) Graphene quantum dots derived from carbon fibers. Nano Lett 12:844–849
22. Stachurski J, Tamariz S, Callsen G, Butté R, Grandjean N (2022) Single photon emission and recombination dynamics in self-assembled GaN/AlN quantum dots. Light Sci Appl 11:114
23. Xia C, Zhu S, Feng T, Yang M, Yang B (2019) Evolution and synthesis of carbon dots: from carbon dots to carbonized polymer dots. Adv Sci 6:1901316
24. Ravi Kumar Y, Deshmukh K, Sadasivuni KK, Khadheer Pasha SK (2020) Graphene quantum dot-based materials for sensing, bio-imaging and energy storage applications: a review. RSC Adv 10:23861
25. Jin SH, Kim DH, Jun GH, Hong SH, Jeon S (2012) Tuning the photoluminescence of graphene quantum dots through the charge transfer effect of functional groups. ACS Nano 7:1239–1245
26. Feng Y, Zhao J, Yan X, Tang F, Xue Q (2014) Enhancement in the fluorescence of graphene quantum dots by hydrazine hydrate reduction. Carbon 66:334–339
27. Luo P, Ji Z, Li C, Shi G (2013) Aryl-modified graphene quantum dots with enhanced photoluminescence and improved pH tolerance. Nanoscale 5:7361–7367
28. Yokoyama A, Yoshida M, Ishii A, Kato YK (2014) Giant Circular dichroism in individual carbon nanotubes induced by extrinsic chirality. Phys Rev X 4:011005
29. Hong G, Diao S, Antaris AL, Dai H (2015) Carbon nanomaterials for biological imaging and nanomedicinal therapy. Chem Rev 115:10816
30. Chua CK, Sofer Z, Šimek P, Jankovský O, Klímová K, Bakardjieva S, Kučková HŠ, Pumera M (2015) Synthesis of strongly fluorescent graphene quantum dots by cage-opening buckminsterfullerene. ACS Nano 9:2548–2555
31. Low T, Avouris P (2014) Graphene Plasmonics for Terahertz to Mid-Infrared Applications. ACS Nano 8:1086–1101

32. Nesakumar N, Srinivasan S, Alwarappan A (2022) Graphene quantum dots: synthesis, properties, and applications to the development of optical and electrochemical sensors for chemical sensing. Microchim Acta 189:258
33. Valappil MO, Pillai VK, Alwarappan S (2017) Spotlighting graphene quantum dots and beyond: synthesis, properties and sensing applications. Appl Mater Today 9:350–371
34. Chen W, Lv G, Hu W, Li D, Chen S, Dai Z (2018) Synthesis and applications of graphene quantum dots: a review. Nanotechnology Rev 7:157
35. Iannazzo D, Pistone A, Salamò M, Galvagno S, Romeo R, Giofré SV, Branca C, Visalli G, Di Pietro A (2017) Graphene quantum dots for cancer targeted drug delivery. Int J Pharm 518:185–192
36. Reina G, González-Domínguez JM, Criado A, Vázquez E, Bianco A, Prato M (2017) Promises, facts and challenges for graphene in biomedical applications. Chem Soc Rev 46:4400–4416
37. Ding H, Zhang F, Zhao C, Lv Y, Ma G, Wei W, Tian Z (2017) Beyond a carrier: graphene quantum dots as a probe for programmatically monitoring anti-cancer drug delivery, release, and response. ACS Appl Mater Interf 9:27396–27401
38. Fan Z, Li S, Yuan F, Fan L (2015) Fluorescent graphene quantum dots for biosensing and bioimaging. RSC Adv 5:19773–19789
39. Zuo W, Tang L, Xiang J, Ji R, Luo L, Rogee L, Lau SP (2017) Functionalization of graphene quantum dots by fluorine: preparation, properties, application, and their mechanisms. Appl Phys Lett 110:221901
40. Zhao XE, Lei C, Gao Y, Gao H, Zhu S, Yang X, You J, Wang H (2017) A ratiometric fluorescent nanosensor for the detection of silver ions using graphene quantum dots. Sens Actua B 253:239–246
41. Liu R, Wu D, Feng X, Müllen K (2011) Bottom-up fabrication of photoluminescent graphene quantum dots with uniform morphology. J Am Chem Soc 133:15221–15223
42. Tian L, Ghosh D, Chen W, Pradhan S, Chang X, Chen S (2009) Nanosized carbon particles from natural gas soot. Chem Mater 21:2803–2809
43. Schwierz F (2010) Graphene transistors. Nat Nanotechnol 5:487–496
44. Nair RR, Blake P, Grigorenko AN, Novoselov KS, Booth TJ, Stauber T, Peres NMR, Geim AK (2008) Fine structure constant defines visual transparency of graphene. Science 320:1308
45. Song L, Shi J, Lu J, Lu C (2015) Structure observation of graphene quantum dots by single-layered formation in layered confinement space. Chem Sci 6:4846–4850
46. Ma X (2021) The use of graphene quantum dots as detection elements in nanomaterials-based sensors for forensic applications, Thesis, KTH Royal Institute of Technology School of Engineering Sciences, Stockholm, Sweden
47. Frigerio C, Ribeiro DSM, Rodrigues SSM, Abreu VLRG, Barbosa JAC, Prior JAV, Marques KL, Santos JLM (2012) Application of quantum dots as analytical tools in automated chemical analysis: a review. Anal Chim Acta 735:9–22
48. Benítez-Martínez S, Valcárcel M (2015) Graphene quantum dots in analytical science. TrAC Trends in Anal Chem 72:93–113
49. Dong Y, Shao J, Chen C, Li H, Wang R, Chi Y, Lin X, Chen G (2012) Blue luminescent graphene quantum dots and graphene oxide prepared by tuning the carbonization degree of citric acid. Carbon 50:4738–4743
50. Pan DY, Zhang JC, Li Z, Wu MH (2010) Hydrothermal route for cutting graphene sheets into blue-luminescent graphene quantum dots. Adv Mater 22:734–738
51. Ye R, Xiang C, Lin J, Peng Z, Huang K, Yan Z, Cook NP, Samuel ELG, Hwang CC, Ruan G, Ceriotti G, Raji ARO, Martı AA, Tour JM (2013) Coal as an abundant source of graphene quantum dots. Nat Commun 4:2943
52. Haque E, Kim J, Malgras V, Reddy KR, Ward AC, You J, Bando Y, Md. Hossain SA, Yamauchi Y (2018) Recent advances in graphene quantum dots: synthesis, properties, and applications. Small Methods 2:1800050
53. Kundu S, Pillai VK (2019) Synthesis and characterization of graphene quantum dots. Phys Sci Rev 5:1–35

54. Valencia AM, Caldas MJ (2017) Single vacancy defect in graphene: Insights into its magnetic properties from theoretical modeling. Phys Rev B 96:125431
55. Li L, Wu G, Yang G, Peng J, Zhao J, Zhu JJ (2013) Focusing on luminescent graphene quantum dots: current status and future perspectives. Nanoscale 5:4015–4039
56. Kim S, Hwang SW, Kim MK, Shin DY, Shin DH, Kim CO, Yang SB, Park JH, Hwang E, Choi SH, Ko G, Sim S, Sone C, Choi HJ, Bae S, Hong BH (2012) Anomalous behaviors of visible luminescence from graphene quantum dots: Interplay between size and shape. ACS Nano 6:8203–8208
57. Tang X, Zhang S, Zhou H, Zhou B, Liu S, Zaixing Y (2020) The role of electrostatic potential polarization in the translocation of graphene quantum dots across membranes. Nanoscale 12:2732–2739
58. Cui Y, Liu L, Shi M, Wang Y, Meng X, Chen Y, Huang Q, Liu C (2024) A review of advances in graphene quantum dots: from preparation and modification methods to application C 10:7
59. Tabish TA, Scotton CJ, Ferguson DCJ, Lin L, van der Veen A, Lowry S, Ali M, Jabeen F, Ali M, Winyard PG, Zhang A (2018) Biocompatibility and toxicity of graphene quantum dots for potential application in photodynamic therapy. Nanomedicine 13:1923–1937
60. Joshi PN, Kundu S, Sanghi SK, Sarkar D (2016) graphene quantum dots - from emergence to nanotheranostic applications. In: System. Sezer AD (ed) Chapter 7 in smart drug delivery. IntechOpen, https://doi.org/10.5772/61932
61. Lu J, Yeo PSE, Gan CK, Wu P, Loh KP (2011) Transforming C60 molecules into graphene quantum dots. Nat Nanotechnol 6:247–252
62. Liu F, Jang MH, Ha HD, Kim JH, Cho YH, Seo TS (2013) Facile synthetic method for pristine graphene quantum dots and graphene oxide quantum dots: origin of blue and green luminescence. Adv Mater 25:3657–3662
63. Daniel Thangadurai T, Manjubaashini N, Nataraj D, Gomes V, Lee YI (2022) A review on graphene quantum dots, an emerging luminescent carbon nanolights: Healthcare and Environmental applications. Mater Sci Eng B 278:115633
64. Ravi PV, Daniel Thangadurai T, Nehru K, Lee YI, Nataraj D, Thomas S, Kalarikkal N, Jose J (2020) Surface and morphology analyses, and voltammetry studies for electrochemical determination of cerium(III) using a graphene nanobud-modified-carbon felt electrode in acidic buffer solution (pH 4.0 $\pm$ 0.05). RSC Adv 10:37409–37418
65. Tabish TA, Hayat H, Abbas A, Narayan RJ (2022) Graphene quantum dots-based electrochemical biosensing platform for early detection of acute myocardial infarction. Biosensors 12:77

Chapter 4
Synthesis of Graphene Quantum Dots and Their Composites

4.1 Introduction

The remarkable physical and chemical properties of graphene, a two-dimensional (2D) nanomaterial containing sp^2 hybridized carbon atoms in a single layer, have drawn a lot of interest. These include good thermal and chemical stability, high electronic mobility of 106 cm 2 V^{-1} s^{-1} [1], high surface area, and superior mechanical flexibility. Studies in physics, chemistry, and materials science will find this to be very appealing and promising. Graphene's zero-bandgap semiconductor status, however, restricts its use in optoelectronic and electronic fields. In pristine graphene, luminescence is absent because of the absence of the bandgap [2]. However, due to the edge effects and quantum confinement of these graphene derivatives, the bandgap can be introduced in GNRs and GQDs [3].

Out of all the graphene derivatives, GQDs are a class of zero-dimensional (0D) materials with typical widths of less than 20 nm [4, 5]. They were regarded as a recently emerging material [6]. The potential for applications in nanoscale optoelectronics is increased by these, which demonstrate distinct quantum confinement and edge effects [7]. Additionally, GQDs exhibit promising properties for nanotechnology applications, including biocompatibility [8], low toxicity, excellent photostability, tunable photoluminescence, nanosecond lifetime, multi-colored emission, excellent dispersibility, high water solubility, and ease of surface grafting [9]. Top-down and bottom-up processes can be used to categorize the current GQD synthesis techniques. According to [10], the synthesis of GQDs is a bottom-up method that necessitates specific organic materials and complex reaction steps, making condition optimization challenging. Consequently, the top-down method which involves breaking up big carbon material blocks into smaller ones is recommended. The method requires a lot of carbon materials, which are inexpensive and simple to obtain. It is also a fairly simple method that makes synthesizing GQDs easy [11].

N. Manjubaashini et al., *Graphene Quantum Dots*, Materials Horizons: From Nature to Nanomaterials, https://doi.org/10.1007/978-981-97-5722-0_4

4.2 Synthesis of GQD Through Various Techniques

GQD synthesis has several advantages, including adjustable optical and electrical characteristics, reduced size and surface area, customized functionalization capacity, compatibility, and adaptability. The development of GQD for a range of industrial, biomedical, and energy-related uses is attracting more attention and funding as a result of the GQD synthesis techniques' ongoing exploration of new avenues and applications. The application, target size, and characteristics of the GQD influence the choice of GQD synthesis techniques. To enhance the effectiveness, scalability, and characteristics of GQDs for diverse technological applications, researchers are still investigating novel approaches. Chemical exfoliation and laser ablation can be performed using top-down methods. Additionally, the following techniques can be used intelligently: chemical vapor deposition, hydrothermal, solvothermal, microwave-assisted technique, electrochemical synthesis, biosynthesis, green synthesis, ultrasonication, and temple-assisted method. This chapter presents a selection of synthesis methodologies with schematics from among the many GQD synthesis techniques; the remaining methodologies have been tabulated (Tables 4.1 and 4.2).

4.2.1 Electrochemical Exfoliation Process

The OH^- ion formation, OH^- ion intercalation, oxygen generation, and GQDs obtained are shown in Fig. 4.1 [45]. An internal view of the graphite rod's core is captured by transmission electron microscopy (TEM). An example of a typical graphite structure is 0.35 nm separating layers [46]. An additional TEM image displays the graphite rod's surface following the electrochemical exfoliation process, highlighting the locations where the GQDs were removed and peeled off. According to Shinde et al. [47], these sites demonstrate that the lattice constant of graphene or GQDs is 0.23 nm or the distance between layers. The distribution of the obtained GQDs produced by the electrochemical exfoliation process is displayed in the final TEM image [48].

4.2.2 Homemade Hydrothermal Process

H K Kondaveeti demonstrates a hydrothermal reactor for a homemade device made from repurposed parts of vintage instruments [49]. A LabVIEW proportional–integral–derivative (PID) controller that controls the amount of electrical power to be delivered and activates a TRIAC as a power switch via the Arduino board allows the user to modify the heating rate, temperature, and residence time (Fig. 4.2). A C program written by hand facilitates communication between the Arduino board

Table 4.1 Analysis of resultant GQDs using various synthesis approaches

Material	Method	Property	Application	Ref. no.
GQD	Microwave-assisted pyrolysis method	Low cellular toxicity, excellent biocompatibility, and high photostability	Fe^{+} ion sensing	[12]
GQD	Oxidative fragmentation	Antioxidant activity	–	[13]
GQD	Wet chemical method	–	Light-emitting devices	[14]
GQD	Pyrolysis	A high quantum yield of 30.9%	Biomedical applications	[15]
GQD	Microwave-assisted synthesis method	Quantum yields up to 35%	White light-emitting diodes and cell-imaging fields	[16]
GQD	Solvothermal method	Strong blue emission at 15% quantum yield	–	[17]
GQD	Hydrothermal method	The lowest limit of detection, superior QY	Optochemical sensors to detect Fe^{3+}	[18]
GQD	Wet chemical method	Quantum yield 20%	Bioimaging, biomedicine, photovoltaics, and optoelectronics	[19]
GQD	Oxidation method	Low cost, rapid, and sustainable	Antioxidant activity	[20]
Biomass-derived GQD	Pyrolysis	Excellent stability, and non-toxic	Mercury detection	[21]

and the LabVIEW interface. The Arduino board gathers the temperature and pressure data as a function of time, transmits it to the LabVIEW interface for graphical display on the computer screen, and saves it in a log file for further data processing [50].

4.2.3 *Pyrolysis Method*

Pyrene (2 g, TCI, purity > 98%) was nitrated into trinitropyrene in hot HNO_3 (160 ml) at 80 °C while refluxing and stirring for 12 h as part of a standard process for the synthesis of OH-GQDs. The mixture was allowed to come to room temperature before being diluted with 1 L of deionized (DI) water and filtered through a 0.22 µm membrane to eliminate the acid. After two hours of ultrasonication (500 W, 40 kHz),

Table 4.2 Summary of various synthesis techniques used for GQD composite preparation

Material	Method	Property	Application	Ref. no.
GQD coupled to Au nanoparticles	Hydrothermal method	–	–	[22]
GQD/Ag nanocomposite	Wet chemical method	Low limit of detection	Glucose detection	[23]
GQD/cobalt ferrite nanocomposite	Carbonization	Superparamagnetic property	Biomedical applications	[24]
Sulfur incorporated **GQD**	Pulsed laser ablation	–	–	[25]
Fluorinated GQD	Hydrothermal method	Upconversion PL	Environmental and energy application	[26]
Nitrogen-doped GQD	Microwave method	–	–	[27]
Nitrogen-doped GQD	Wet chemical method	Low detection limit	–	[28]
GQD-stabilized AuNPs	Wet chemical method	Excellent catalytic performance	Industrial wastewater treatment	[29]
Nitrogen-doped GQD	Pyrolysis method	Good biocompatibility, high quantum yield of 12%	Copper ion detection in living cells	[30]
Amine-functionalized GQD	Hydrothermal method	Quantum yield of 51%	Immunosensing of Ferritin	[31]
Carboxyl-functionalized GQD	Microwave irradiation conditions	Simple procedure, excellent yields	–	[32]
EDA-GQD	Wet chemical method	Low toxicity, improve and modulate the photoluminescence of GQD	Bioimaging applications	[33]
Boron and sulfur co-doped GQD	Thermal pyrolysis	High selectivity, low cost	Detection of dopamine	[34]
GQD-ferrocene (FC)-HA formulation	Wet chemical and freeze-drying	Targeted delivery, imaging, and cancer-specific treatment capabilities	Nanotherapeutic approach	[35]
GQD-modified Ag_3PO_4	Wet chemical method	–	–	[36]
His-GQD/LDH	Microwave method	–	Supercapacitor	[37]
GQDs-SiNP nanocomposite	Ultrasonic aided synthesis	Selectivity, reproducibility, long-term stability	Detection of glutathione	[38]

(continued)

Table 4.2 (continued)

Material	Method	Property	Application	Ref. no.
Lignin sulfonate/graphene quantum dots composites	Wet chemical method	Excellent sensitivity and selectivity	Detection of Fe^{3+} ions	[39]
GQDs@ZrO_2	Hydrothermal method	Enhanced biosensing parameters	Electrochemical Immunosensing	[40]
Sulfur-doped GQD	pyrolysis	selectivity, simplicity, economics, and high sensitivity	Detection of quercetin	[41]
HFHPB-GQD hexafluorohydroxypropanyl benzene	Wet chemical method	High sensitivity	Molecular detection	[42]
Nitrogen and phosphorus co-doped GQD	Carbonization	high fluorescence quantum yield, strong two-photon upconversion, high photostability, and good biocompatibility	Cellular imaging	[43]
NS-GQD	Wet chemical method	High PL quantum yield	Deep-tissue imaging	[44]

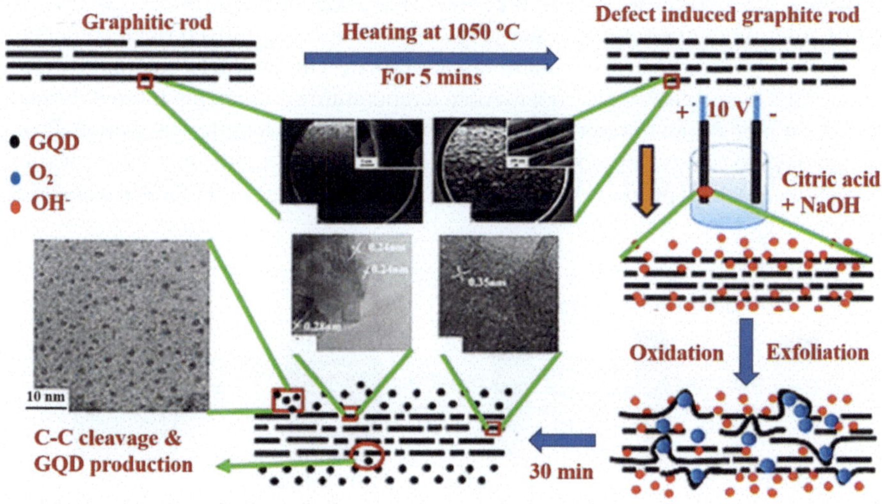

Fig. 4.1 Electrochemical exfoliation of defect-induced graphite rod for the production of GQD

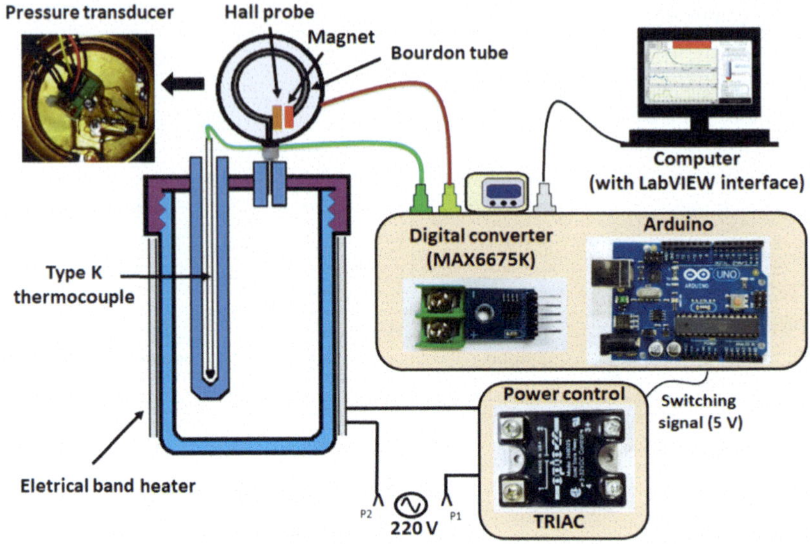

Fig. 4.2 Working scheme of the homemade hydrothermal reactor

the yellow 1,3,6-trinitropyrene (3.0 g, 90% yield) was distributed in a 0.6 l, 0.2 M NaOH solution containing DI water (Fig. 4.3). After being moved to an autoclave, the suspension was heated for ten hours at 200 °C. The product containing water-soluble GQDs was allowed to cool to room temperature before being filtered through a 0.22 μm membrane to remove the insoluble carbon product. It was then dialyzed for two days to remove the sodium salt and unfused small molecules. For structural characterization and property measurement, the purified black OH-GQDs were dried at 80 °C [51].

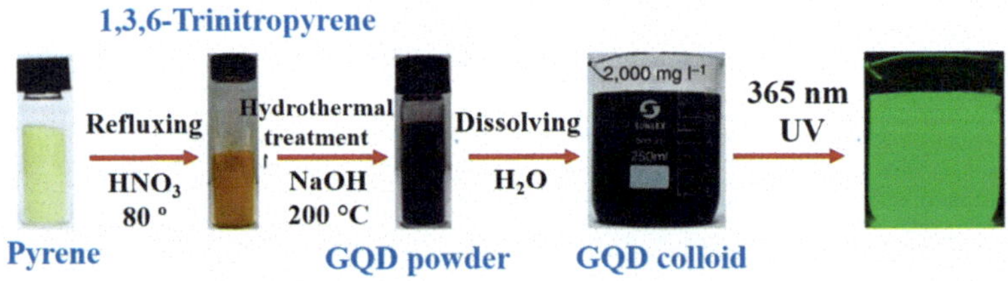

Fig. 4.3 Schematic for the preparation of GQDs

4.2.4 Hydrothermal Method

A single, quick, and efficient hydrothermal process uses H_2O_2 to help produce GQDs from GO (Fig. 4.4a). According to [52], the as-synthesized GQDs from glucose molecules demonstrated many superior qualities, including their highly water-dispersible, uniformly distributed size (4–8 nm), and excellent optical properties (Fig. 4.4b). Excitation-dependent photoluminescent GQDs, GNSs with various photoluminescence quantum yields (PLQYs), and CNSs lacking any photoluminescence were obtained by varying the hydrothermal reaction time [53]. AFM, or atomic force microscopy, reveals single or double layers. Transmission electron microscopy (TEM) images revealed the high PLQY of 22.7% for the reaction time of 8 h and the sizes of the GQDs as <10 nm for the reaction time of 2 h and the homogeneous carbon nanospheres (CNSs~560 nm) for the reaction time of 30 h [2]. A hydrothermal one-pot synthesis process was created to produce graphene oxide quantum dots (GOQDs). H_2O_2 is used as the oxidant and toner waste is the precursor (Fig. 4.4c). Strong acids or difficult purification procedures are not needed for the 4 h synthesis, which also doesn't produce any metal ions that are bad for the environment [54]. T Renbing describes a straightforward, one-step hydrothermal process that uses GO and hydrogen peroxide to react in an environment of high pressure and high temperature (HPHT) to create GQDs (Fig. 4.4d). Under HPHT conditions, hydroxyl radicals (-OH) were produced using hydrogen peroxide, acting as chemical "scissors" to break down GO sheets. It is important to note that as the reaction progresses, different PLs are obtained by different sizes of GQDs [55].

4.2.5 Exfoliation Method

A novel method for creating GQD from GO with tuneable physical properties uses supercritical water and p-tetrasulfonic acid calix[4]arene (SCX4) [56].

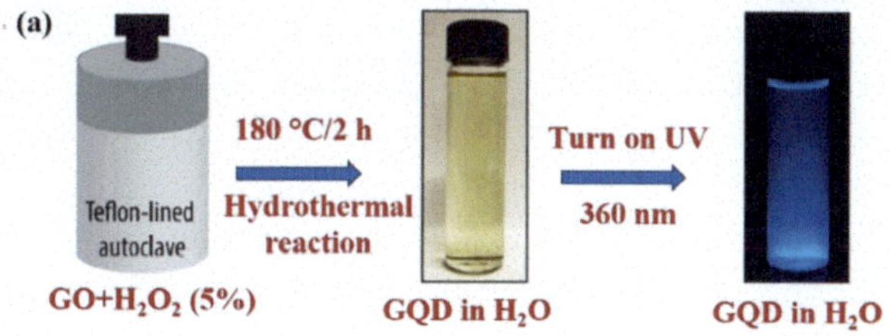

Fig. 4.4 **a** Formation of GQDs from GO. Reprinted with permission from @ ACS Appl Bio Mater 1, 452–461 (2018). **b** Schematic diagram of GQD from glucose molecules using a hydrothermal process. **c** Preparation of GOQDs from waste toner; **d** multi-color GQD were prepared by reacting GO with H_2O_2 in hydrothermal setup

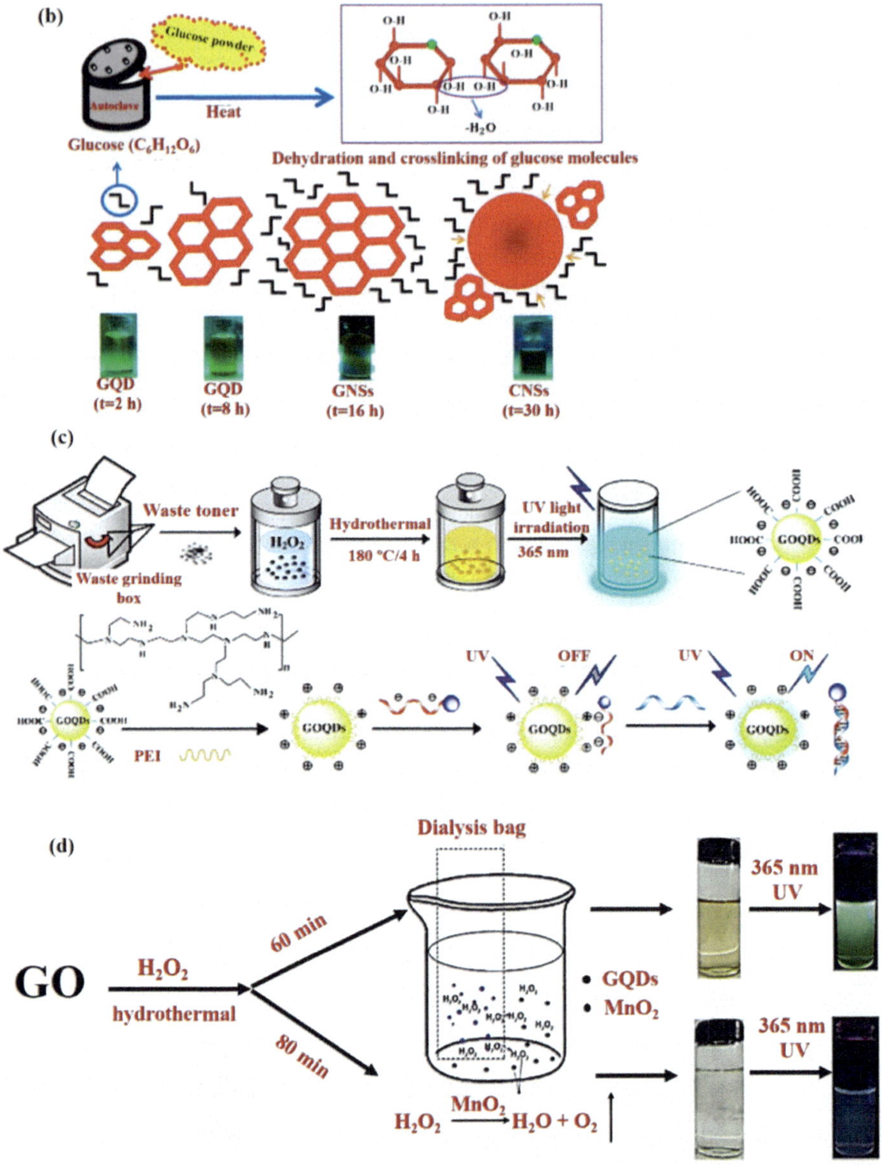

Fig. 4.4 (continued)

This process reduces the amount of hazardous or toxic reagents used, limits the use of complex and time-consuming processes, and can be used to reduce reaction times to a matter of seconds (Fig. 4.5a). During this procedure, a low-temperature KOH solution was pumped to combine with a pre-mixed GO and SCX4 aqueous solution at a T-junction ("T"). Following this mixture's exposure to supercritical water (450

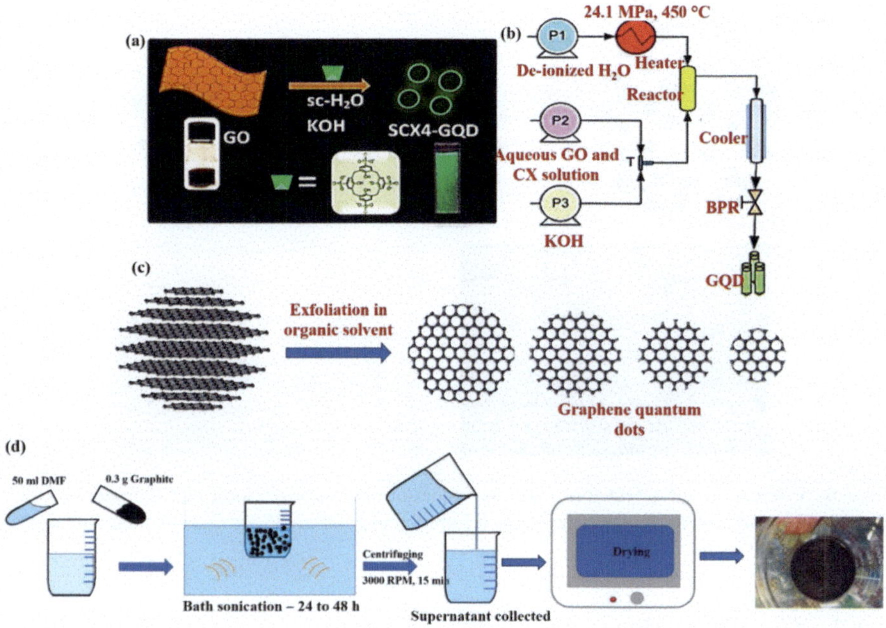

Fig. 4.5 **a** Synthesis of GQD from GO using Calix[4]arene tetrasulfonic acid (SCX4), **b** CHFS reactor utilized for the synthesis of GQDs, **c** synthesis of GQD by exfoliation technique, and **d** pictorial representation of GQDs. Reprinted with permission from @ Materials Today: Proceedings (2023)

°C, 24.1 MPa) in a current "Reactor", GQD formation took place [57]. Sonication was used in the liquid phase exfoliation (LPE) process to create the synthesized GQDs (Fig. 4.5b). The optical characteristics show that GQDs have an energy gap of roughly 3.086 eV and are light blue luminescent. The hydrophobicity of the prepared GQDs may be attributed to the amide groups' uniform functionalization (Fig. 4.5c), which greatly improves moisture stability [58].

4.2.6 Oxidation Method

Owing to their well-defined dimensions, fullerene C_{60} is used as a starting material to produce very small GQD (~2–3 nm). Fullerene C_{60} underwent the oxidation, cage-opening, and fragmentation processes when treated with a strong acid and chemical oxidant mixture. According to [59], the quantum dots showed strong luminescence properties, with the highest intensity at 460 nm under 340 nm excitation wavelength. They remained fully dispersed in aqueous suspension (Fig. 4.6a–b).

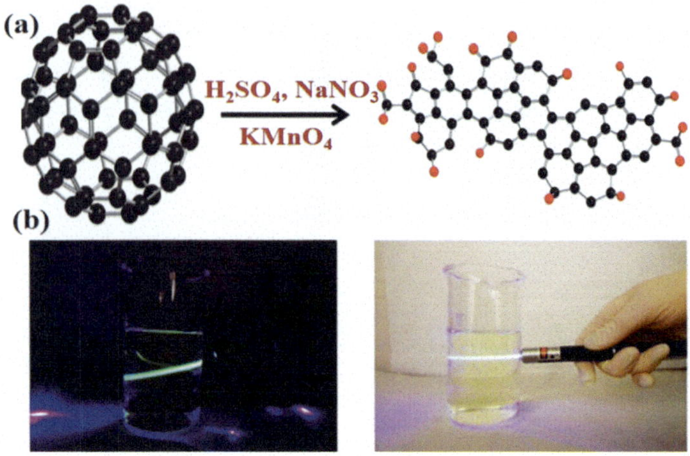

Fig. 4.6 **a** Schematic of GQD formation by oxidation and cage-opening of fullerene C_{60} with strong acid and chemical oxidant treatment; **b** luminescence of graphene QDs excited with a blue laser pointer (405 nm)

4.2.7 Green Synthesis Method

Weifeng C presents a green and efficient hydrothermal one-pot method for the synthesis of GQD. The natural polymer starch is employed as a novel precursor in the synthesis of GQDs for the first time (Fig. 4.7). The only reactants are starch and water; no strong acids or other oxidizers are used, and no metal impurities are added. Furthermore, the only products are carbide precipitate, water, and GQDs. Centrifugation makes it easy to separate the precipitate from the GQD solution, so complex post-processing is not needed. According to [60], the reaction happens quickly and the entire synthetic process takes about 120 min. Using only D-fructose powder as precursors and hydrochloric acid in the presence of deionization, blue–green–yellow photoluminescent GQDs were successfully synthesized by the dehydration method at a specific fixed temperature. When a GQD is functionalized by GQD-OOH, its bandgap drops from 3.4 to 1.51 eV [61].

4.2.8 Electrochemical Approach for GQD Synthesis

Direct electrochemical preparation of GQDs with a consistent 3–5 nm size from graphene oxide (GO) using non-aqueous propylene carbonate and $LiClO_4$ media. Through prolonged oxidation of GO at 1.05 V for various times, followed by reduction, we were able to obtain discrete GQD particles (Fig. 4.8a). Unzipping through the GO defect sites may be the likely mechanism underlying the formation of GQDs [62].

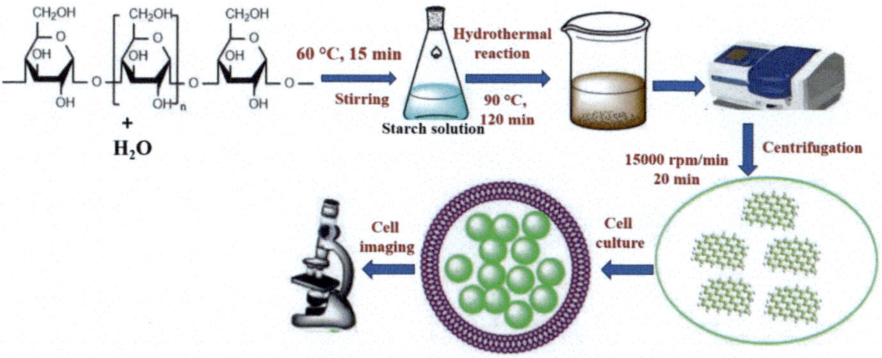

Fig. 4.7 Schematic illustration of the green route of GQDs synthesis

Simultaneous, easy, and simple GQD preparation is shown. GQDs show a homogeneous and uniform particle distribution in addition to stunning lattice fringe patterns (Fig. 4.8b. Additionally, these fluorescent GQDs are investigated for bio-imaging applications; all of the GQDs investigated in this work exhibit low cytotoxicity, good biocompatibility, and bright, intense fluorescence emission [63].

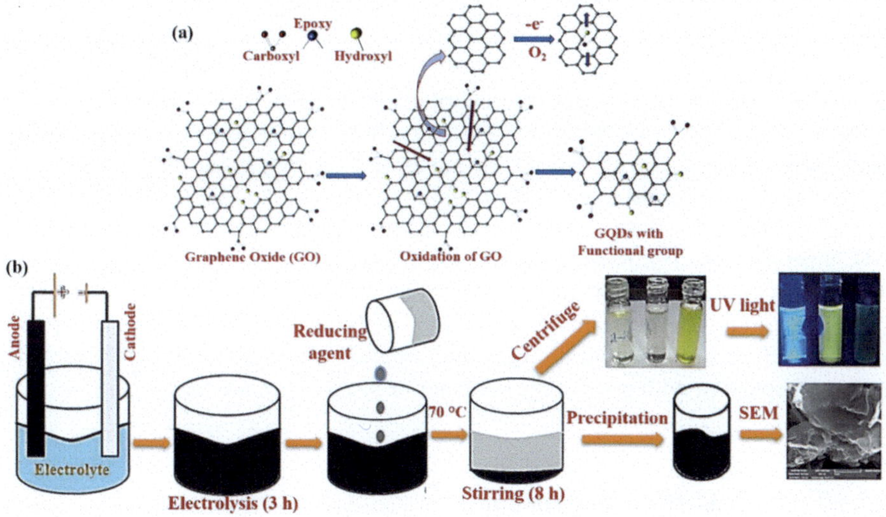

Fig. 4.8 **a** Preparation of GQD from GO using an electrochemical technique; **b** visual depiction of the suggested approach for the concurrent preparation of GQD

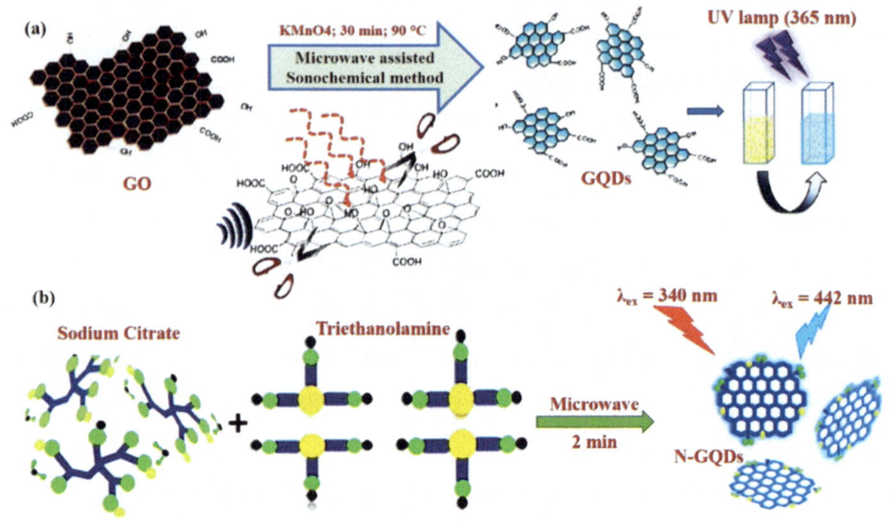

Fig. 4.9 **a** A sonochemical combined microwave technique for GQD synthesis; **b** doped GQD formation by microwave synthesis

4.2.9 Microwave Technique

A quick, environmentally friendly, one-pot synthesis of size-tunable GQDs from GO using a sonochemical technique alternates between microwave heating bursts of 90 °C. In just 30 min, under acid-free conditions, the GQDs were created by oxidatively cutting GO with $KMnO_4$ acting as an oxidizing agent (Fig. 4.9a).

High quality and good QY (23.8%), high product yield (>75%), and lower cytotoxicity (tested up to 1000 μg/mL) were all displayed by the synthesized GQDs [64]. Nitrogen-doped graphene quantum dots (N-GQDs) with high fluorescence were produced by heating sodium citrate and triethanolamine in a microwave (Fig. 4.9b). The as-prepared N-GQDs demonstrated good stability and pH-sensitive characteristics, as well as significant bright blue fluorescence with a QY of 8% and excellent uniform dispersion with an average diameter of roughly 5.6 nm [65].

4.2.10 Ball-Milling Technique

Meng-Jey Y describes a low-cost ball-milling technique for creating GQDs by combining sodium carbonate as the intercalation agent with carbon nanocapsules (20 nm) as the carbon source (Fig. 4.10). The findings demonstrate that the suggested exfoliation process yields GQDs with diameters ranging from 1.0 to 4.0 nm as well as a very small number of grafted functional groups that glow blue when exposed

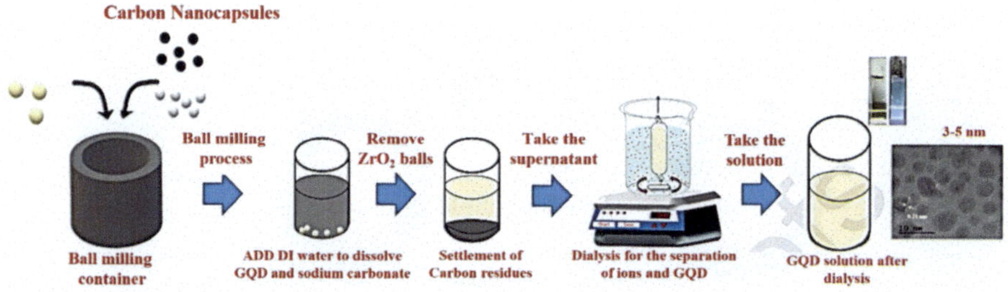

Fig. 4.10 Formation of GQDs by ball-milling technique. Reprinted with permission from @ Inorganic Chemistry Communications 119, 108,061 (2020)

to UV light. About 5% is the QY. According to Meng-Jey et al. [66], the suggested method has a lot of potential for producing GQDs from inexpensive raw materials.

The numerous parameters that were examined for the as-prepared GQDs that are synthesized using different techniques are compiled in Table 4.1.

4.3 Preparation of GQD Composites

This work presents a novel supramolecular hybrid comprising mercaptopyridine-substituted zinc phthalocyanine (Pc)-Au@Ag nano-alloys and polyethyleneimine-functionalized GQD (PEI-GQDs), as reported by Achadu (Fig. 4.11a). Using PEI-GQDs as the fluorescence switching signal probe, the hybrid was utilized as a dual fluorescence nanoprobe for the detection of Hg^{2+} and biothiol in an aqueous solution [67]. Using wet chemical sensing, a functional hybrid of GQDs and cobalt pyrene-derivatized phthalocyanine (CoPc) was presented as a sensitive nanoprobe (Fig. 4.11b). Upon non-covalent (π–π stacking) hybrid formation with CoPc, the fluorescence of GQDs was quenched through a potential energy transfer pathway [68].

Using ammonia as a nitrogen source and high-softening point asphalt as a precursor, R Wang et al. demonstrated the synthesis of amino-functionalized GQDs (af-GQDs) with yellow fluorescence. The mechanism of fluorescence enhancement by nitrogen-containing functional groups is shown in Fig. 4.12a. Following the amino-hydrothermal treatment, the photoluminescence QY of the resulting GQDs increased from 4.7 to 13.8% [69]. J Shen has developed an easy and efficient method for synthesizing graphene quantum dots using glucan as a precursor for cell imaging and as a deoxidizer. Without the need for additional reductants, GQD and Au (III) salts are successfully mixed to create AuNPs (Fig. 4.12b). These GQDs have a far weaker reducing driving force than powerful reducing agents like $NaBH_4$. The average size of the as-synthesized AuNPs is 15 nm, which is significantly larger [70]. Citric acid and tris(hydroxymethyl) 10 aminomethane are pyrolyzed in one step to create highly fluorescent nitrogen-doped graphene quantum dots (N-GQDs), as demonstrated by L

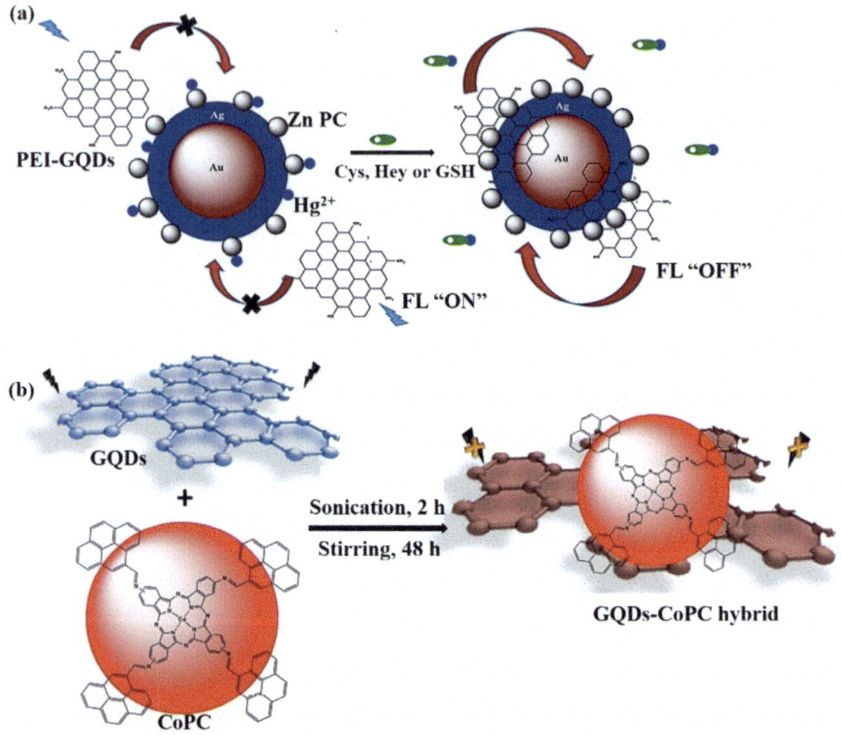

Fig. 4.11 **a** Formation scheme for GQD-ZnPC-Au@Ag nanoparticles for sensing. Reprinted with permission from @ Dyes and Pigments 145, 189–201 (2017). **b** GQD-CoPc supramolecular hybrid synthesis process. Reprinted with permission from @ Dyes and Pigments 160, 328–335 (2019)

Lin's simple bottom-up technique. Under 365 nm UV light excitation, the produced N-GQDs showed strong blue fluorescence with a high QY of 59.2% [71].

Ngoc ANT created a straightforward, one-step synthesis process for producing a gold nanoparticle (AuNPs) @sulfur-doped GQD (Au@S-GQD) with tunable size for the sensitive and targeted detection of 4-nitrophenol in water and wastewater with a complex matrix (Fig. 4.13a). The mean diameter of two to six-layered S-GQD is 4.0 ± 0.5 nm, and their particle sizes range from 1 to 7 nm. To create Au@S-GQD nanocomposites and the 4-aminophenol absorption signal at 307 nm, the S-GQD acts as a reducing agent for both Au ions and 4-nitrophenol reduction [72]. To create GQD–europium (III) complex composites, Y Liu first non-covalently bonded GQDs to the chelating ligands 1,10-phenanthroline (Phen) and dibenzoylmethane (DBM), and then coordinated to Eu (III) (Fig. 4.13b). The resultant composites exhibit high color purity red fluorescence and are well soluble in water. The composites exhibit fluorescence in aqueous solutions when excited at 365 nm, with a peak at 613 nm and a QY of 15.5% [73]. By directly pyrolyzing citric acid, Pavithra R created a surfactant-free GQD solution, which is an equimolar concentration of DAN in

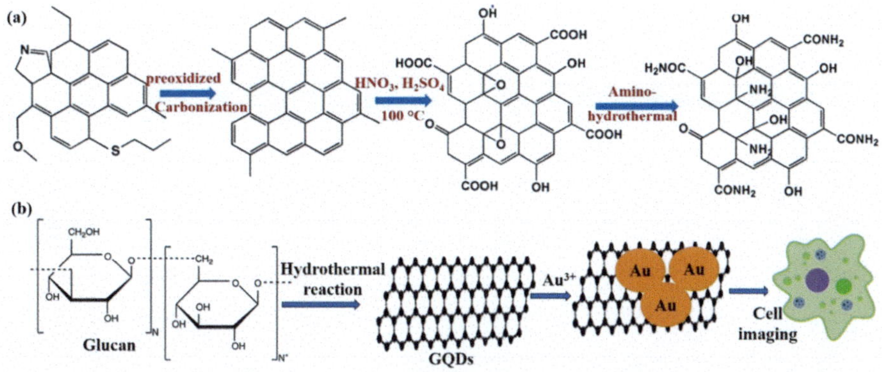

Fig. 4.12 **a** Illustration of amino-functionalized GQD by pyrolysis method. Reprinted with permission from @ Applied Surface Science 467–468, 446–455 (2019). **b** Scheme of GQD-Au composite using hydrothermal method

methanol under the right circumstances (Fig. 4.13c). The formation of DAN-GQDs at room temperature was confirmed by the appearance of a brown color [74].

Wang T used straightforward cross-linking to create a nitrogen-doped graphene quantum dots (NGQDs)/poly(vinyl alcohol) (PVA) composite membrane (NGQDs-PVA) with strong adsorption selectivity and fluorescence quenching for Ag^+ (Fig. 4.14a). The material displayed quantum dot fluorescence characteristics, as demonstrated by the uniform distribution of NGQDs in PVA as revealed by the SEM characterization [75]. L Linbo described a single-step thermolysis method that used glycine as a carbon and nitrogen source to create highly luminescent nitrogen-doped and amino acid functionalized GQD (NA-GQDs) (Fig. 4.14b). With a QY of about 16% and maximum excitation and emission wavelengths of 380 and 450 nm, respectively, the NA-GQDs exhibit an excitation wavelength-dependent fluorescence [76]. GQDs/GO hybrid composites with visible-range emission are prepared in one pot using heteroatom doping and hydroxyl-radical-induced GO degradation. Hydroxyl radicals result from the dissociation of H_2O_2 mediated by NH_4OH and thiourea. Small-sized GO sheets and GQDs are produced when hydroxyl radicals are applied to GO. These particles then self-assemble to form GQDs/GO through strong π–π interactions. For instance, GO reacts with a mixture of NH_4OH and H_2O_2 for 40, 120, and 270 min to produce GQDs that emit yellow, green, and blue light. On the other hand, GO is incubated with a mixture of thiourea and H_2O_2 to produce GQDs that emit red light (R-GQDs/GO) [77].

The various techniques used for the synthesis of GQDs-based composite are summarized in Table 4.2.

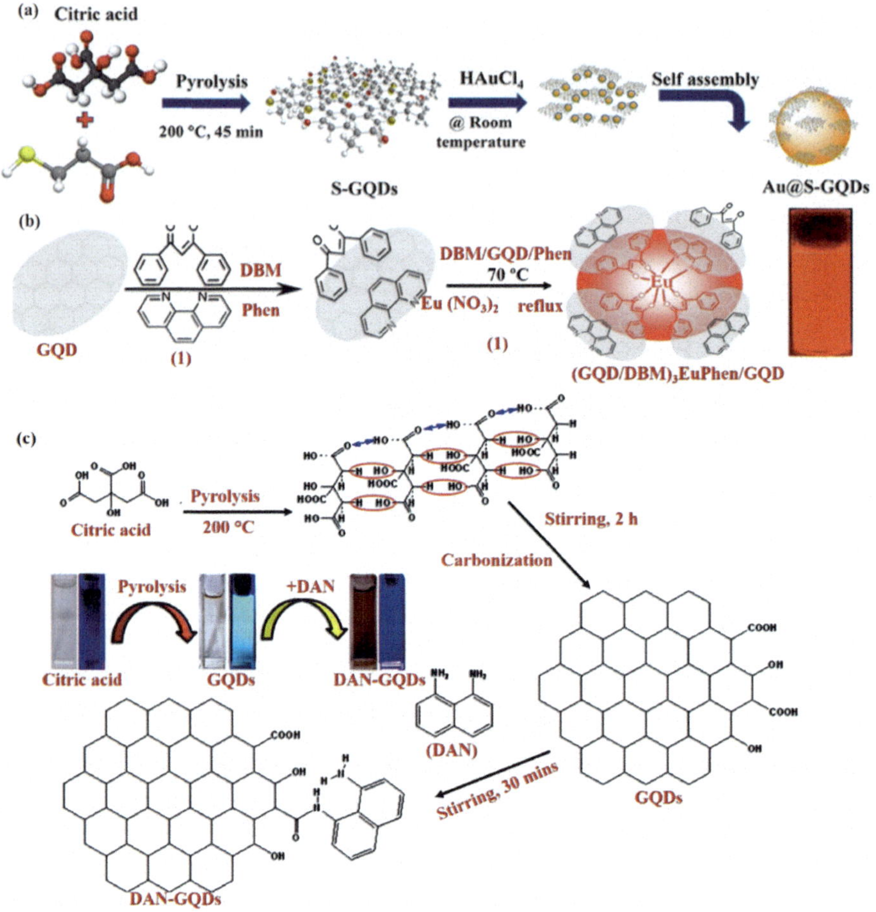

Fig. 4.13 **a** Gold@Sulfur-Doped GQD nanocomposites by pyrolysis method, **b** red fluorescent GQD–europium complex composites, and **c** synthesis of surfactant-free DAN-GQDs by pyrolysis technique

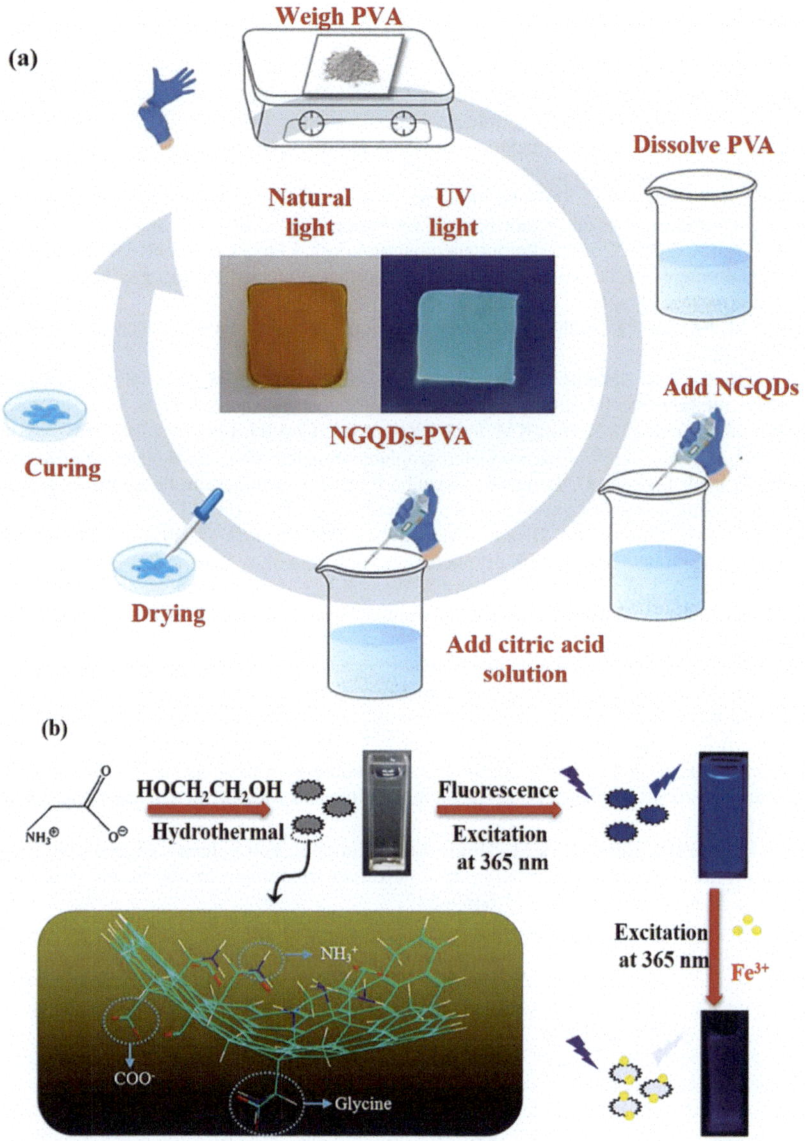

Fig. 4.14 **a** Flow diagram for the preparation of NGQDs-PVA composite membrane, and **b** synthesis of high fluorescent GQDs

References

1. Long MQ, Tang L, Wang D, Wang L, Shuai Z (2009) Theoretical predictions of size-dependent carrier mobility and polarity in graphene. J Am Chem Soc 131:17728–17729
2. Kalluri A, Debnath D, Dharmadhikari B, Patra P (2018) Graphene quantum dots: synthesis and applications. methods in enzymology. Elsevier
3. Ritter KA, Lyding JW (2009) The influence of edge structure on the electronic properties of graphene quantum dots and nanoribbons. Nat Mater 8:235–242
4. Li LS, Yan X (2010) Colloidal graphene quantum dots. J Phys Chem Lett 1:2572–2576
5. Pan D, Zhang J, Li Z, Wu M (2010) Hydrothermal route for cutting graphene sheets into blue-luminescent graphene quantum dots. Adv Mater 22:734–738
6. Bacon M, Bradley SJ, Nann T (2014) Graphene quantum dots. Part Part Syst Characterization 31:415–428
7. Ponomarenko LA, Schedin F, Katsnelson MI, Yang R, Hill EW, Novoselov KS, Geim AK (2008) Chaotic Dirac billiard in graphene quantum dots. Science 320:356–358
8. Nurunnabi M, Khatun Z, Huh KM, Park SY, Lee DY, Cho KJ, Lee YK (2013) In vivo biodistribution and toxicology of carboxylated graphene quantum dots. ACS Nano 7:6858–6867
9. Farka Z, Juřík T, Kovář D, Trnková L, Skládal P (2017) Nanoparticle-based immunochemical biosensors and assays: recent advances and challenges. Chem Rev 117:9973–10042
10. Sweetman MJ, Hickey SM, Brooks DA, Hayball JD, Plush SE (2019) A practical guide to prepare and synthetically modify graphene quantum dots. Adv Func Mater 29:1808740
11. Zhao C, Song X, Liu Y, Fu Y, Ye L, Wang N, Wang F, Li L, Mohammdniaei M, Zhang M, Zhang Q, Liu J (2020) Synthesis of graphene quantum dots and their applications in drug delivery. J Nanobiotechnol 18:142
12. Zhang C, Cui Y, Song L, Liu X, Hu Z (2016) Microwave-assisted one-pot synthesis of graphene quantum dots as highly sensitive fluorescent probes for detection of iron ions and pH value. Talanta 150:54–60
13. El-Hnayn R, Canabady-Rochelle L, Desmarets C, Balan L, Rinnert H, Joubert O, Medjahdi G, Ben Ouada H, Schneider R (2020) One-step synthesis of diamine-functionalized graphene quantum dots from graphene oxide and their chelating and antioxidant activities. Nanomaterials 10:104
14. Yoon H, Kim HS, Kim J, Park M, Kim B, Lee S, Kang K, Yoo S, Jeon S (2020) Blue graphene quantum dots with high color purity by controlling subdomain formation for light-emitting devices. ACS Appl Nano Mater 3:6469–6477
15. Eun LN, Jong-Min J, San LH, Yoon LS, Oh CS (2020) Ultraviolet/blue light emitting high-quality graphene quantum dots and their biocompatibility. Carbon 170:213–219
16. Li W, Li M, Liu Y, Pan D, Li Z, Wang L, Wu M (2018) Three minute ultrarapid microwave-assisted synthesis of bright fluorescent graphene quantum dots for live cell staining and white LEDs. ACS Appl Nano Mater 1:1623–1630
17. Tian R, Zhong S, Wu J, Jiang W, Shen Y, Jiang W, Wang T (2016) Solvothermal method to prepare graphene quantum dots by hydrogen peroxide. Optical Mater 60:204–208
18. Facure MHM, Schneider R, Mercante LA, Correa DS (2022) Rational hydrothermal synthesis of graphene quantum dots with optimized luminescent properties for sensing applications. Materials Today Chemistry 23:100755
19. Ye R, Xiang C, Lin J et al (2013) Coal as an abundant source of graphene quantum dots. Nat Commun 4:2943
20. Hu X, Xiao-Yan M, Tian J, Huang Z (2020) Rapid and facile synthesis of graphene quantum dots with high antioxidant activity. Inorg Chem Commun 122:108288
21. Sarah R, Yingfen W, Xiao L, Rahul S, Jacob B, Xu W, Qianli RC, Nuri O, Jin Z, Xiaodong H, Colin C, Antonio V, Julia XZ (2021) Synthesis of highly near-infrared fluorescent graphene quantum dots using biomass-derived materials for in vitro cell imaging and metal ion detection. ACS Appl Mater Interfaces 13:43952–43962
22. Ibarra D, Kharissova O, Gomez I (2023) Synthesis of graphene quantum dots coupled to au nanoparticles: a facile and versatile route using different carbon sources. J Carbon Res 9:45

23. Bich QTT, Huu NN, Thuy Duong DTN, My NNT, Hong Thien TDV, Vu TLH (2021) Rapid and simple synthesis of graphene quantum dots/ag nanocomposites and its application for glucose detection by photoluminescence spectroscopy. Int J Sci Eng Sci 5:1–5
24. Shilpa R, Sathishkumar M, Kothurkar NK, Senthilkumar R (2018) Synthesis and characterization of graphene quantum dots/cobalt ferrite nanocomposite. IOP Conf Ser: Mater Sci Eng 310:012139
25. Kang S, Jeong YK, Jung KH, Son Y, Kim WR, Ryu JH, Kim KM (2020) One-step synthesis of sulfur-incorporated graphene quantum dots using pulsed laser ablation for enhancing optical properties. Opt Express 28:21659–21667
26. Feng Q, Cao Q, Li M, Liu F, Tang N, Du Y (2013) Synthesis and photoluminescence of fluorinated graphene quantum dots. Appl Phys Lett 102:013111
27. Nguyen KY, Le XH, Dao NT, Pham NT, Vu NT, Thi HH, Nguyen NH, Pham TN (2019) Microwave-assisted synthesis of graphene quantum dots and nitrogen-doped graphene quantum dots: Raman characterization and their optical properties. Adv Natural Sci: Nanosci Nanotech 10:025005
28. Li J, Ying W, Fengkai Y, Jianpo Z, Fangli Z (2019) The synthesis and application of nitrogen-doped graphene quantum dots on brilliant blue detection. J Nanomaterials: Article ID 1471728
29. Chen W, Shen J, Chen S, Yan J, Zhang N, Zheng K, Liu X (2019) Synthesis of graphene quantum dot-stabilized gold nanoparticles and their application. RSC Adv 9:21215–21219
30. Liu X, Sun B (2023) One-pot synthesis of nitrogen-doped graphene quantum dots and their applications in bioimaging and detecting copper ions in living cells. ACS Omega 8:27333–27343
31. Garg M, Vishwakarma N, Sharma AL, Singh S (2021) Amine-functionalized graphene quantum dots for fluorescence-based immunosensing of ferritin. ACS Appl Nano Mater 4:7416–7425
32. Shomali A, Valizadeh H, Banan A, Mohammad-Rezaei R (2015) Efficient synthesis of Xanthene derivatives using carboxyl functionalized graphene quantum dots as an acidic nano-catalyst under microwave irradiation. RSC Adv 5:88202–88208
33. Qian Z, Ma J, Shan X, Shao L, Zhou J, Chen J, Feng H (2013) Surface functionalization of graphene quantum dots with small organic molecules from photoluminescence modulation to bioimaging applications: an experimental and theoretical investigation. RSC Adv 3:14571–14579
34. Chatterjee M, Nath P, Kadian S, Kumar A, Kumar V, Roy P, Manik G, Satapathi S (2022) Highly sensitive and selective detection of dopamine with boron and sulfur co-doped graphene quantum dots. Sci Rep 12:9061
35. Elizabeth C, Md Tanvir H, Gonzalez-Rodriguez R, Truly T, Lee BH, Green KN, Akkaraju G, Naumov AV (2021) Graphene quantum dot formulation for cancer imaging and redox-based drug delivery. Nanomedicine: Nanotechnol, Biol Med 37:102408
36. Chen Z, Zhao Z, Liu J, Zhu X, Wang L, Xu Y, Xu H, Li H (2018) Graphene quantum dots modified Ag_3PO_4 for facile synthesis and the enhanced photocatalytic performance. J Chinese Adv Mater Society 6:255–269
37. Qiu H, Sun X, An S, Lan D, Cui J, Zhang Y, He W (2020) Microwave synthesis of histidine-functionalized graphene quantum dots/Ni-Co LDH with flower ball structure for supercapacitor. J Collo Interf Sci 567:264–273
38. Kaimal KR, Vinoth VV, Salunke SAS, Hector V, Viswanathan MR, Belqasem A, Sambandam A (2022) Highly sensitive and selective detection of glutathione using ultrasonic aided synthesis of graphene quantum dots embedded over amine-functionalized silica nanoparticles. Ultrason Sonochem 82:105868
39. Xu L, Mao W, Huang J, Li S, Huang K, Li M, Xia J, Chen Q (2016) Economical, green route to highly fluorescence intensity carbon materials based on lignin sulfonate/graphene quantum dots composites: Application as excellent fluorescent sensing platform for detection of Fe^{3+} ions. Sens Actua B: Chem 230:54–60
40. Gupta PK, Chauhan D, Khan ZH, Solanki PR (2020) ZrO_2 Nanoflowers decorated with graphene quantum dots for electrochemical immunosensing. ACS Appl Nano Mater 3:2506–2516

41. Sachin K, Gaurav M (2020) Sulfur-doped graphene quantum dots as a potential sensitive fluorescent probe for the detection of quercetin. Food Chem 317:126457
42. Hwang E, Hwang H, Shin Y, Yoon Y, Lee H, Yang J, Bak S, Lee H (2016) Chemically modulated graphene quantum dot for tuning the photoluminescence as novel sensory probe. Sci Rep 6:39448
43. Arundithi A, Yue W, Parimal R, Mahasin Alam SK, Aung T, Ming L, Jie Z, Jie C, Handong S, Peng C (2015) Nitrogen and phosphorus co-doped graphene quantum dots: synthesis from adenosine triphosphate, optical properties, and cellular imaging. Nanoscale 7:8159–8165
44. Vijay Kumar S, Soumyodeep D, Shubhayan B, Pooria L, Yogambha R, Gurvinder S, Hala Z, Prem Ballabh B (2022) Probing heteroatoms co-doped graphene quantum dots for energy transfer and 2-photon assisted applications. J Photochem. Photobio A: Chem 423:113618
45. Satyaprakash A, Sudhanshu M, Dhirendra B (2017) Electrochemical method to prepare graphene quantum dots and graphene oxide quantum dots. ACS Omega 2:8343–8353
46. Lee SH, Seo SD, Jin YH, Shim HW, Kim DW (2010) A graphite foil electrode covered with electrochemically exfoliated graphene nanosheets. Electrochem Commun 12:1419–1422
47. Shinde DB, Pillai VK (2013) Electrochemical resolution of multiple redox events for graphene quantum dots. Angew Chem Int Ed 52:2482–2485
48. Shinde DB, Majumder M, Pillai VK (2014) Counter-ion dependent, longitudinal unzipping of multi-walled carbon nanotubes to highly conductive and transparent graphene nanoribbons. Sci Rep 4:4363
49. Kondaveeti HK, Kumaravelu NK, Vanambathina SD, Mathe SE, Vappangi SA (2021) Systematic literature review on prototyping with Arduino: Applications, challenges, advantages, and limitations. Comput Sci Rev 40:100364
50. Darabian LM, Costa TLG, Cipriano DF, Cremasco CW, Schettino MA Jr, Freitas JCC (2022) Synthesis of graphene quantum dots by a simple hydrothermal route using graphite recycled from spent Li-Ion batteries. J Carbon Res 8:48
51. Wang L, Wang Y, Xu T, Liao H, Yao C, Liu Y, Li Z, Chen Z, Pan D, Sun L, Wu M (2014) Gram-scale synthesis of single-crystalline graphene quantum dots with superior optical properties. Nat Commun 5:5357
52. Arnab H, Godoy-Gallardo M, Ashley J, Feng X, Zhou T, Hosta-Rigau L, Sun Y (2018) One-pot green synthesis of biocompatible graphene quantum dots and their cell uptake studies. ACS Appl Bio Mater 1:452–461
53. Behzadi F, Saievar-Iranizad E, Bayat A (2019) One-step synthesis of graphene quantum dots, graphene nanosheets and carbon nanospheres: investigation of photoluminescence properties. Mater Res Express 6:105615
54. Xu Q, Gong Y, Zhang Z, Miao Y, Li D, Yan G (2019) Preparation of graphene oxide quantum dots from waste toner, and their application to a fluorometric DNA hybridization assay. Microchim Acta 186:483
55. Renbing T, Suting Z, Juan W, Wei J, Tianhe W (2016) Facile hydrothermal method to prepare graphene quantum dots from graphene oxide with different photoluminescence. RSC Adv 6:40422–40426
56. Suela K, John A, Nicholas PP, Morgan DJ, Paolo C, Heil T, Saha B (2017) Rapid synthesis of graphene quantum dots using a continuous hydrothermal flow synthesis approach. RSC Adv 7:14716
57. Liu F, Min-Ho J, Dong HH, Je-Hyung K, Yong-Hoon C, Seok ST (2013) Facile synthetic method for pristine graphene quantum dots and graphene oxide quantum dots: Origin of blue and green luminescence. Adv Mater 25:3657–3662
58. Aleena W, Abhishek N, Sobha A (2023) Synthesis of graphene quantum dots by liquid phase exfoliation method and its applications. Mater Today: Proc. https://doi.org/10.1016/j.matpr.2023.04.362
59. Kiang CC, Zdenek S, Simek P, Ondrej J, Katerina K, Snejana B, Hrdlickova KS, Martin P (2015) Synthesis of strongly fluorescent graphene quantum dots by cage-opening buckminsterfullerene. ACS Nano 9:2548–2555

60. Weifeng C, Dejiang L, Li T, Wei X, Tianyuan W, Weimin H, Yulin H, Shaona C, Jianfeng C, Zhongxu D (2018) Synthesis of graphene quantum dots from natural polymer starch for cell imaging. Green Chem 20:4438–4442
61. Al Ghamdi HS, Al-Ghamdi AA (2023) Opening band gap of multi-color graphene quantum dots using D-fructose as a green precursor. Alex Eng J 79:155–163
62. Hemen K, Palaparthy VS, Shojaei BM, Aslam M (2020) Electrochemical synthesis of graphene quantum dots from graphene oxide at room temperature and its soil moisture sensing properties. Carbon 165:9–17
63. Muthurasu A, Dhandapani P, Ganesh V (2016) Facile and simultaneous synthesis of graphene quantum dots and reduced graphene oxide for bio-imaging and supercapacitor applications. New J Chem 40:9111–9124
64. Nair RV, Thankam TR, Vandana S, Hanif M, Mingdong D, Saju P (2017) Rapid, acid-free synthesis of high-quality graphene quantum dots for aggregation induced sensing of metal ions and bioimaging. ACS Omega 2:8051–8061
65. Ren Q, Ga L, Ai J (2019) Rapid synthesis of highly fluorescent nitrogen-doped graphene quantum dots for effective detection of ferric ions and as fluorescent ink. ACS Omega 4:15842–15848
66. Meng-Jey Y, Min-Yu J, Meng-Chih C, Hung-Chun T, Yuan-Yao L (2020) Formation of graphene quantum dots by a ball-milling technique using carbon nanocapsules and sodium carbonate. Inorg Chem Commun 119:108061
67. Achadu OJ, Nyokong T (2017) Graphene quantum dots anchored onto mercapto pyridine-substituted zinc phthalocyanine-Au@Ag nanoparticle hybrid: application as fluorescence "off-on-off" sensor for Hg^{2+} and biothiols. Dyes Pigm 145:189–201
68. Achadu OJ, Nyokong T (2019) Fluorescence "turn-ON" nanosensor for cyanide ion using supramolecular hybrid of graphene quantum dots and cobalt pyrene-derivatized phthalocyanine. Dyes Pigm 160:328–335
69. Wang R, Huailin F, Wei J, Guosong N, Shijie Q (2019) Amino-functionalized graphene quantum dots prepared using high-softening point asphalt and their application in Fe^{3+} detection. Appl Surf Sci 467–468:446–455
70. Shen J, Chen W, Liu (2021) Facile synthesis of graphene quantum dots from glucan and their application as a deoxidizer and in cell imaging. J Chem Res 45:242–247
71. Lin L, Rong N, Lu S, Song X, Zhong Y, Yan J, Wang Y, Chen X (2015) A facile synthesis of highly luminescent nitrogen-doped graphene quantum dots for the detection of 2,4,6-trinitrophenol in aqueous solution. Nanoscale 7:1872–1878
72. Ngoc ANT, Rueyan D (2018) One-step synthesis of size-tunable gold@sulfur-doped graphene quantum dot nanocomposites for highly selective and sensitive detection of nanomolar 4-nitrophenol in aqueous solutions with complex matrix. ACS Appl Nano Mate. 1:2153–2163
73. Liu Y, Zhou S, Fan L, Shixin Z, Hong F (2016) Synthesis of red fluorescent graphene quantum dot-europium complex composites as a viable bioimaging platform. Microchim Acta 183:2605–2613
74. Pavithra R, Daniel Thangadurai T, Nataraj D, Senthilkumar K, Manonmani G, Nandakumar K, Sabu T, Praveen G (2019) Graphene nanobuds: a new second-generation phosgene sensor with ultralow detection limit in aqueous solution. ACS Appl Mater Interf 11:19339–19349
75. Wang T, Yang F, Zhang L, Tang Z, Liu W, Zhong L, He Z, Chai S (2022) Fluorescence quenching and highly selective adsorption of Ag+ using N-doped graphene quantum dots/poly(vinyl alcohol) composite membrane. Ind Eng Chem Res 61:18090–18099
76. Linbo L, Lin L, Chao W, Kangyu L, Ruohua Z, Hong Q, Yuqing L (2015) Synthesis of nitrogen-doped and amino acid-functionalized graphene quantum dots from glycine, and their application to the fluorometric determination of ferric ion. Microchim Acta 182:763–770
77. Ke CC, Yang YC, Tseng WL (2016) Synthesis of blue-, green-, yellow-, and red-emitting graphene-quantum-dot-based nanomaterials with excitation-independent emission. Part Part Syst Charact 33:132–139

Chapter 5
Biomass-Based Graphene Quantum Dots

5.1 Introduction

Graphene quantum dots (GQDs) are characterized as zero-dimensional graphene sheet fragments with a maximum lateral size of 100 nm. The intriguing physicochemical properties of graphene-like quantum dots (GQDs) have attracted a lot of interest from researchers [1]. Typically, graphene clusters' nanometer-sized sp^2 domains are broken up to create GQDs. GQDs have a nonzero bandgap, enriched edge defects, strong photoluminescence, and high aqueous solubility due to quantum confinement effects at the nanoscale. Higher electron transfer rates at the edges, outstanding chemical stability, minimal cytotoxicity, and ease of fabrication are further benefits of GQDs [2]. Furthermore, GQDs' surface and electric properties can be changed to suit specific applications by adjusting their size and edge structure. GQDs have the potential to be used in many different applications, such as biosensing, photocatalysis, fluorescence, moisture sensing, and others, due to their unique properties and nanoscale structure [3].

5.2 Preparation of GQDs

Synthesis of GQDs from biomass sources, which in a futuristic sense adds to the growing interest and holds numerous merits. Using biomass to prepare GQDs lessens the need for conventional carbon precursors and encourages sustainability; possibly lowering the total cost of production when compared to the use of more costly precursors; and less of an influence on the environment than conventional techniques that might use dangerous chemicals. Size turnability, optical characteristics, and surface functionalization are encouraged. It is possible to produce on a large scale, implement a waste management system, and enhance the efficiency of the waste-to-wealth ratio.

N. Manjubaashini et al., *Graphene Quantum Dots*, Materials Horizons: From Nature to Nanomaterials, https://doi.org/10.1007/978-981-97-5722-0_5

GQDs are made by combining "bottom-up" and "top-down" methods. In the top-down method, bulk carbon materials like graphene, carbon black, etc., are usually broken into nanoscale GQDs by hydro or solvothermal treatment, chemical or electrochemical exfoliation, and microwave or ultrasonication. The top-down approach is ideal for mass production because it is simple to use and has a large supply of precursor materials [4]. On the other hand, non-selective chemical cutting leaves little control over the size and shape of the finished product. However, the bottom-up strategy depends on the gradual conversion of small precursor molecules into nanoscale graphene quantum dots (GQDs) employing procedures such as carbonization, pyrolysis, chemical vapor deposition, etc. [5].

5.3 Preparation of GQDs from Biomass Waste

Rich carbon source materials like graphene, graphene oxide, graphene nanoribbons (GNRs), graphite, carbon nanotubes (CNTs), carbon fibers (CFs), and fullerenes (C60), as well as other biomaterials like cellulose, lignin, durian, and carbon-based polymers, are the first materials used in the top-down method [6]. Novel approaches for producing GQDs from biomass sources have been investigated by researchers. Large-scale GQD production has been accomplished by using biomass waste, such as lignin. This strategy provides a different, affordable, and adaptable way to make large-area graphene. As a result, the GQD derived from biomass has enormous potential to transform several industries. For example, its low toxicity makes it safer for use in a variety of applications [7]. Their high-quality quantum dots can be achieved through photo-induced energy transfer, which makes them more commercially viable. Additionally, because GQDs are compatible with biological systems, they can be used for imaging and diagnostics, and their ability to dissolve well in a variety of solvents enhances their usability. However, the majority of the techniques have certain drawbacks, including low yield, a high density of defects, the need for dangerous chemicals, intricate synthetic protocols, or expensive equipment [8].

5.4 Importance and Uses of Biomass Energy Resources

One flexible energy source is biomass. Trees, wood waste, wood chips, corn, rice hulls, peanut shells, sugar cane, grass cuttings, leaves, manure, sewage, and municipal solid waste are examples of organic matter that is used as a source of biomass energy (Fig. 5.1). Woodstoves have been used and are being used by people all over the world for cooking and building heating, making biomass one of the most widely used energy sources [9]. The World Bank estimates that half of the world's population cooks with wood, accounting for 50–60% of energy in developing countries in Asia and 70–90% in developing countries in Africa [10].

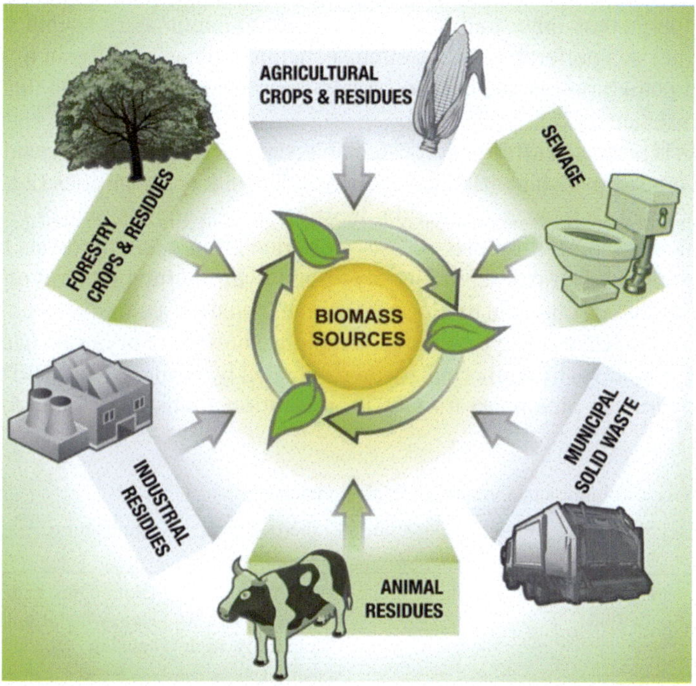

Fig. 5.1 Different sources of biomass [16]

Manures, slurries, grass silage, and animal bedding like chicken litter are examples of animal waste [11]. Along with food wastes from various food materials during the processing stage to remove components that are inedible or not required, such as peel/skin, shells, husks, cores, pips/stones, fish heads, pulp from juice, and oil extraction, it also includes gases and liquids recovered from the decomposition of nonfossilized and biodegradable organic materials [12, 13]. The use of less expensive, non-toxic, and renewable reinforcing fillers in polymer composites made of sisal, jute, pineapple leaves, banana stems, oil palm, rice hulls, and sugar palm is being investigated using biomass residues from agricultural wastes [14, 15].

5.5 Significance of Biomass-Based GQD Synthesis

Strong acids or organic solvents are frequently required for the preparation of GQD, and there are still significant obstacles to overcome in the green production of these materials using sustainable methods [17]. As a result, eco-friendly synthetic approaches that require little post-processing and have simple separation need to be developed, using natural renewable raw materials as precursors [18, 19]. Because it is a more environmentally friendly method and makes biological entities more readily available than other traditional approaches to nanomaterial synthesis, bioinspired synthesis is preferable [20]. Cellulose, hemicellulose, and lignin make up the majority of plant biomass, the latter is largely responsible for structural integrity and chemical resistance [21].

Biomass is a natural, affordable, sustainable, dependable, and eco-friendly source of carbon, but it can also be used to produce GQDs in large quantities [22, 23]. The Biomass-derived GQD was admired by most of the researchers, which leads to the newer implementation of biomass GQDs from rice grain using pyrolysis technique by [24], fenugreek leaf extract via pyrolysis and hydrothermal method reported by Roy et al. [25], wood charcoal by electrochemical oxidation technique done by Nirala et al. [26], Neem leaves by hydrothermal treatment shown by Suryawanshi et al. [27], and coffee ground via hydrothermal method revealed by Wang et al. [28]. Thus, most of the used biomass precursors for GQDs preparation are listed in Table 5.1.

Table 5.1 Various biomass precursors for GQD preparation

Biomass used	Synthesis method	Quantum yield	Properties	Application	Ref. no.
Coffee grounds	Hydrothermal method	33%	Highly fluorescence, low cytotoxicity, high sensitivity	Bioimaging and Heavy metal ion sensing	[29]
Mangifera indica (mango) leaves	Microwave-assisted method	–	Biocompatible, scalable, photo-stable	Nano-thermometry	[30]
Rice husk	Hydrothermal ethod	–	Excellent biocompatibility	Fluorescent bio-probes	[31]
Spent ground coffee	Hydrothermal method	–	Active compound and electrochemically active	Glucose sensor	[32]
Spent tea leaves	Pyrolysis method	–	Size-dependent optical properties and considerable surface grafting	Photocatalytic, bioimaging, and sensing	[33]
Spent tea	Pyrolysis method	0.23	High yield, high selectivity, and cost-effectiveness	Fe^{3+} sensing	[34]
Spent tea leaves	Hydrothermal method	21%	Highly selective, high quality, inexpensive, and sustainable	Optical sensing probe	[35]
Honey and Mangosteen	Carbonization and pyrolysis	69%	Molecular switching ability	Pesticide Detection	[36]
Corn powder	Hydrothermal method	–	–	–	[37]

5.6 Formation of GQDs from Different Biomass Wastes

High-grade fluorescent GQDs are synthesized from cellulose-rich tender coconut water. According to Jayan et al. 2023, this is a novel method for producing value-added nanoproducts (GQDs) based on tender coconut water (Fig. 5.2a). The research was done on nitrogen-doped graphene quantum dots (N-GQDs) using ethylenediamine (EDA) as the nitrogen source and Pennisetum purpureum (elephant grass) as the carbon precursor (Fig. 5.2b) [38]. Simple, large-scale, one-step, affordable, catalyst- and template-free hydrothermal synthesis of highly crystalline, sustainable S-GQDs (Fig. 5.2c) from second-generation (2G) biowastes [39]. A straightforward, low-cost, and ecologically benign process for hydrothermally transforming cellulose nanocrystals (Bf-CNCs) derived from bamboo timber waste (Bf) into potent blue luminescent graphene quantum dots (Bf-GQDs) (Fig. 5.3). The effect of pH and particle size on the fluorescence properties of Bf-GQDs is also investigated, as is the most likely mechanism by which Bf-GQDs form from Bf-CNCs [40]. GQD is efficiently synthesized by microwave pyrolysis and an Opuntia sp. Extract (Fig. 5.4a). We can detect phosphates and phytic acid, as well as enhance fluorescence quenching at pH = 8.8, by employing this green synthesis method to produce nanostructures with a narrow size distribution and good dispersion [41].

Biowaste is used to synthesize GQDs, which are then converted into amine-terminated GQDs (Am-GQDs), which are more photoluminescent intense, and

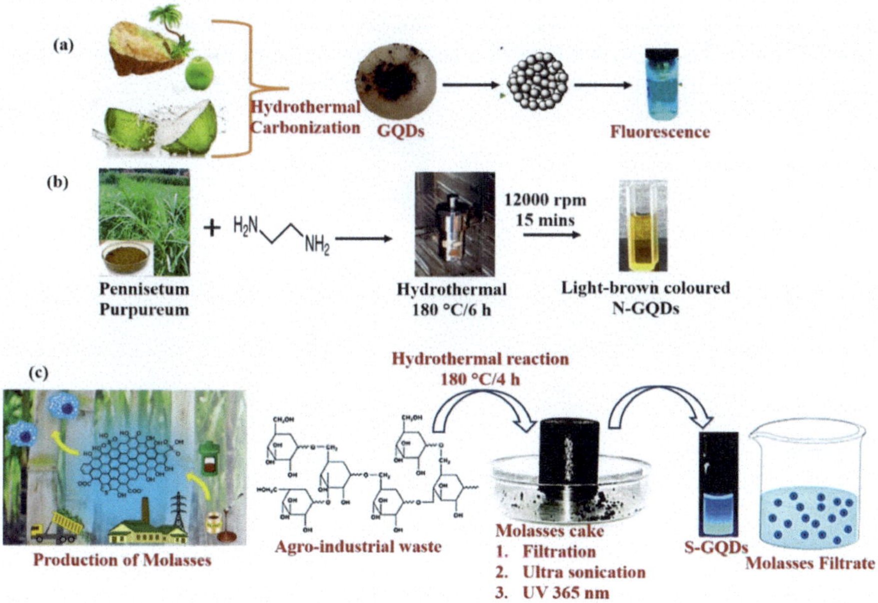

Fig. 5.2 Schematic illustration of biomass-derived GQDs; **a** coconut water-based GQD [16], **b** one-step synthesis of nitrogen-doped GQD [38], and **c** single crystalline sulfur-doped GQD [39]

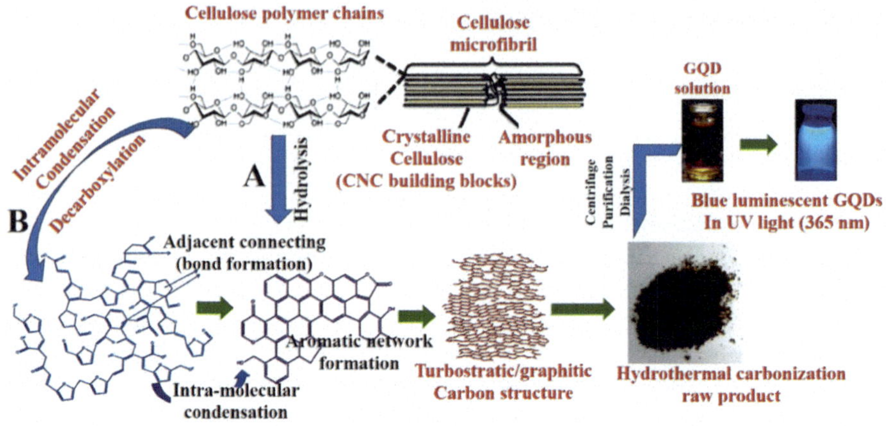

Fig. 5.3 Green synthesis of blue fluorescent GQD. Reprinted with permission from @ Current Applied Physics 20, 1226–1236 (2020)

dispersible than GQDs [42] (Fig. 5.4b). Various weight percentages of graphene (0.25, 0.5, and 1% wt. graphene) were mixed with polyester resin to create nanocomposite samples [43]. Using rice as a carbon source, a straightforward environmentally friendly process yields monodisperse GQDs with sizes ranging from 2 to 6.5 nm (Fig. 5.4c). As the size of the GQDs increases from 2–6.5 nm, the quantum confinement effect results in a redshift (blue to cyan) in the photoluminescence emission spectra [44]. With an average size of 3.9 nm, GQDs were produced using lemon extract [43] (Fig. 5.5a), and rice husk biomass as the raw materials in a straightforward one-step, one-pot hydrothermal process [45] (Fig. 5.5b).

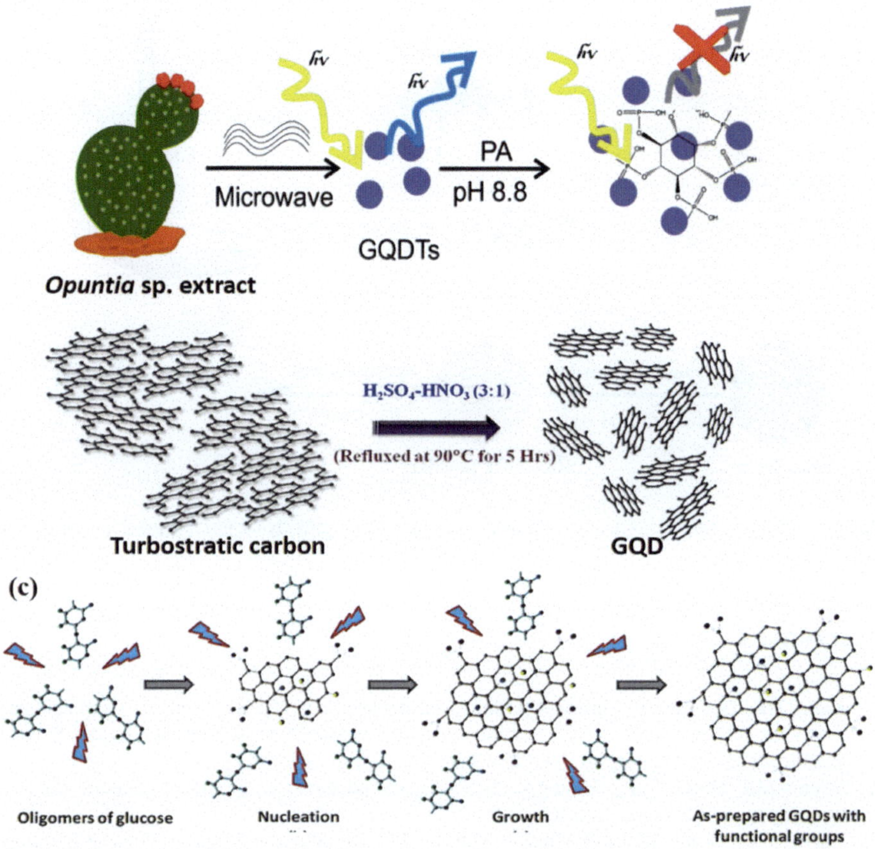

Fig. 5.4 **a** *Opuntia* sp. extract-based pH-dependent GQD. Reprinted with permission from @ Sensing and biosensing research 32, 100,412 (2021), **b** GQD from biomass Turbostratic carbon [42], and **c** GQD synthesis from rice-based biomass [42]

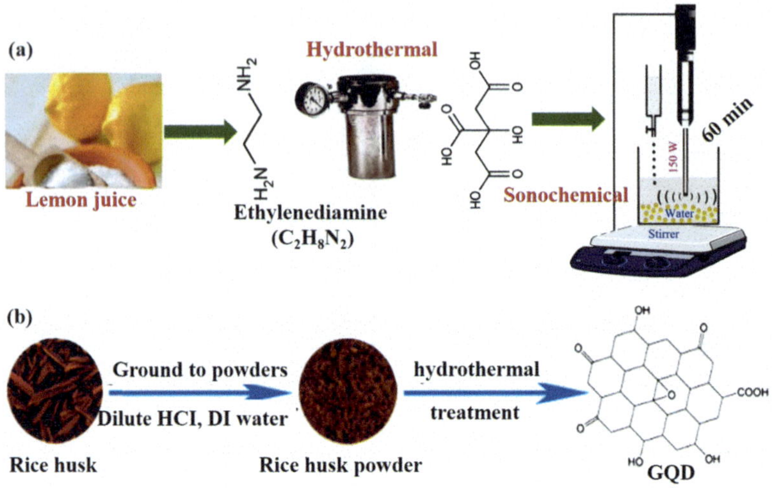

Fig. 5.5 **a** GQDs formation using lemon extract and EDA [43], **b** GQDs synthesized from rice husk biomass [45]

References

1. Kalita HK, Dehingia B, Ghosh R, Ojha A, Kushwaha AK (2023) Chapter 7: Graphene and graphene derivatives for wastewater treatment. In: Sillanpää M, Khadir A, Gurung K (eds) Resource recovery in industrial waste waters. Elsevier, pp 143–163
2. Younis MR, He G, Lin J, Huang P (2020) Recent advances on graphene quantum dots for bioimaging applications. Front Chem 8:424
3. Chen W, Lv G, Hu W, Li D, Chen S, Dai Z (2018) Synthesis and applications of graphene quantum dots: a review. Nanotechnol Rev 7:157–185
4. Ankarao K, Bhushan D, Debika D, Prabir P, Challa VK (2023) Advances in structural modifications and properties of graphene quantum dots for biomedical applications. ACS Omega 24:21358–21376
5. Balkanloo PG, Sharifi KM, Marjani AP (2023) Graphene quantum dots: synthesis, characterization, and application in wastewater treatment: a review. Mater Adv 4:4272–4293
6. Ding Z, Li F, Wen J, Wang X, Sun R (2018) Gram-scale synthesis of single-crystalline graphene quantum dots derived from lignin biomass. Green Chem 20:1383–1390
7. Abbas A, T. Mariana LT, Phan AN (2018) Biomass-waste derived graphene quantum dots and their applications. Carbon 140:77–99
8. Abbas A, Abbas S, Tabish TA, Bull SJ, Phan AN, Lim TM (2021) Role of precursor microstructure in the development of graphene quantum dots from biomass. J Environ Chem Eng 9:106154
9. Jawaid M, Paridah MT, Saba N (2017) Chapter 1: introduction to biomass and its composites. In: Jawaid M, Md Tahir P, Saba N (eds) Composites science and engineering, lignocellulosic fibre and biomass-based composite materials. Woodhead Publishing Series, pp 1–11
10. Singh J (1996) Economic and environmental aspects of bio-energy productivity, in biomass for energy and the environment. In: Chartier P, Ferrero GL, Henius UM, Hultberg S, Sachau J, Wiinblad M (eds) Pergamon. pp 1865–1868
11. Zafar S (2022) An Introduction to Biomass Energy, BioENERGY CONSULT
12. Bracmort KS, Gorte RW (2009) Biomass: comparison of definitions in legislation. Congressional Research Service, Library of Congress, Washington DC, p 12

13. Huber GW, Iborra S, Corma A (2006) Synthesis of transportation fuels from biomass: chemistry, catalysts, and engineering. Chem Rev 106:4044–4098
14. Vo LT, Navard P (2016) Treatments of plant biomass for cementitious building materials-a review. Constr Build Mater 121:161–176
15. Ibrahim MS, Sapuan SM, Faieza AA (2012) Mechanical and thermal properties of composites from unsaturated polyester filled with oil palm ash. J Mech Eng Sci 2:181–186
16. Jitha SJ, Sajitha SJ (2023) Synthesis of fluorescent graphene quantum dots from tender coconut water. Available at SSRN: https://doi.org/10.2139/ssrn.4362687
17. Iravani S, Varma RS (2020) Green synthesis, biomedical and biotechnological applications of carbon and graphene quantum dots. A review. Environ Chem Lett 18:703–727
18. Chen W, Shen J, Lv G, Li D, Hu Y, Zhou C, Liu X, Dai Z (2019) Green Synthesis of Graphene Quantum Dots from Cotton Cellulose. ChemistrySelect 4:2898–2902
19. Saha JK, Dutta AA (2022) Review of graphene: material synthesis from biomass sources. Waste Biomass Valor 13:1385–1429
20. Irshad A, Sarwar N, Sadia H, Riaz M, Sharif S, Shahid M, Khan J (2020) Silver nanoparticles: synthesis and characterization by using glucans extracted from *Pleurotus ostreatus*. Appl Nanosci 10:3205–3214
21. Wong SS, Shu R, Zhang J, Liu H, Yan N (2020) Downstream processing of lignin derived feedstock into end products. Chem Soc Rev 49:5510–5560
22. Pooja K, Rohit M, Saurabh B, Harrasi AA, Syam M, Mansi Y, Sunita D, Seema R, Sudhir Kumar K, Arasu T (2023) Comprehensive review on synthesis, applications, and challenges of graphene quantum dots. Article ID 2832964
23. Seyyed MM, Seyyed AH, Masoomeh YK, Darwin K, Ahmad G, Vahid R, Navid O, Wei-Hung C (2022) Recent advances in inflammatory diagnosis with graphene quantum dots enhanced SERS detection. Biosensors 12:461
24. Kalita H, Mohapatra J, Pradhan L, Mitra A, Bahadur D, Aslam M (2016) Efficient synthesis of rice-based graphene quantum dots and their fluorescent properties. RSC Adv 6:23518–23524
25. Roy P, Periasamy AP, Chuang C, Liou YR, Chen YF, Joly J, Liang CT, Change HT (2014) Plant leaf-derived graphene quantum dots and applications for white LEDs. New J Chem 38:4946–4951
26. Nirala NR, Khandelwal G, Kumar B, Prakash R, Kumar V (2017) One step electro-oxidative preparation of graphene quantum dots from wood charcoal as a peroxidase mimetic. Talanta 173:36–43
27. Suryawanshi A, Biswal M, Mhamane D, Gokhale R, Patil S, Guin D, Ogale S (2014) Large scale synthesis of graphene quantum dots (GQDs) from waste biomass and their use as an efficient and selective photoluminescence on–off–on probe for Ag^{+} ions. Nanoscale 6:11664–11670
28. Wang L, Li W, Wu B, Li Z, Wang S, Liu Y, Pan D, Wu M (2016) Facile synthesis of fluorescent graphene quantum dots from coffee grounds for bioimaging and sensing. Chem Eng J 300:75–82
29. Wang L, Li W, Wu B, Li Z, Wang S, Liu Y, Pan D, Wu M (2016) Facile synthesis of fluorescent graphene quantum dots from coffee grounds for bioimaging and sensing. Chem Eng J 300:75–82
30. Mukesh Kumar K, Mukeshchand T, Gurung RB, Rohit S (2017) Graphene quantum dots from mangifera indica: application in near-infrared bioimaging and intracellular nanothermometry. ACS Sust Chem Eng 5:1382–1391
31. Wang Z, Yu J, Zhang X, Li N, Liu B, Li Y, Wang Y, Wang W, Li Y, Zhang L, Dissanayake S, Suib SL, Sun L (2016) Large scale and controllable synthesis of graphene quantum dots from rice husk biomass: a comprehensive utilization strategy. ACS Appl Mater Interf 8:1434–1439
32. Filimon H, Meng-Hsuan C, Yi-Fang H, Shao-Yu W, Shyankay J (2022) Boron-Doped Graphene Quantum Dots (BGQDs) from spent coffee ground for glucose sensor. Hindawi Adv Mater Sci Eng. Article ID 4574772, 11 pages
33. Abbas A, Liang Q, Abbas S, Liaqat M, Rubab S, Tabish TA (2022) Eco-Friendly sustainable synthesis of graphene quantum dots from biowaste as a highly selective sensor. Nanomaterials 12:3696

34. Abbas A, Tabish TA, Bull SJ, Lim TM, Phan AN (2020) High yield synthesis of graphene quantum dots from biomass waste as a highly selective probe for Fe^{3+} sensing. Sci Rep 10:21262
35. Abbas A, Rubab S, Rehman A, Irfan S, Sharif HMA, Liang Q, Tabish TA (2023) One-step green synthesis of biomass-derived graphene quantum dots as a highly selective optical sensing probe. Materials Today Chemistry 30:101555
36. Lekshmi L, Swathy US, Keerthi V, Haleema S, Sumalekshmy S (2023) Graphene quantum dots derived from honey and mangostin as sustainable materials to construct fluorescence turn-on molecular switches for pesticide detection. ChemistrySelect 8:e202204869
37. Teymourini H, Salavati-Niasari M, Amiri O, Safardoust-Hojaghan H (2017) Synthesis of graphene quantum dots from corn powder and their application in reduce charge recombination and increase free charge carriers. J Mol Liquids 242:447–455
38. Sari FD, Chotimah, Roto, I. Kartini I (2023) Highly fluorescent nitrogen-doped graphene quantum dots (NGQDs) synthesized from Pennisetum purpureum for selective and sensitive detection of Fe^{3+} ions. Mater Res Express 10:075603
39. Sujata S, Apoorv G, Adeeba S, Rohan B, Arun Kumar S, Deepa S, Sandip C, Sandeep Kumar G, Sourav C, Biswarup B, Vinod K, Satyendra Kumar R, Malay Kishore D, Monalisa M (2018) Sustainable synthesis of single crystalline sulphur-doped graphene quantum dots for bioimaging and beyond. Green Chem 20:4245
40. Tade RS, Patil PO (2020) Green synthesis of fluorescent graphene quantum dots and its application in selective curcumin detection. Current Appl Phys 20:1226–1236
41. Centeno L, Romero-García J, Alvarado-Canche C, Gallardo-Vega C, Telles-Padilla G, Barriga-Castro ED, Cabrera-Alvarez EN, Ledezma-Perez A, de Leon A (2021) Green synthesis of graphene quantum dots from Opuntia sp. extract and their application in phytic acid detection. Sens Bio-Sens Res 32:100412
42. Anil S, Mandakini B, Dattakumar M, Rohan G, Shankar P, Debanjan G, Satishchandra O (2014) Large scale synthesis of graphene quantum dots (GQDs) from waste biomass and their use as an efficient and selective photoluminescence on-off-on probe for Ag^{+} ions. Nanoscale 6:11664–11670
43. Safari M, de Sousa RA, Salamat-Talab M, J. Joudaki J, D. Ghanbari D, A. Bakhtiari A (2021) Mechanical properties of green synthesized graphene nano-composite samples. Appl Sci 11: 4846
44. Hemen K, Jeotikanta M, Lina P, Arijit M, Dhirendra B, Mohammed A (2016) Efficient synthesis of rice-based graphene quantum dots and their fluorescent properties. RSC Adv 6:23518–23524
45. Wang W, Wang Z, Liu J, Peng Y, Yu X, Wang W, Zhang Z, Sun L (2018) One-pot facile synthesis of graphene quantum Dots from rice husks for Fe^{3+} sensing. Ind Eng Chem Res 57:9144–9150

Chapter 6
Characterization Techniques for Graphene Quantum Dots

6.1 Introduction

Several experimental methods have been used to produce differentiable properties on GQD materials. The following characterization techniques and methods are primarily used for the investigation of GQDs: (i) phase and lattice parameter identification, (ii) distribution of particle sizes in a liquid media and to investigate colloidal nature, (iii) optical characteristics and absorbance to determine surface plasmon resonance, (iv) light emitting properties, (v) surface changes and chemical interactions by analyzing molecule vibrations and functional groups, (vi) information on molecular vibrations, crystal structures, chemical composition, and flaws, (vii) size, shape, and morphology, crystal lattice configurations and defects at atomic scales, and (viii) surface topography.

a. X-ray diffraction (XRD),
b. Dynamic Light Scattering (DLS),
c. UV–Visible spectroscopy,
d. Photoluminescence spectroscopy,
e. Fourier transform infrared (FTIR) spectroscopy,
f. Raman spectroscopy,
g. High-resolution transmission electron microscopy (HRTEM),
h. Field emission scanning electron microscopy (FESEM),
i. X-ray photoelectron spectroscopy (XPS),
j. X-ray absorption fine structure (XAFS) spectroscopy, and
k. Atomic force microscopy (AFM).

N. Manjubaashini et al., *Graphene Quantum Dots*, Materials Horizons: From Nature to Nanomaterials, https://doi.org/10.1007/978-981-97-5722-0_6

6.2 X-Ray Diffraction (XRD)

XRD is an effective non-destructive method for crystalline material characterization. In addition to preferred crystal orientations and structures, it offers details on phases, average grain size, strain, crystallinity, and crystal defects. A monochromatic X-ray beam dispersed at different angles from each pair of lattice planes in a sample generates constructive interference that results in XRD peaks (Fig. 6.1b) [1]. The distribution of atoms within the lattice determines the peak intensities. As a result, the fingerprint of periodic atomic configurations in a particular material is represented by the XRD pattern [2]. When Bragg's law is met, the incident rays' interaction with the sample results in constructive interference

$$n\lambda = 2d\ \sin \tag{6.1}$$

where n—integer, λ—wavelength of the X-rays, d—interplanar spacing from diffraction, and θ—diffraction angle (Fig. 6.1a). This law connects the lattice spacing and diffraction angle of a crystalline sample to the wavelength of electromagnetic radiation.

Due to the random orientation of the powdered material, all potential diffraction directions of the lattice should be achieved by scanning the sample through a range of 2θ angles [3]. Since every compound has a distinct set of d-spacings, conversion of the diffraction peaks to d-spacings enables compound identification. This is usually accomplished by comparing the d-spacings with industry-standard reference patterns. The most common use of XRD is the detection of unidentified minerals and inorganic substances. Studies in geology, environmental science, material science, engineering, and biology depend on the determination of unknown solids. Additionally, the technique includes measuring sample purity, determining unit cell dimensions, characterizing crystalline materials, and identifying fine-grained minerals that are challenging to identify visually, such as clays and mixed layer clays [4].

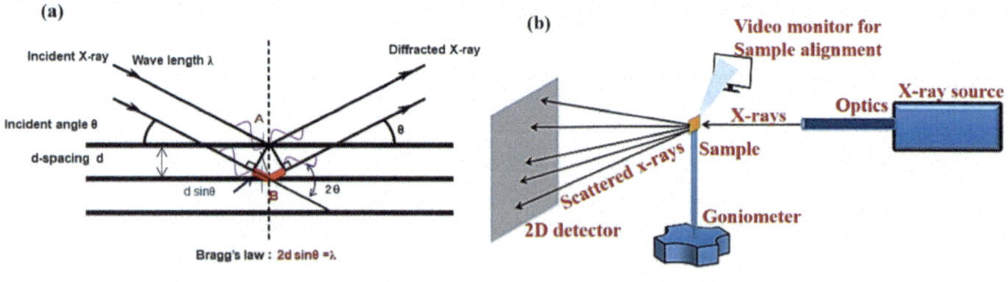

Fig. 6.1 **a** Schematic XRD principle, **b** XRD instrumentation

6.3 Dynamic Light Scattering (DLS)

DLS, also known as quasi-elastic light scattering or photon correlation spectroscopy (PCS), is used to measure the size of the nanoparticles in the colloidal suspension. By using a laser to scatter light through a colloidal solution and analyzing the change in scattered light intensity over time, DLS determines the hydrodynamic size of particles [5]. The Brownian movement theory underlies DLS's operation (Fig. 6.2). Larger particles will move more slowly, and smaller particles will move faster. A molecule's hydrodynamic radius (RH) is measured via DLS. Particles' hydrodynamic diameter and Brownian motion are correlated [6].

According to Falke et al. [7], the DLS instrument will provide a correlation function that is mathematically related to particle size and its time-dependent light scattering capability. Smaller particles will disperse more quickly than bigger ones. DLS has been used to assess formulation stability, estimate the size of particles in dispersing colloidal materials, and identify aggregation or agglomeration. According to Maskos et al. [8], DLS is the best method for identifying and quantifying the aggregation condition of nanoparticles. Using a helium-neon laser as the light source, the DLS measurements employ scattering angles of 90 or 173°, meaning that the detector is positioned at a back angle of 173° and a right angle of 90° to the incident light [9]. In Brownian motion, the individual particles are renewed into particle size, which will be calculated by the Stokes–Einstein equation:

$$D = \frac{kT}{6\pi \eta Rh} \tag{6.2}$$

where D—diffusion coefficient, k—Boltzmann's constant, T—temperature, η—solvent viscosity, and Rh—rheodynamic radius [7]. Compared to previous techniques, the dynamic light-scattering method has several advantages. For instance, studies can be carried out using a broad range of concentrations, temperatures, and sample buffers [10]. DLS is another non-invasive method that uses a small amount of sample and quickly yields accurate preparation quality assessments. The use of

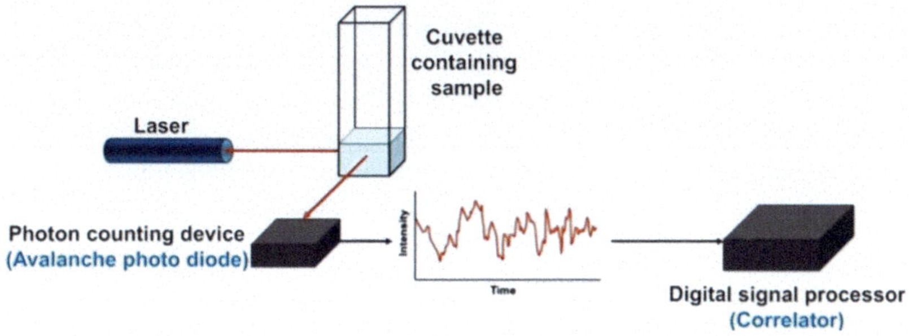

Fig. 6.2 Instrumentation and components of DLS

DLS to investigate the homogeneity of proteins and their complexes is discussed in the sections that follow [11].

6.4 UV–Visible Spectroscopy

The analytical method known as UV-visible spectroscopy counts the number of distinct UV or visible light wavelengths that a sample absorbs or transmits when compared to a reference or blank sample. The composition of the sample has an impact on this attribute, which may reveal information about the contents and concentrations of the sample [12]. Since light is used in this spectroscopy approach, let's start by talking about what light is made of. The wavelength of light and its energy content are inversely correlated. As a result, light has a greater energy at shorter wavelengths and less at longer wavelengths. Some measure of energy will promote the electrons to a higher energy level for detection via absorption [13].

To move electrons to a higher energy state, a substance's electrons in various bonding settings need distinct amounts of energy. This explains why various compounds absorb light at different wavelengths. Visible light has a wavelength range that humans can perceive, ranging from around 380 nm, or violet, to 780 nm, or red [14]. Up to about 100 nm, the wavelengths of UV radiation are shorter than those of visible light. Since light may therefore be characterized by its wavelength, UV-visible spectroscopy can be used to analyze or identify various compounds by identifying the precise wavelengths that correlate to maximal absorbance. When using two lamps in an instrument, the visible light is usually produced by a tungsten or halogen lamp, while the UV light is often produced by a deuterium lamp. The light source in the device must switch during measurement since two separate light sources are required to scan both the UV and visible wavelengths [15]. In actuality, this transition usually takes place in the range between 300 and 350 nm, when both light sources emit light similarly and the change may be accomplished more seamlessly.

Data from UV-visible spectroscopy can be displayed as a wavelength-dependent graph of transmittance, optical density, or absorbance. On the other hand, a graph with absorbance on the vertical y-axis and wavelength on the horizontal x-axis is more commonly used to display the data. If a linear relationship can be found using a measured set of standard solutions containing the same material, Beer–Lambert's law is particularly helpful for determining the concentration of a substance [16]. The mathematical correlations between absorbance, transmittance, light intensities measured in the device, and Beer–Lambert's law are displayed in Equation 6.3.

$$A = \varepsilon Lc = \log_{10}\left(\frac{I_0}{I}\right) = \log_{10}\left(\frac{1}{T}\right) = -\log_{10}(T) \tag{6.3}$$

Thus, the method is non-destructive, the sample can be processed or analyzed further, or it can be reused [17]. Rapid measurement enables simple integration

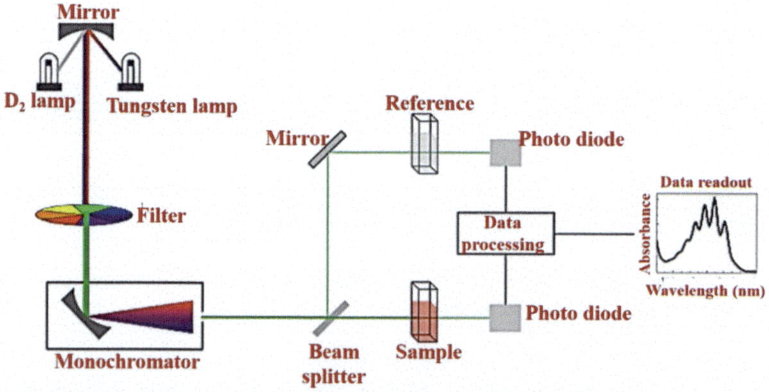

Fig. 6.3 Instrumentation and working procedure of UV-visible spectroscopy

into experimental methods. The instruments are simple to operate and don't require any user education before usage. Again, this means that not much processing is needed for data analysis, which means less user training is needed. The apparatus can be accessed by many laboratories due to its low acquisition and operating costs (Fig. 6.3). UV-visible spectroscopy has applications in bacterial culture, beverage analysis, pharmaceutical analysis, and DNA and RNA analysis [18].

6.5 Fluorescence Spectroscopy

Luminescence is a group of methods known as spectroscopy used to quantify the fluorescence that materials release into the ultraviolet, visible, or other electromagnetic radiation spectrum [19]. Fluorescence is the result of a molecule emitting electromagnetic radiation at a wavelength where it has a dipole moment of transition [20]. Photons are accelerated to an excited singlet state by the ground state molecule's excitation energy and then decay to the lowest vibrational energy level of the state. As the molecule's energy levels settle back to their ground state, photons are produced (Fig. 6.4b). In any fluorescence apparatus, a light source, sample holder, and detector are essential components. Additionally, the wavelength of the incident radiation should be selectable, and the detector signal must be accurately adjustable and displayed for analytical purposes [21].

Fluorescence spectroscopy is highly sensitive, making it the perfect tool for detecting minute amounts of contaminants such as pollutants, biological molecules, and more (Fig. 6.4a). Fluorescence spectroscopy can be used to selectively detect single molecules or small groups of molecules by using the appropriate excitation and emission wavelengths [22]. Because fluorescence spectroscopy has a low background noise, it can be used to identify faint fluorescence signals. Fluorescence spectroscopy can detect many molecules in parallel, a technique known as multiplexing,

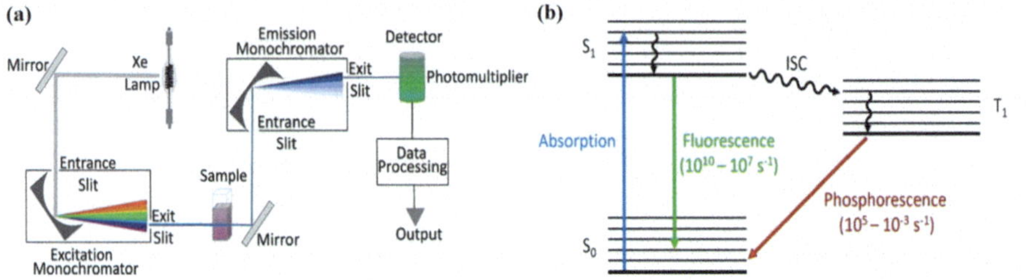

Fig. 6.4 **a** Working and components of fluorescence spectroscopy, **b** formation and Jablonski diagram of fluorescence and phosphorescence

by using multiple fluorescent labels. Fluorescence spectroscopy analysis does not involve sample destruction or manipulation [23]. Fluorescence spectroscopy can be used *in vivo* to image biomolecules in motion by using fluorescent labels that have been targeted to particular cells or tissues. Many different kinds of scientists employ fluorescence spectroscopy since it is inexpensive and simple to use. Fluorescence spectroscopy's versatility can be seen in its prospective applications in a wide range of domains, including chemistry, biology, medicine, environmental monitoring, and industrial analysis [24].

The application includes:

1. Biomolecule detectionFluorescence spectroscopy is a quick and non-invasive technique; it is commonly used in the industry for contamination assessment. Groundwater contamination from hydraulic fracturing for gas exploration can contain organic compounds; this technology has been used to find these molecules [25].
2. Chemical binding—When biological fluids come into touch with nanoparticles, proteins, and other biomolecules attach to and coat the particles. The interactions of nanoparticles with the protein corona impact their safety in vivo [26].
3. Pharmaceutical analysis—Spectrophotometric techniques are frequently used by the pharmaceutical industry to analyze medications. One such example is a review of co-formulated cholesterol-treatment pills [27].
4. Agricultural crop detection—Spectroscopic methods are widely used in agriculture, such as in crop variety identification [28].

6.6 Fourier Transform Infrared (FTIR) Spectroscopy

Substances that absorb in the mid-infrared (400–4000 cm^{-1}) can be measured using FTIR. The fundamental idea behind the interaction between matter and radiation is that absorption can take place at inherent energy levels that are unique to each molecule. Certain frequencies will be absorbed when mid-infrared radiation passes through a molecule if they coincide with the molecule's transition between vibrational states [29]. As such, both qualitative and quantitative examination can be done

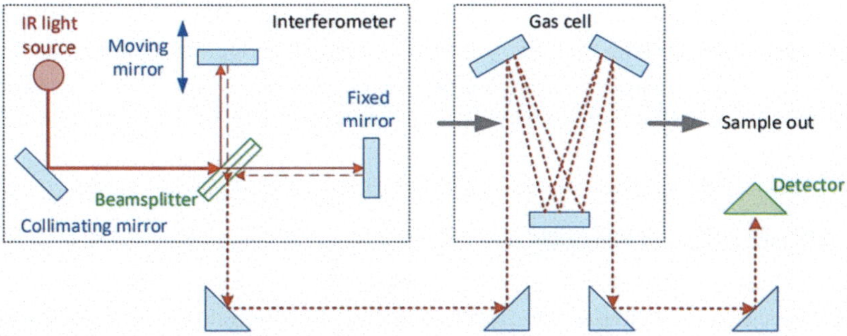

Fig. 6.5 Principle and operation of FTIR technique

with FTIR. However, because diatomic gases like oxygen (O_2), nitrogen (N_2), and hydrogen (H_2) do not have absorbance bands in the infrared area of the electromagnetic spectrum due to their vibration not creating a dipole moment, they cannot be detected by this method. Noble gases like helium (He) and argon (Ar) are also inaccessible. Other chemicals are undetectable in low quantities because they absorb very little radiation. An optical device known as an interferometer is the brains behind any FTIR instrument. The Michelson interferometer is the most traditional and widely used variety. Typically, a heated ceramic serves as the infrared source [30]. Light is collected from the source and its rays are made parallel by a collimating mirror. The half of the light that strikes a beamsplitter that is transmitted and the half that is reflected is measured in KBr. A portion of the light that is transmitted goes to a mirror that is fixed, and another portion goes to a mirror that is moving [31]. The two mirrors reflect the light to the beamsplitter, where it is combined again into a single beam of light (Fig. 6.5). After interacting with the gas cell's exhaust, this light beam hits the detector. To achieve a long optical path length with the smallest possible cell volume, a multireflection cell is utilized [32].

For the detection of a wide range of samples in various sectors, FTIR is a crucial tool. This approach can be used to analyze a wide range of substances, including resins, adhesives, paints, coatings, polymers, metal oxides, and numerous medications. For example, FTIR can be used to evaluate and identify various natural samples, including gummy materials, solids, liquids, and solutions. This method can be used to identify a wide variety of chemical and inorganic compounds [33]. It is possible to test polymers and polymer blends as well as indirectly verify the presence of trace organic pollutants on the surfaces of different materials. It is possible to carry out routine qualitative and quantitative analyses. Thin film, coatings, adhesion boosters, coupling agents, and adhesives can all be readily evaluated with the aid of FTIR adhesives [34]. Using FTIR in conjunction with optical microscopy, scanning electron microscopy (SEM)/energy dispersive X-ray analysis (EDX), X-ray photoelectron spectroscopy (XPS), and secondary ion mass spectrometry (SIMS) techniques, we can examine stains and surface blemishes left over from cleaning and degreasing procedures [35]. This method was used to identify the composition of rubbers and filled rubbers. It can also be used to identify low-density polyethylene (LDPE) and high-density polyethylene (HDPE) and to measure the degrees of crystallinity in polymers [36]. Comparative chain lengths in organic compounds can likewise be distinguished using this technique [37].

6.7 Raman Spectroscopy

Infrared spectroscopy and Raman spectroscopy are now thought to work best together as a molecular structure investigation tool. The Raman effect, discovered in 1928 by Indian physicist Chandrasekhara Venkata Raman, is the foundation of Raman spectroscopy. According to quantum mechanics, a molecule may be advanced to a virtual state of higher energy when photons interact with it [38]. There are several conceivable outcomes from this higher energy state. One such result could be that the molecule produces a photon with a different energy, which lowers its vibrational energy level to a level distinct from its initial state (Fig. 6.6).

The Raman Shift is the term used to describe the difference in energy between the incident photon and the scattered photon [39]. The molecule will be excited to some virtual state between the two states if the incident photon's energy is insufficient to excite it from the ground state to the lowest electronic state. However, the excited molecule's electron must quickly return to the ground state because it cannot remain in the virtual state for very long [40]. The intensity of the scattered light (shorter wavelength) is inversely proportional to the fourth power of wavelength (higher wavelength) is known as rayleigh scattering [41]. On the other hand, the Raman scattering is determined by the possible number of electrons to go to a virtual state that is different from where it is excited, giving rise to an energy difference between the emitted photon and the incident photon, and therefore, a shift in wavelength for the emitted photon as compared to the incident photon [42, 43]. Raman

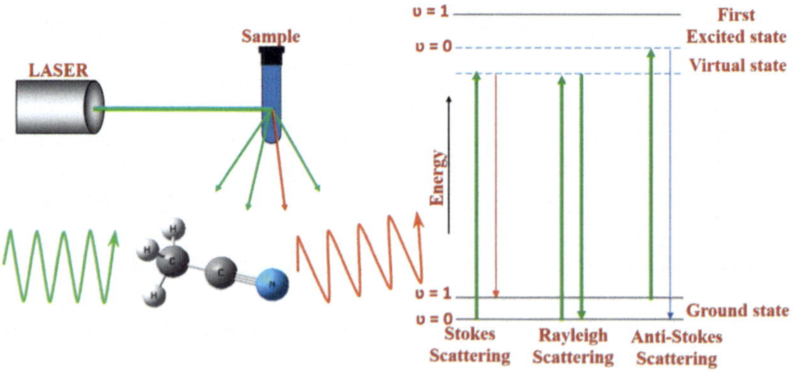

Fig. 6.6 Working procedure and scattering of Raman spectroscopy

Spectroscopy is frequently employed in food analysis, including structural analysis, impurity content, fruit and vegetable carotenoids measurement, and distinguishability [44]. Additionally, the structural examination of agricultural products and the distinction between transgenic and regular crops are discussed. Additionally, in the pharmaceutical sector, the quantification of active ingredients, the identification of minor ingredients or raw materials, and the biocompatibility of a material [45].

6.8 Transmission Electron Microscopy (TEM)

TEM is a technique that involves passing an electron beam through a specimen that is 100 nm thin and interacts with itself as it does so. According to [46], the two key characteristics of the TEM are its great lateral spatial resolution and its capacity to offer the sample's diffraction as well as an image. This method is utilized to extract the samples' whole morphological, crystallographic, atomic structural, chemical composition, bonding, electronic structure, and coordination number. Extremely tiny objects are necessary for high-resolution (HR) TEM imaging,these particles may only slightly alter the electron wave's phase while largely maintaining its amplitude. For objects classified as weak-phase, phase contrast can only be achieved through the generation of an extra phase shift between the unscathed and scattered portions of the object wave. Such a phase shift is produced in traditional TEM by making use of the imaging system's lens aberrations.

A TEM is a device, artifact, or tool designed especially for the analysis and display of samples or samples submitted in dimensions between 1 μm and 1 nm. According to Cristian Fabian Escalante Sierra [47], this type of electron microscope can disclose extremely sophisticated levels of detail that are not visible with a traditional light microscope. A high-energy electron beam is sent through a solid using the TEM technique to image the interior structure of the material. This configuration is comparable to a simple transmission-illuminated optical microscope (Fig. 6.7b). When it comes to the exterior morphology of solids, TEM photographs are typically more challenging to understand than SEM images. Although the fundamentals of TEM imaging remain unchanged, over time, numerous additional facilities, primarily analytical ones, have been added to a basic TEM microscope [48]. The electron cannon, which is positioned at the top of the TEM, generates the electron that passes through the vacuum tube and helps to make the image using a high-voltage electron beam. The ultra-thin specimen is subjected to a tiny electron beam, which is focused using an electromagnetic lens [49]. The fluorescent screen at the base of the microscope is struck by the electrons that are transmitted. Based on its density, the specimen's various components' image is displayed on the screen (Fig. 6.7a) [50].

Life sciences, nanotechnology, medical, biological, and material research, forensic analysis, gemmology, metallurgy, industry, and education are just a few of the many domains where TEM finds its perfect use. Information on topography, morphology, composition, and crystal structure is provided by TEMs. The pictures

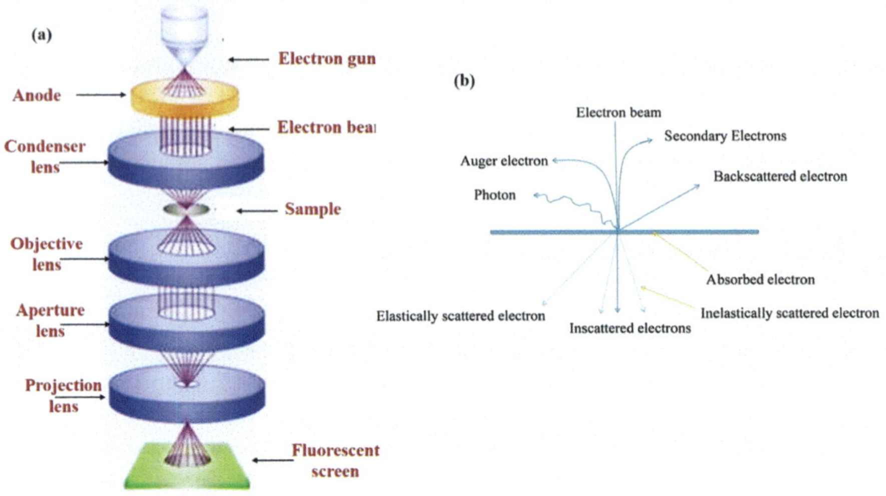

Fig. 6.7 **a** Working and instrumentation of TEM analysis, **b** principle operation of TEM technique

enable the scientists to examine materials at the molecular level, enabling the examination of texture and structure. In addition to industrial uses, this information is helpful in the study of metals and crystals. According to Priyanka and Om Prakash [51], TEMs are useful in semiconductor investigation as well as the creation and manufacturing of silicon and computer chips.

6.9 Field Emission Scanning Electron Microscopy (FESEM)

The foundation of FESEM's development is its high-resolution imaging technology and various contrasting techniques, which work toward a thorough characterization of specimens. Without requiring any prior preparation, surface-sensitive and non-conductive sample imaging is one of the many uses for FESEM. Various attachments for elemental analysis are also included with FESEM [52]. The scanning electron microscope uses emitted electrons rather than transmitted ones as the TEM does. The scanning electron microscope operates using the kinetic energy concept, which generates signals based on electron contact (Fig. 6.8).

These electrons, which are utilized to see photons and crystalline components, are secondary electrons, diffracted backscattered electrons, and backscattered electrons. An image is created using backscattered and secondary electrons [53]. While the backscattered electrons reveal differences in the elements' composition, the secondary electrons released from the material serve as the main means of identifying its topography and shape. SEM offers the benefit of Their interfaces are

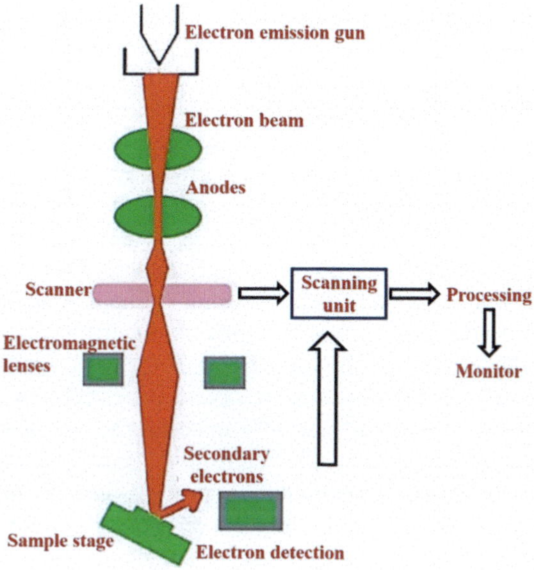

Fig. 6.8 Working procedure and instrumentation of FESEM analysis

user-friendly and they are simple to use. They are employed to examine solid object surfaces in a range of industrial settings. Certain contemporary SEMs can produce digital data that is portable. Data from the SEM can be easily obtained in a short time (5 min) [54, 55].

6.10 Atomic Force Microscopy (AFM) Spectroscopy

AFM holds the high-resolution type of scanning probe microscopy that offers a 3D profile of the surface on a nanoscale. It was measured by the forces between a sharp probe (less than 10 nm) and the surface at extremely short distances. The resolution of AFM has been demonstrated to be on the order of fractions of a nanometer. There is a flexible cantilever supporting the probe. According to Xiangying et al. [56], the AFM tip lightly touches the surface and measures the little force that exists between the probe and the surface. A piezoelectric ceramic scanning device for sample scanning, a micro-cantilever with a probe, a micro-cantilever motion detection device, a feedback loop for monitoring micro-cantilever motion, and a computer-controlled image acquisition, display, and processing system make up the AFM system. By identifying extremely weak interatomic interactions between the sample surface and the probe tip, AFM analyzes the surface structure and characteristics of the sample [57]. The concept of operation involves fixing one end of a micro-cantilever, which is very sensitive to small forces, and bringing the other end of the cantilever, which is inside the probe, close to the sample. After this is done, there is a very faint force between the probe's tip atom and the sample surface atoms that could be either repulsive or attracted. The micro cantilever's deformation and motion state are affected by the force's magnitude. The user can receive surface structure information with nanoscale resolution by using sensors to detect these changes and obtain force distribution information during the scanning process [58].

The AFM scanner has three directions of motion: *X*, *Y*, and *Z*. The vertical *Z*-direction is usually restricted to a few microns, although the *X* and *Y* directions exhibit variable distance traveled depending on the scanner. Concurrently, the position of the piezoelectric ceramic scanning tube in the *Z*-direction and the *X*–*Y* plane can be reconstructed to acquire the morphology of the scanning area sample [59]. Samples such as cells and tissues have their Young's modulus values fitted with a linear ft by Hertz model,a lower Young's modulus suggests that the sample is more prone to deformation. The quantitative force between the tip and the sample is reflected in the force-distance curve obtained by AFM measurement. The elastic modulus, adhesion, and stiffness of a material can be quantitatively measured using force-displacement curves derived from the nano-indentation of the micro areas measured by AFM using the appropriate probes outfitted with specialized software (Fig. 6.9). During the test, each force curve is preserved for subsequent quantitative analysis [60]. Different modes of AFM include; contact mode, Intermittent mode (Tapping mode), and non-contact mode [61].

The ability to scan tunneling microscopy (STM) and AFM devices to investigate samples in ambient conditions, liquids, and ultrahigh vacuums is a unique characteristic [62]. Because AFM can photograph non-conducting materials, biological systems can benefit greatly from its use [63]. Grain size analysis, pinholes or other defects in oxide growth, substrate roughness analysis, step formation in thin film epitaxial deposition, and phase mode are all highly sensitive to changes in material properties, such as adhesion, surface stiffness, and elasticity [64, 65]. By comparing the tip-sample forces curves for various materials, one can investigate Young's modulus ratio and learn more about the behavior of the material during

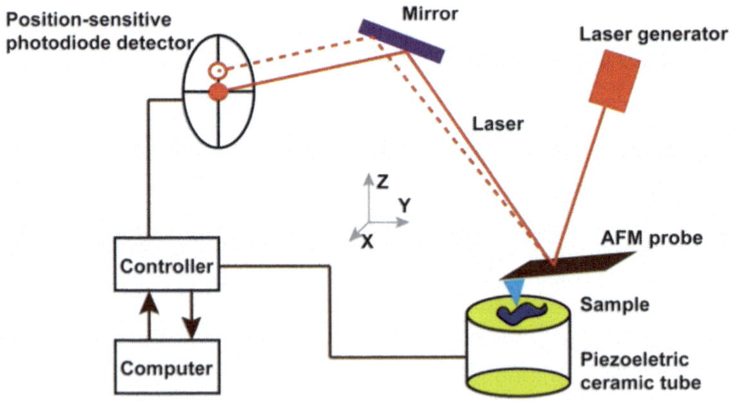

Fig. 6.9 Operating procedure of AFM technique

indentation at extremely small loads. We can investigate structural changes by using in situ AFM analysis as the temperature varies [66].

6.11 X-ray Absorption Fine Structure (XAFS)

The intricacies of how an atom absorbs X-rays at energies close to and above its core-level binding energies are referred to as XAFS. To be more precise, XAFS is the variation in an atom's X-ray absorption likelihood brought about by its physical and chemical composition. According to Evans [67], XAFS spectra are particularly sensitive to the formal oxidation state, coordination chemistry, and the atoms' distances, coordination numbers, and species that directly surround the chosen element. Owing to this dependence, XAFS is widely used in a variety of scientific domains, such as biology, environmental science, catalyst research, and material science. It offers a useful and reasonably easy method to ascertain the chemical state and local atomic structure for a chosen atomic species (Fig. 6.10b). As an atomic probe, XAFS can be applied in a range of systems and sample settings, with few restrictions on the shape of the samples that can be investigated [68].

Measurements of X-ray absorption are quite simple as long as you have a powerful and adjustable X-ray source (Fig. 6.10c). Generally speaking, this refers to the application of synchrotron radiation, and the evolution and history of XAFS closely

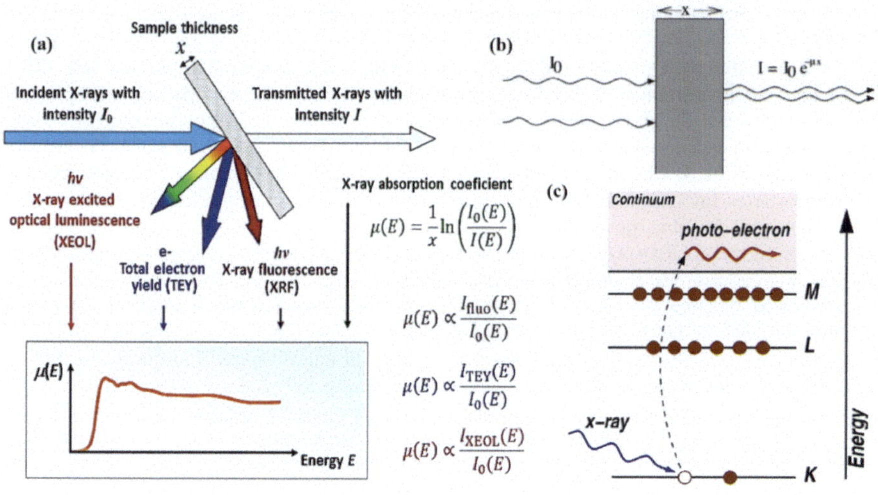

Fig. 6.10 **a** XAFS schematic and principle of operation, **b** procedure of an x-ray absorption measurement in transmission mode, and **c** formation of photoelectric effect by absorbing x-rays and promoting out atom from core-level of electrons

resemble that of synchrotron sources. Even though there are a few inherent limitations to the XAFS technique itself, the features of synchrotron sources and experimental stations dictate what energy ranges, beam sizes, and intensities are available (Fig. 6.10a). As a result, there are often practical experimental limits on the XAFS measurements that can be done at a particular station [69]. Generally speaking, there are two regimes in the X-ray absorption spectrum: extended X-ray absorption fine-structure spectroscopy (EXAFS) and X-ray absorption near-edge spectroscopy (XANES). Even if their physical origins are identical, the interpretation benefits from this distinction. While the EXAFS is utilized to ascertain the distances, coordination number, and species of the absorbing atom's neighbors, the XANES is highly sensitive to the formal oxidation state and coordination chemistry of the absorbing atom [70].

References

1. Bunaciu AA, Udristioiu EG, Aboul-Enein HY (2015) X-Ray diffraction: instrumentation and applications. Critical Rev Anal Chem 45:289–299
2. Ujwala AJ, Lagad CE, Ingole RK (2023) A review on application and importance of analytical methods for metallic preparations (incinerated ash) W.S.R. to XRF and XRD. World J Pharmaceu Med Res 9:47–54
3. Ribis S (2022) Fundamental principles of X-ray powder diffraction and applications. Pharm Anal Acta 13:1000688
4. Ali A, Chiang YW, Santos RM (2022) X-Ray diffraction techniques for mineral characterization: a review for engineers of the fundamentals, applications, and research directions. Minerals 12:205
5. Nidhi R, Rahul M, Dnyaneshwar K, Youngren-Ortiz SR, Chougule MB, Tekade RK (2019) Chapter 10: importance of physicochemical characterization of nanoparticles, in pharmaceutical product development. In: Tekade RK (ed) In advances in pharmaceutical product development and research, basic fundamentals of drug delivery. Academic Press, pp 369–400
6. Surendra N (2013) Chapter 3: Tools and techniques for physicochemical characterization of nanoparticles. In: Surendra N (ed) In Woodhead Publishing series in biomedicine, gene therapy. Woodhead Publishing, pp 43–63
7. Falke S, Betzel C (2019) Dynamic light scattering (DLS): principles, perspectives, applications to biological samples. Radiat Bioanal 8:173–193
8. Maskos M, Stauber RH (2017) Chapter 3.2:1 characterization of nanoparticles in biological environments. In: Ducheyne P(ed) Comprehensive biomaterials II. Elsevier, pp 467–481
9. Xiaoli Z, Tao G (2019) Chapter 10—spectrometry, in nano-inspired biosensors for protein assay with clinical applications. In: Li G (ed) Elsevier, pp 237–264
10. Rita S, Nishi SS, Jyotika D, Kamal G, Rajan S (2018) Dynamic light scattering (DLS) technique, principle, theoretical considerations and applications. In: Centre of advanced faculty training in dairy processing, nanotechnological and biochemical techniques for assessing the quality and safety of milk and milk products. ICAR-NDRI, pp 135–137
11. Naveen J, Deshmukh GP, Rekha RM (2019) Dynamic light scattering: advantages and applications. Acta Scientific Nutritional Health 3:50–52
12. Picollo M, Aceto M, Vitorino T (2019) UV-Visible spectroscopy. Phys Sci Rev 4:4
13. Diffey BL (2002) Sources and measurement of ultraviolet radiation. Methods 28:4–13
14. Liu PF, Avramova LV, Park C (2009) Revisiting absorbance at 230 nm as a protein unfolding probe. Anal Biochem 389:165–170

15. Domingo C, Saurina J (2012) An overview of the analytical characterization of nanostructured drug delivery systems: Towards green and sustainable pharmaceuticals: a review. Anal Chim Acta 744:8–22
16. Stevenson K, McVey AF, Clark IBN, Swain PS, Pilizota T (2016) General calibration of microbial growth in microplate readers. Sci Rep 6:38828
17. Patel MUM, Demir-Cakan R, Morcrette M, Tarascon JM, Gaberscek M, Dominko R (2013) Li-S battery analyzed by UV/Vis in operando mode. Chemsuschem 6:1177–1181
18. Behzadi S, Ghasemi F, Ghalkhani M et al (2015) Determination of nanoparticles using UV-visible spectra. Nanoscale 7:5134–5139
19. Lakowicz JR (2006) Principles of fluorescence spectroscopy. Springer, New York, NY, p 954
20. Albani J (2007) Principles and applications of fluorescence spectroscopy. Blackwell Science, Oxford, Ames, Iowa
21. Ardui S, Ameur A, Vermeesch JR, Hestand MS (2018) Single-molecule real-time (SMRT) sequencing comes of age: applications and utilities for medical diagnostics. Nucleic Acids Res 46:2159–2168
22. Diaz-Garcia ME, Badia-Laino R (2005) Fluorescence, overview, in encyclopaedia of analytical science, 2nd edn. In: Worsfold P, Townshend A, Poole C (eds) Elsevier, pp 97–106
23. Senesi N, D'Orazio V (2005) Fluorescence spectroscopy. In: Hillel D (ed) Encyclopaedia of soils in the environment. Elsevier, pp 35–52
24. Ewa S, Igor K, Sikorski M (2019) Chapter 19: Fluorescence spectroscopy and imaging instruments for food quality evaluation. In: Jian Z, Xichang W (eds) Food science, technology and nutrition, evaluation technologies for food quality. Woodhead Publishing, pp 491–533
25. Aldo R, Catia C, Costanza M, Gianna F (2010) Fluorescence spectroscopy: a powerful technique for the non-invasive characterization of artwork. Acc Chem Res 43:837–846
26. Markus S, Johan H, Jorg E (2011) Handbook of fluorescence spectroscopy and imaging: from single molecules to ensembles. Wiley-VCH Verlag GmbH & Co KGaA. https://doi.org/10.1002/9783527633500
27. Shahzad A, Kohler G, Knapp M et al (2009) Emerging applications of fluorescence spectroscopy in the medical microbiology field. J Transl Med 7:99
28. Kommu N (2014) Applications of fluorescence spectroscopy. J Chem Pharmaceu Sci 5:18–21
29. Barouch G, Michael C (2021) Fourier transform infrared (FTIR) spectroscopy for measurements of vehicle exhaust emissions: a review. Appl Sci 11:7416
30. Peak D (2005) Fourier transform infrared spectroscopy. In: Hillel D (ed) Encyclopedia of soils in the environment. Elsevier, pp. 80–85
31. Tranter GE (2017) FTIR spectroscopy of aqueous solutions. In: Lindon JC, Tranter GE, Koppenaal DW (eds) Encyclopedia of spectroscopy and spectrometry, 3rd edn. Academic Press, pp 762–769
32. Shuming W, Jian L, Jiushuai D (2021) Chapter 3: methods for the detection and composition study of fluid inclusions. In: Wen S, Liu J, Deng JJ (eds) Fluid inclusion effect in flotation of sulfide minerals. Elsevier, pp 27–68
33. Shahid Ali K, Sher Bahadar K, Latif Ullah K, Aliya F, Kalsoom A, Asiri AM (2018) Chapter 9: Fourier transform infrared spectroscopy: fundamentals and application, in functional groups and nanomaterials characterization. In: Sharma K (ed) Handbook of materials characterization. Springer International Publishing, pp 317–344
34. Berthomieu C, Hienerwadel R (2009) Fourier transform infrared (FTIR) spectroscopy. Photosynth Res 101:157–170
35. Kosa G, Shapaval V, Kohler A et al (2017) FTIR spectroscopy as a unified method for simultaneous analysis of intra- and extracellular metabolites in high-throughput screening of microbial bioprocesses. Microb Cell Fact 16:195
36. Mester L, Govyadinov AA, Chen S et al (2020) Subsurface chemical nanoidentification by nano-FTIR spectroscopy. Nat Commun 11:3359
37. Zapata F, Lopez-Fernandez A, Ortega-Ojeda F, Quintanilla G, Garcia-Ruiz C, Montalvo G (2021) Introducing ATR-FTIR spectroscopy through analysis of acetaminophen drugs: practical lessons for interdisciplinary and progressive learning for undergraduate students. J Chem Educ 98:2675–2686

38. Sayantan N (2021). Raman Spectroscopy. https://doi.org/10.13140/RG.2.2.24191.33445
39. Kolesov BA (2022) An introduction to Raman spectroscopy. Cambridge Scholars Publishing
40. de Oliveira Penido CAF, Pacheco MTT, Lednev IK, Silveira L (2016) Raman spectroscopy in forensic analysis: identification of cocaine and other illegal drugs of abuse. J Raman Spectroscopy 47:28–38
41. Dresselhaus MS, Dresselhaus G, Jorio A, Filho AGS, Saito R (2002) Raman spectroscopy on isolated single wall carbon nanotubes. Carbon 40:2043–2061
42. Abdelsalam M (2011) Theoretical studies of Raman scattering. Universitets service US-AB
43. Radzol ARM, Lee KY, Mansor W, Yahaya SR (2012) Nano-Scale characterization of surface enhanced raman spectroscopic substrates. Procedia Eng 41:867–873
44. Anamika TM (2019) Raman spectroscopy: an experimental tool to probe molecular level processes. Gyankosh: Interdiscip J 2:34–47
45. Samir K, Prabhat K, Anamika D, Shakher PC (2020) Surface-enhanced Raman scattering: introduction and applications, recent advances in nanophotonics - fundamentals and Applications. IntechOpen
46. Hettler S, Dries M, Zeelen J, Oster M, Schröder RR, Gerthsen D (2016) High-resolution transmission electron microscopy with an electrostatic Zach phase plate. New J Phys 18:053005
47. Cristian Fabian Escalante Sierra (2019) Fundamentals of transmission electron microscopy, the technique with the best resolution in the world
48. Liu BX, Lai WS, Zhang ZJ (2001) Solid-state crystal-to-amorphous transition in metal-metal multilayers and its thermodynamic and atomistic modeling. Adv Phys 50:367–429
49. Sid AS, Zoubeida M, Abdeldjalil O, Sylvain T, Cedric M, Sergio M (2015) Nonparametric denoising methods based on contourlet transform with sharp frequency localization: application to low exposure time electron microscopy images. Entropy 17:3461–3478
50. Nabi U, Raina Aman Q, Sami U, Saraf K (2022) Application and importance of scanning and transmission electron microscopes in science and technology. Sec Nat Math Biotech Sci MASA 42–43:75–85
51. Priyanka, Om Prakash C (2018) Uses of transmission electron microscope in microscopy and its advantages and disadvantages. Int J Curr Microbiol.App.Sci 7:743–747
52. Gnanamoorthy P, Karthikeyan V, Ashok Prabu V (2014) Field emission scanning electron microscopy (FESEM) characterization of the porous silica nanoparticulate structure of marine diatoms. J Porous Mater 21:225–233
53. Alyamani A, Lemine OM (2012) FE-SEM Characterization of Some Nanomaterial, in Scanning Electron Microscopy. Editor(s) Kazmiruk V, InTechOpen. https://doi.org/10.5772/34361
54. Erdman N, Bell DC, Reichelt R (2019) Scanning electron microscopy. In: Hawkes PW, Spence JCH (eds) Springer handbook of microscopy. Springer
55. Anshida M, Shaji LK, Nair AK, Nandakumar L (2018) Chapter 12: morphological characterization of nanomaterials.In: Mohan BS, Samuel OO, Nandakumar K, Sabu T (eds) Micro and nano technologies, characterization of nanomaterials. Woodhead Publishing, pp 335–364
56. Xiangying D, Fang X, Xiayu L, Xiang XB, Zheng L, Xu W, Can G, Xiaoling L, Yong L, Guiyuan L, Wei X, Zhaoyang Z (2018) Application of atomic force microscopy in cancer research. J Nanobiotechnol 16:102
57. Wang YA, Li XL, Mo YZ, Fan CM, Tang L, Xiong F et al (2018) Effects of tumor metabolic microenvironment on regulatory T cells. Mol Cancer 17:168
58. Dufrene YF, Pelling AE (2013) Force nanoscopy of cell mechanics and cell adhesion. Nanoscale 5:4094–4104
59. Fan C, Tang Y, Wang J, Wang Y, Xiong F, Zhang S et al (2018) Long non-coding RNA LOC284454 promotes migration and invasion of nasopharyngeal carcinoma via modulating the Rho/Rac signaling pathway. Carcinogenesis 40:380–391
60. Ling-Zhi C, Weidong Z, Shuang S, Cai S (2019) Lab on a tip: Applications of functional atomic force microscopy for the study of electrical properties in biology. Acta Biomater 99:33–52
61. Xia F, Youcef-Toumi K (2022) Review: advanced atomic force microscopy modes for biomedical research. Biosensors 12:1116

62. Sun W (2018) Principles of Atomic Force Microscopy, in Atomic Force Microscopy in Molecular and Cell Biology, Editor(s) Cai J, Springer, Singapore
63. Robins LH, Brubaker MD, Tung RC, Killgore JP (2020) Isomorphic contact resonance force microscopy and piezoresponse force microscopy of an AlN thin film: demonstration of a new contact resonance technique. Nano Futures 4:025003
64. Killgore JP, Deolia A, Robins LH, Murray TW (2019) Experimental reconstruction of the contact resonance shape factor for quantification and amplification of bias-induced strain in atomic force microscopy. Appl Phys Lett 114:133108
65. MacDonald GA, DelRio FW, Killgore JP (2018) Higher-eigenmode piezo response force microscopy: a path towards increased sensitivity and the elimination of electrostatic artifacts. Nano Futures 2:015005
66. Wu Y, Fang Y, Fan Z, Liu C (2021) Accurate morphology characterization using atomic force microscopy via vertical drift correction and illusory slope elimination. Microsc Microanal 27:1366–1374
67. Evans J (2018) X-Ray absorption spectroscopy for the chemical and materials sciences, 1st edn. Wiley & Sons, pp 224
68. Newville N (2014) Fundamentals of XAFS. Rev Mineral Geochem 78:33–74
69. Shrivastava BD (2012) X-ray absorption fine structure (XAFS) spectroscopy using synchrotron radiation. J Phys: Conf Ser 365:012002
70. Abhijeet G, Shrivastava BD, Nigam HL (2013) X-Ray absorption fine structure (XAFS) spectroscopy – a review. Proc Indian Natn Sci Acad 79:921–966

Chapter 7
Photophysics of Graphene Quantum Dots and Their Composites

7.1 Introduction

Graphene is an interesting 2D honeycomb carbon structure with very good electrical and thermal conductivity, strong chemical stability, and high mechanical strength [1–6]. It can be prepared in different forms, such as sheets [7], nanoribbons [8], and graphene quantum dots (GQDs [9], by various methods. The size and shape, in turn, determine the properties and applications of graphene. For example, GQDs and graphene nanoribbons have photoluminescence properties [9, 10], whereas graphene sheets do not due to their metallic character [4]. While GQDs are finding applications in biology as biomarkers [11, 12], graphene sheets are being used in optoelectronic device fabrication [13]. As far as preparation is concerned, graphene sheets are obtained by exfoliating graphite using various methods, which include Bordie's and Hummer's methods [14 15, 16, 17, 18, 19]. The chemical vapor deposition method is also used to prepare pristine graphene sheets [13]. In contrast, GQDs are obtained by top-down and bottom-up methods [9, 20]. A pristine graphene sheet is a zero-bandgap material due to the convergence of the valence band maximum and conduction band minimum at the Dirac point [21].

Photo-excited electron–hole pairs will undergo a non-radiative relaxation process through the continuum band states toward the band edge states, and therefore, photoluminescence is not possible. If a graphene sheet has to be used as an active layer in device fabrication, similar to other semiconducting materials, its bandgap must be opened. However, opening up the bandgap in graphene sheets is a challenging task, because of the nature of the carbon network. In this sp^2 hybridized network, each carbon atom is connected to three other carbon atoms on the 2D plane through their 2 s, $2p_x$, and $2p_y$ (sigma bond) state electrons. The fourth valence electron of a carbon atom which was left out in the 2D network formation is $2p_z$ electrons and it is available in the third dimension to make the π bond network. In graphene sheets, larger numbers of pi bonds are there, and when they are all well interconnected, then a band of continuum energy states is produced in the valence and conduction

N. Manjubaashini et al., *Graphene Quantum Dots*, Materials Horizons: From Nature to Nanomaterials, https://doi.org/10.1007/978-981-97-5722-0_7

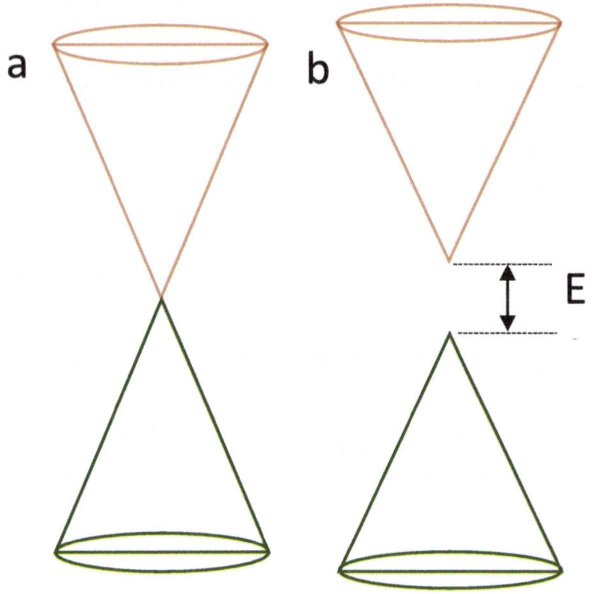

Fig. 7.1 Schematic illustration of the band structure of **a** pristine graphene and **b** semiconducting graphene with sp^2 domains

bands as shown in Fig. 7.1a. This, in addition to zero effective mass value for charge carriers in the graphene, brings zero-bandgap value to the graphene sheet, as reported elsewhere [22].

The energy band gap in graphene is possible through the formation discontinuity in the pi-bond network, such that part of the sp^2 network is isolated into a nanoscale domain, and as a consequence, separated molecular-like energy levels can be seen from these domains. Such an energy level distribution has a non-zero energy band gap in the graphene as represented in Fig. 7.1b. But the challenge is about the mechanism of introducing a discontinuity in the pi-bond network, in a way such that several localized sp^2 domains of similar size are formed in the graphene network. If it is so then the graphene sheet becomes a graphene QD interconnected solid. Efforts have been taken to see the quantum behavior in graphene sheets. For example, a surface oxidation/reduction process can help to form such islands of GQDs [23]. In this mechanism, oxygen played a major role in isolating sp^2 domains through the formation of a sp^3 hybridized carbon network around sp^2 islands. Even it was possible to see a weak blue emission from such graphene sheets [24]. In a recent research report, the titanium metal layer was coated over the as-grown graphene sheet to introduce a disorder-assisted discontinuity in the sp^2 carbon network to see the quantum behavior. In the molecular-like energy levels of the graphene domain, the photo-excited electron–hole pairs can undergo radiative recombination, resulting in an intrinsic emission. Such an intrinsic emission is an indication that the energy band gap has been opened in the graphene sheets. Similarly, H. Yoon et al., have formed oxidation-controlled graphene quantum dots from graphite oxide (prepared by the modified Hummers) and studied the intrinsic light emission property [25].

N. Amirhasan et al., have used plasma treatment on micromechanically exfoliated single-layer graphene to induce intrinsic properties and see a broad green light emission [26]. In these modified graphene sheets, a single-layer configuration is necessary to avoid the highly possible interlayer PL quenching effect [27].

From the overall observations, it is clear that there is difficulty in obtaining a strong intrinsic emission yield from a graphene sheet sample. Reports are showing a weak emission from graphene sheet samples and it is due to surface-attached molecules with excitation-energy-dependent behavior [28, 29]. The best way to obtain a strong emission yield from a graphene system is by preparing the samples in size-reduced form as QDs (Fig. 7.2). In such GQDs of dimensions 5–10 nm, the presence of 2–3 nm sized sp^2 domains can show quantum size-effect-induced intrinsic emission. Mostly blue emission is reported from GQDs ranging from 1.8 to 7 nm [9]. The GQDs are also well known for the blue emission from the edge effects (free zigzag sites/carbene states) [20] and $n \rightarrow \pi^*$ transitions from C = O and C = N bond [9, 30]. The emissive sites of GQDs and the optical transitions corresponding to these emissive sites are represented in Fig. 7.3. For example, carbene-like states and the $n \rightarrow \pi^*$ transition due to the absorption at 350 nm always show Stokes-shifted emissions at 450 nm, which matches with the size-effect-induced emissions at 450 nm. Hence, it is always difficult to obtain a true size-effect-induced strong intrinsic emission from GQDs. Few reports claim to have obtained size-effect-induced emissions from GQDs [31]. Others have mostly attributed the reason for the emissions to $n \rightarrow \pi^*$ transitions of C = O/C = N. Hence, there is an open challenge in graphene research to obtain high-yield photoluminescence that is purely from an intrinsic nature arising from the structure and not involving any functional groups.

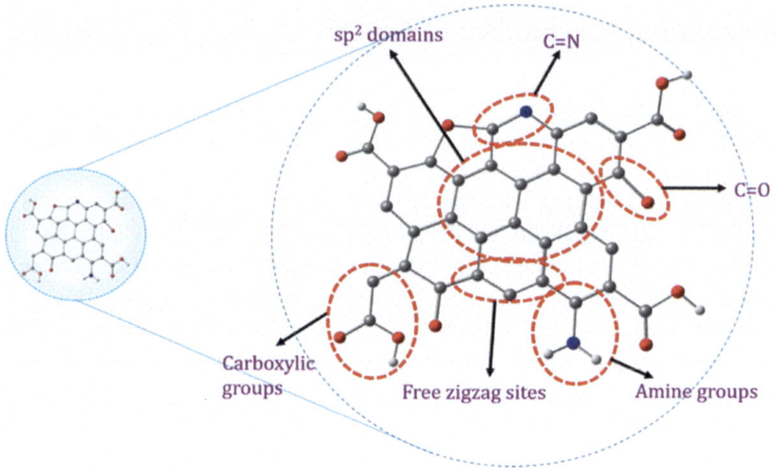

Fig. 7.2 Schematic illustration of different origins of emission in GQDs

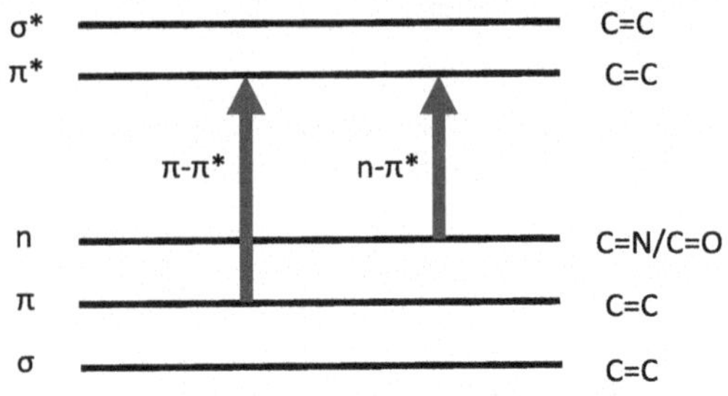

Fig. 7.3 Schematic representation of optical transitions in GQDs and other graphene systems

7.2 Challenges

The main challenges in the field of graphene sheet research are:

(i) Preparing graphene systems with intrinsic optical absorption and emission properties.
(ii) Inducing semiconducting property by band gap opening.

It is attempted to prepare a semiconducting graphene sheet with induced intrinsic optical absorption and emission properties. Able to prepare graphene quantum dot interconnected composite-type solid sheets by adding the zinc metal precursor to the hydrothermal reaction. The metal-mediated interconnection process resulted in the removal of edge-attached functional groups in GQDs, and consequently, the contribution of functional groups was suppressed effectively.

7.3 Formation Mechanism of GQDs

More extensively, the GQDs will be prepared by the incorporation of citric acid and urea molecules to improve the graphitic quality and enhance the luminescence effect of the quantum dots. When the citric acid and urea were subjected to the hydrothermal treatment (180 °C for 3 h in a muffle furnace), the reaction between citric acid and urea forms the citric acid amide. Furthermore, de-ammoniation and dehydration take place between the amino, carboxyl, and hydroxyl groups of the intermediate compound, and as a result, a nanosheet structure is formed. Aggregations of these citric acid amide condensates yield the GQDs. The GQD samples were collected and stored for further characterization. Figure 7.4 schematically represents the formation mechanism.

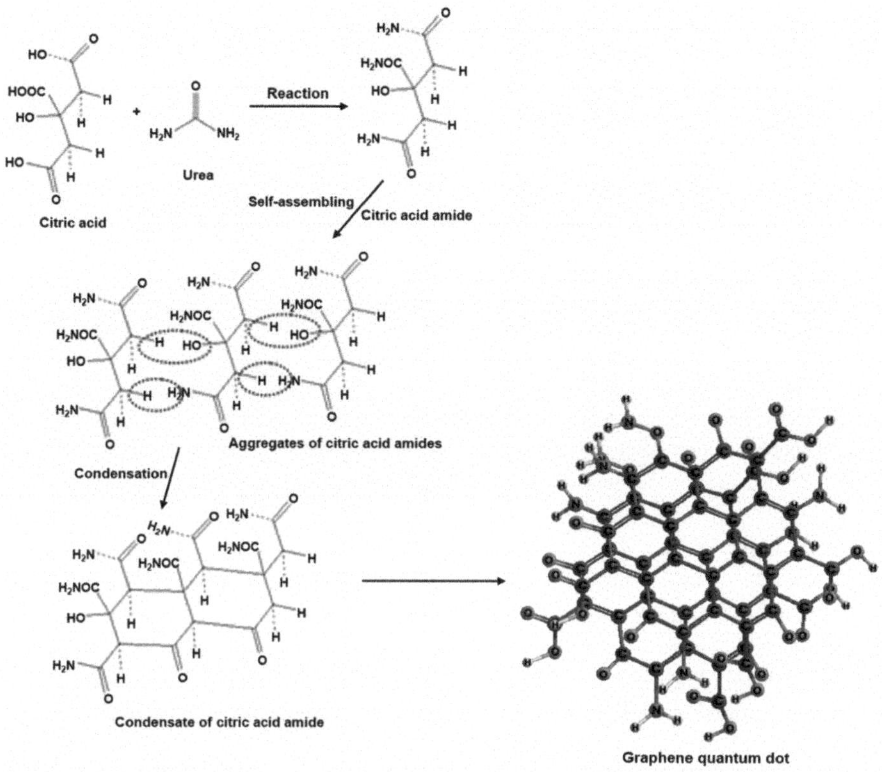

Fig. 7.4 Possible formation mechanism of graphene quantum dots by the hydrothermal treatment of citric acid and urea

7.4 HRTEM Analysis of GQDs

The hydrothermally prepared GQD sample was subjected to High-resolution transmission electron microscopy analysis. The image of GQDs given in Fig. 7.5a shows spherically shaped GQD morphology with an average size value of 5 nm (Fig. 7.5b). The corresponding selected area electron diffraction (SAED) pattern depicts the polycrystalline nature of the as-prepared GQDs (inset of Fig. 7.5b).

7.5 Optical Properties of GQDs

As-prepared GQDs were studied by UV–visible absorption and photoluminescence emission spectroscopy to investigate the optical properties. UV–Vis absorption spectrum (Fig. 7.6a) has shown two strong absorption bands at 234 nm and 334 nm, respectively, corresponding to $\pi \rightarrow \pi^*$ transition of the $C = C$ bond and $n \rightarrow \pi^*$

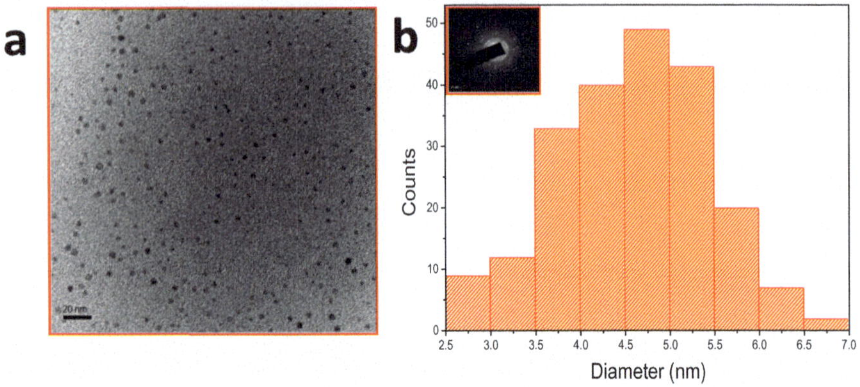

Fig. 7.5 **a** HRTEM image and **b** size distribution chart of graphene quantum dots (GQDs) showing spherical-shaped quantum dots with an average size of 5 nm. The scale bar represents the measure of 20 nm in **a**

transition of the C = O/C = N bond [9], [30]. The PL emission spectrum from GQDs (shown in Fig. 7.6b) has shown a strong emission band at the blue region (440 nm) corresponding to an excitation wavelength of 350 nm. Figure 7.7a is an excitation spectrum corresponding to the emission at 440 nm, exhibiting a strong peak at 350 nm. Figure 7.7b is the emission spectra from GQDs recorded at the different excitation wavelengths. Since there is no shift in the emission peak position, it indicates the excitation-independent emission property of the GQD samples. The as-prepared GQDs have shown strong UV–Vis absorption and PL emission properties. The PL emission from the GQD sample is the contribution of both intrinsic (originating from edge states) and extrinsic (originating from sp^2 domains) origin.

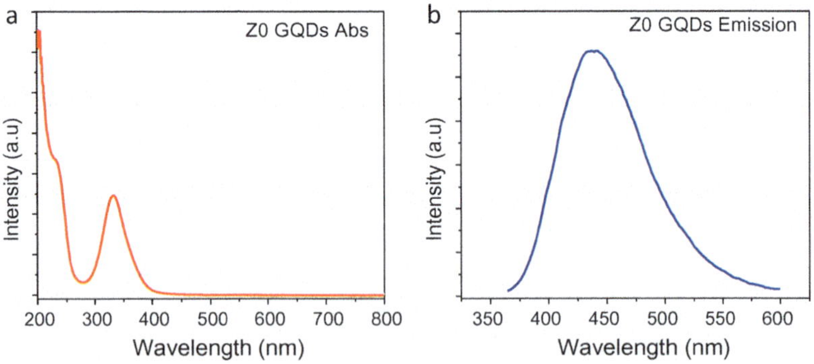

Fig. 7.6 **a** UV–Visible absorption spectrum of GQDs, showing absorption bands at 234 nm and 334 nm corresponding to $\pi \rightarrow \pi^*$ and $n \rightarrow \pi^*$ transitions, **b** PL emission spectrum of GQDs, which shows a strong emission band at 440 nm

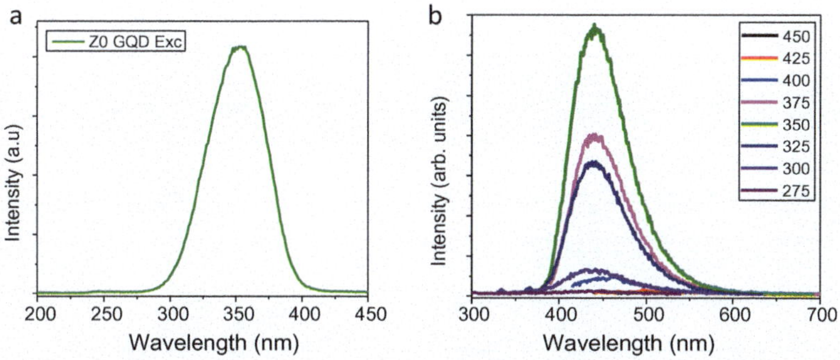

Fig. 7.7 **a** The excitation spectrum of GQDs shows an excitation peak centered at 350 nm, **b** PL emission spectra of GQDs, which show the excitation-independent emission behavior of the corresponding sample

7.6 Preparation of Zn-Graphene Nanostructures

The GQD interconnected composite-type solid sheets were obtained by adding the zinc chloride into the citric acid and urea solution. The stirring process was continued for an additional 10 min. Different concentrations of Zinc chloride viz 0.1, 0.3, 0.5, 0.7, and 0.9 mol were used and the respective samples were named Z1, Z3, Z5, Z7, and Z9 [36].

7.7 Electron Microscopy Analysis

It has been observed from HRTEM analysis that there is an increase in the size of graphene with the increase in the zinc metal concentration. Figure 7.8a shows the HRTEM images of graphene with 0.1 mol and 0.3 mol of zinc metal concentrations. The Z1 and Z3 samples respectively show 10 and 300 nm size distribution. When the metal concentration was increased to 0.5 (Z5) and 0.7 mol (Z7) well-grown micron-sized composite-type graphene sheets were obtained (Fig. 7.8b). It was further noted that all the raw samples prepared had the presence of GQDs as well, in the size range of 5 nm. From the results, it was found that the average size of the composite-type graphene sheet increases with the increase in the metal concentration (Fig. 7.9a–d).

7.8 Crystal Twins in the Composite-Type Graphene Sheets

Figure 7.10a-d shows the SAED patterns GQDs (Z1) and GQD interconnected composite solid sheets (Z3, Z5, and Z7). The GQD composite sheet samples with higher metal concentrations have produced a single crystalline SAED pattern with

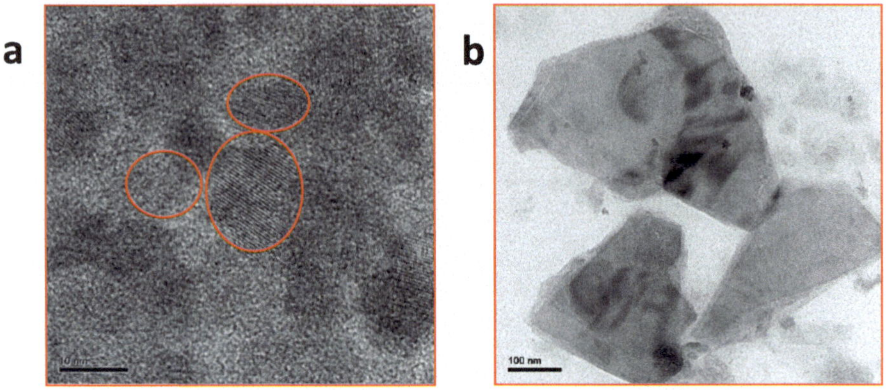

Fig. 7.8 HRTEM images of **a** spherical-shaped GQDs with a size around 10 nm were recorded from the Z1 sample and **b** slab-like GQD solid sheets in the 100 nm size range were recorded from the Z3 sample. The scale bar represents the measure of 10 nm in **a** and 100 nm in **b**

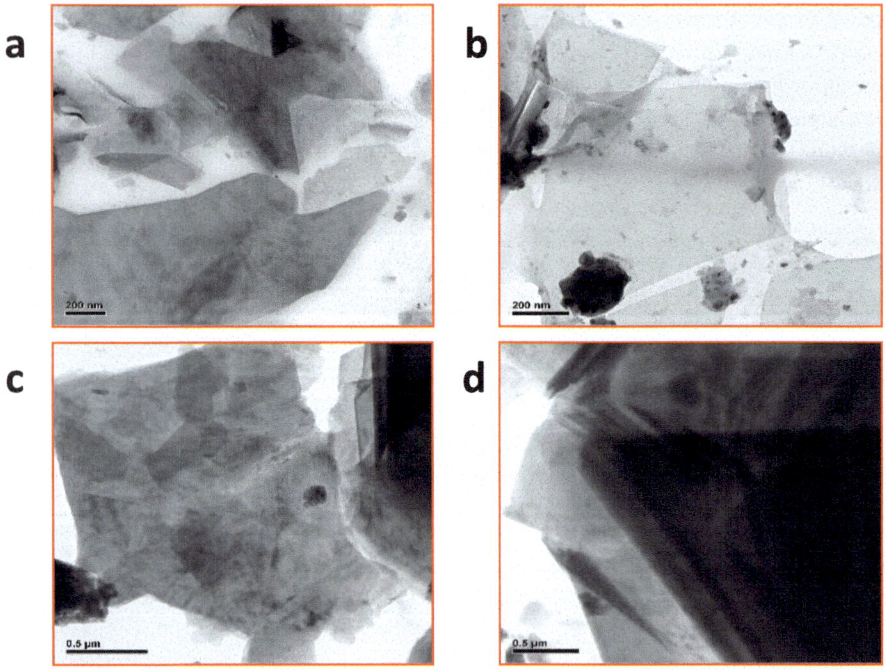

Fig. 7.9 HRTEM images of well-grown micro-sized GQD solid sheets were prepared using **a**, **b** 0.5 mol and **c**, **d** 0.7 mol metal concentrations. The scale bar represents the measure of 200 nm in **a** and **b** and 0.5 μm in **c** and **d**

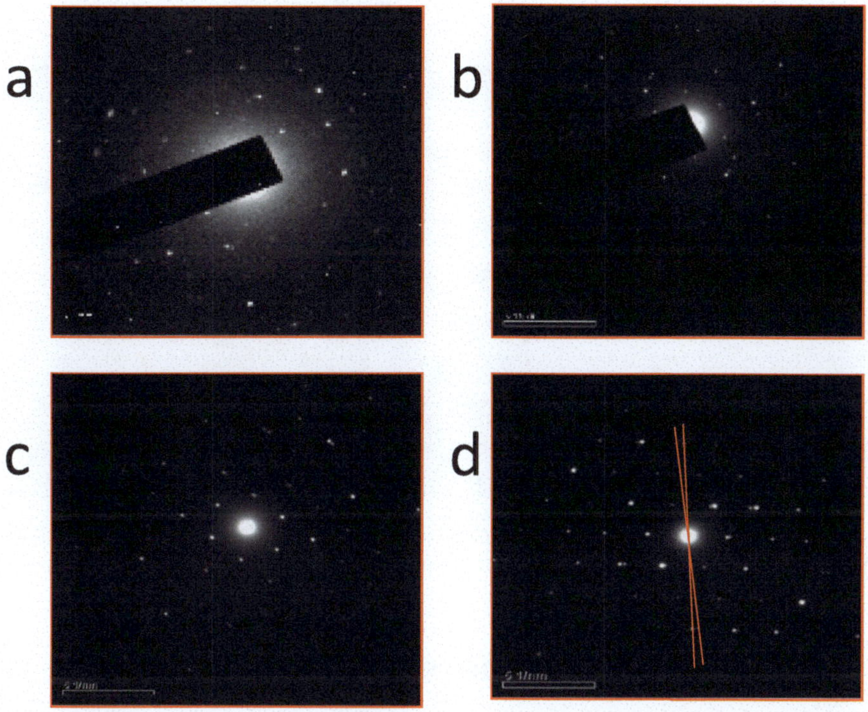

Fig. 7.10 Selected Area Electron Diffraction patterns recorded from **a** Z1 sample, **b** Z3 sample, **c** Z5 sample, and **d** Z7 sample. The SAED patterns show the polycrystalline nature of Z1 and Z3 samples, and the single crystalline nature of Z5 and Z7 samples. The double diffraction spots indicate the presence of a crystal twinning effect in the GQD interconnected solid sheet samples

double diffraction spots. The double diffraction spots indicate the presence of crystal twinning in the sheets. It is believed that crystal twinning has been introduced during the growth of the composite sheet. The zinc metal atom binds to the edges of individual GQDs and helps to grow larger structures. In this process, there is a chance for a slight orientation mismatch between the interconnecting dots leading to the twinning effect in the graphene solid sheets.

Moiré patterns in graphene occur due to the rotational or stacking misalignment of graphene layers one over the other (Fig. 7.11). The observed moiré pattern is a definitive indication of the formation of the layered structure. Moiré patterns have definite periodicity and it is because of this reason, that moiré patterns possibly form only in the layered structure, which is stacked with the van der Waals distance. Hence, the presence of a Moiré pattern in the samples gives direct evidence of the stacked graphene layers. Moiré patterns can be recorded using an aberration-corrected transmission electron microscope, which can form an optical-like projection of the electron beam transmitted by the misoriented graphene lattices in the beam path.

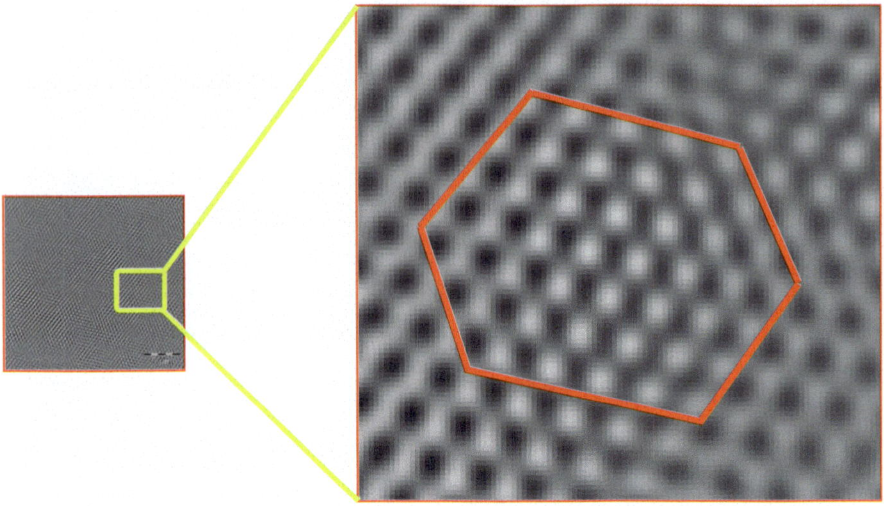

Fig. 7.11 Moiré pattern recorded from the GQD solid sheet sample (Z7). The observed pattern is similar to the reported moiré patterns created for rotation of 4°. The scale bar represents the measure of 5 nm

7.9 Raman Spectroscopy Analysis

Raman analysis was performed on the quantum dots composite sheets to determine their structural conformation. The D and G bands are displayed in Fig. 7.12a at 1350 and 1580 cm^{-1}, respectively [17]. The dominant G band in the Raman spectra of the metal-GQD composite samples (Z1-Z7) suggests the presence of graphene-like molecules within the GQDs. Because of the metal incorporation in the graphene sheet, the spectra corresponding to samples Z3–Z7 exhibit supremacy in the D band because of the sp^3-type disorder network. Moreover, substantial sheets are visible in the 2D band of GQD (Fig. 7.12b), indicating the development of a graphene sheet of the composite type. Furthermore, the Raman examination verified that the samples (Z1-Z7) had the atomic type zinc incorporated into the graphene sheet without experiencing any movement in the G band location. The findings so indicate that atomic zinc, or zinc metal, is dispersed throughout the graphene network. The D band's intensity rises with increasing zinc concentration because the structural disarray becomes more improvised. With an increase in the solid sheet dimension, the unsaturated carbon atoms in the network strengthen the D band by forming a sp^3-like disorder with oxygen atoms. The examination of Raman mapping on the separated solid sheet (Fig. 7.13a–d) from the Z7 sample shows the distribution of D and G bands throughout the flake's dimensions. The graphitic quality of the sample is shown by the ratio between the D and G bands. Extended graphene lattice development is visible in lattice-resolved HRTEM images of the sheets generated with 0.5 mol (Fig. 7.14a, b) and 0.7 mol (Fig. 7.14c, d) of metal concentration. Furthermore, Fig. 7.14d illustrates how two GQDs are coupled to each other.

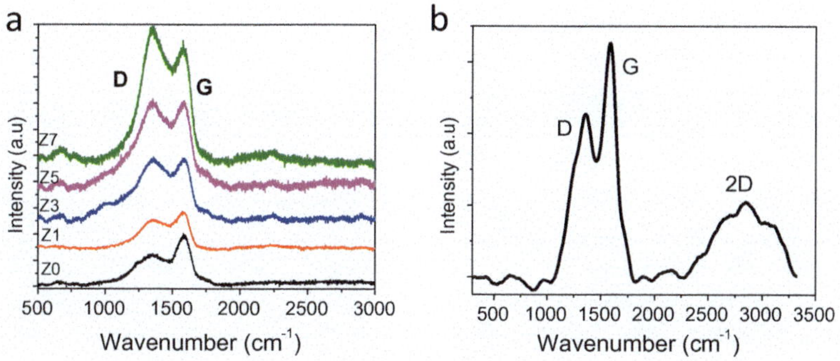

Fig. 7.12 Raman spectra **a** the raw samples, **b** the GQD solid sheet separated from the Z7 sample

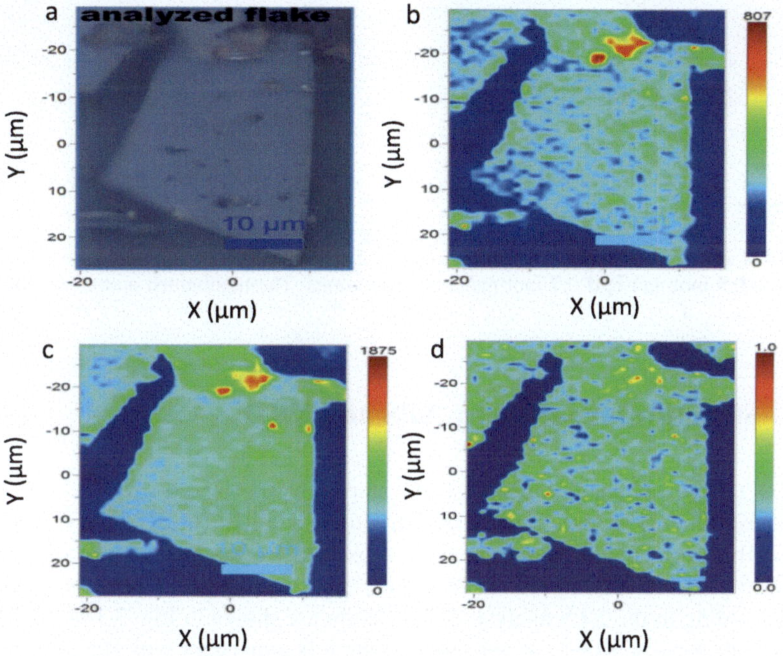

Fig. 7.13 **a** Optical microscope image of solid sheet separated from Z7 sample, Raman mapping of **b** D band, **c** G band, and **d** ratio between D and G bands. While the Red spots in the mapping of the ratio between D and G bands indicate the equal presence of D and G bands, the yellow and green spots indicate the G band dominance

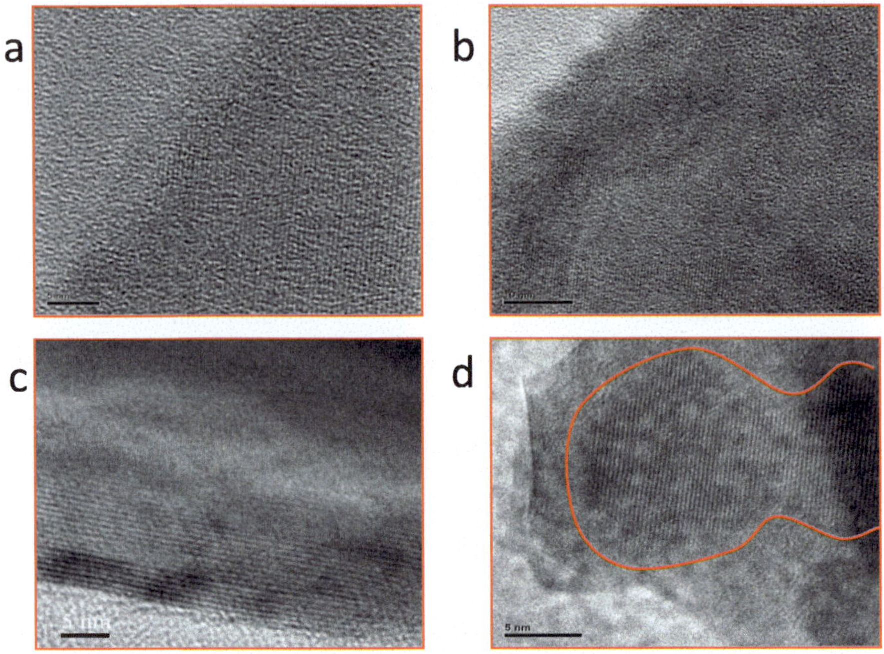

Fig. 7.14 Lattice resolution HRTEM images of well-grown micro-sized GQD solid sheets prepared using **a**, **b** 0.5 mol and **c**, **d** 0.7 mol metal concentrations. The highlighted area in **d** shows the interconnection of two GQDs. The scale bar represents the measure of 5 nm in **a**, **c**, **d** and 10 nm in **b**

7.10 Structural Analysis by XPS and XRD

The GQD and graphene composite sheet samples were subjected to XPS and XRD tests to examine the sample purity and structural quality. The XPS survey spectra showed the presence of strong peaks for oxygen, nitrogen, and carbon species (Fig. 7.15). All of the samples have a more significant C = C peak at 284.6 eV [32], according to the XPS analysis (C1s spectra are shown in Fig. 7.16a, b). Moreover, a faint peak at 288.0 eV was noted, which is consistent with the presence of C = O groups in the samples. Good-quality graphene structures made using other techniques are characterized by a low oxygen-to-carbon ratio (0.23–0.28) [33]. The nitrogen peak's strength remained constant in all the samples due to an ideal urea concentration. The percentage of nitrogen dopants was roughly 10%. Table 7.1 lists the elements for the formation of graphene composite sheets and GQDs. Additionally, the broad symmetric peak in the N1s XPS spectra (Fig. 7.17a) indicates that nitrogen is mostly found in the graphitic form in all the samples. The large signal at 531.2 eV in the O1s XPS spectra displayed in Fig. 7.17b is indicated as C = O groups.

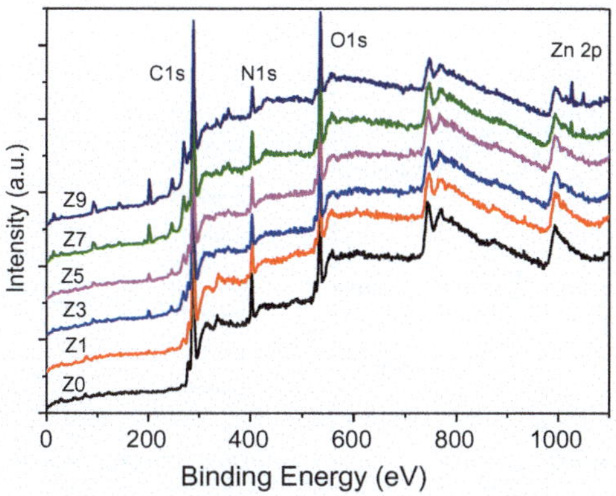

Fig. 7.15 XPS survey spectra of GQDs and GQD solid sheets showing the presence of C, N, O, and Zn elements in the as-prepared samples

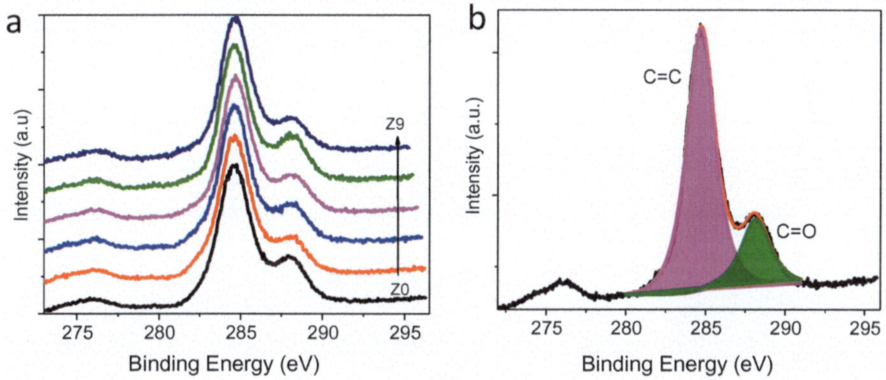

Fig. 7.16 **a** C1s high-resolution XPS spectra of GQDs and GQD solid sheets, **b** representative deconvolution of C1s spectra

Table 7.1 Elemental composition of GQDs and GQD solid sheets calculated from XPS analysis

Sample	C, at%	N, at%	O, %	Zn, %
Z0	68.20	10.98	20.82	0.0
Z1	73.16	9.27	17.57	0.0
Z3	71.46	9.54	19.00	0.3
Z5	72.63	9.44	17.93	0.4
Z7	71.75	10.05	18.20	0.6
Z9	73.27	9.45	17.28	0.9

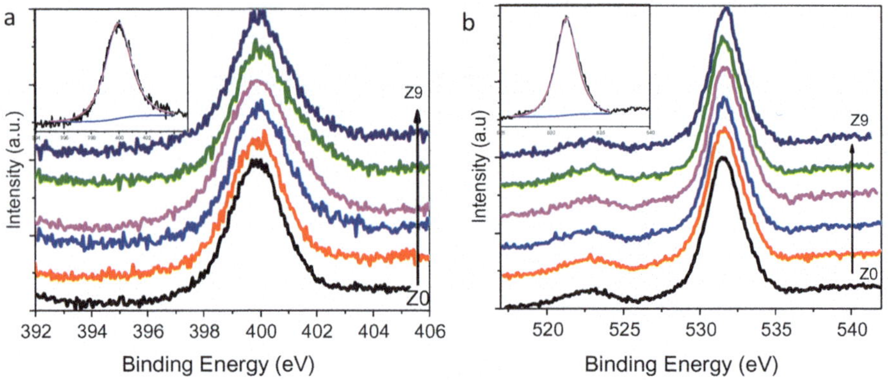

Fig. 7.17 **a** N1s high-resolution XPS spectra of GQDs and GQD solid sheets **b** O1s high-resolution XPS spectra of GQDs and GQD solid sheets. Insets of **a** and **b** show the representative deconvolution of corresponding high-resolution XPS spectra

The presence of metal was confirmed by the Zn2p XPS spectra (Fig. 7.18), which correspond to the zinc component at 1022.1 eV (Zn2p3/2). The peak strength increased as the metal concentration in the sample increased. According to Table 7.1, the Z3, Z5, Z7, and Z9 samples had computed Zn metal contents of 0.3, 0.4, 0.6, and 0.9%, respectively.

X-ray diffraction (Fig. 7.19) was used to confirm the structure of graphene, and the results showed that the Z0, Z1, and Z3 samples have interlayer spacing values of 0.346 nm, which is physically similar to graphite. Because oxygen functional groups were added between the layers of the larger-sized sheet samples, the d spacing value for the Z5, Z7, and Z9 samples increased somewhat to 0.36 nm [34]. The rise in d spacing value that occurs when oxygen functional groups are introduced between graphene layers is schematically depicted in Fig. 7.20.

Fig. 7.18 Zn 2p High-resolution XPS spectra of GQD solid sheets showing the increase of metal content in the as-prepared solid sheet samples

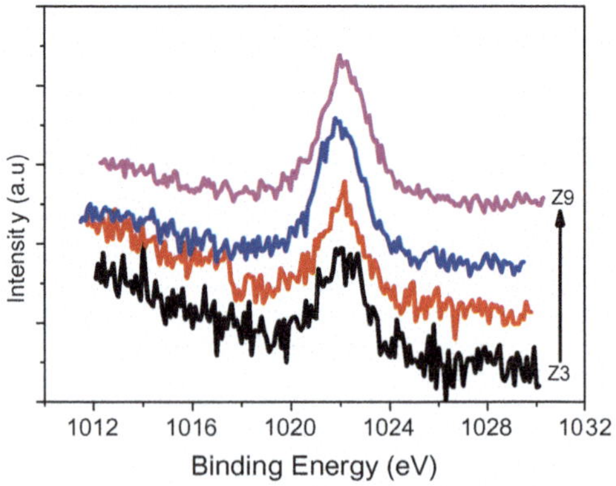

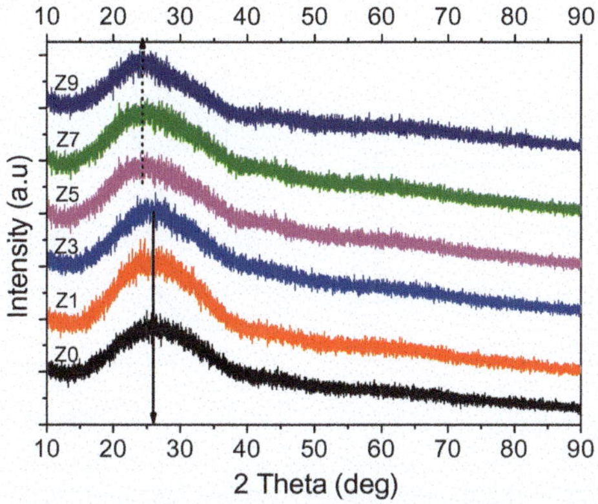

Fig. 7.19 X-ray diffraction pattern of GQDs and GQD solid sheet structures

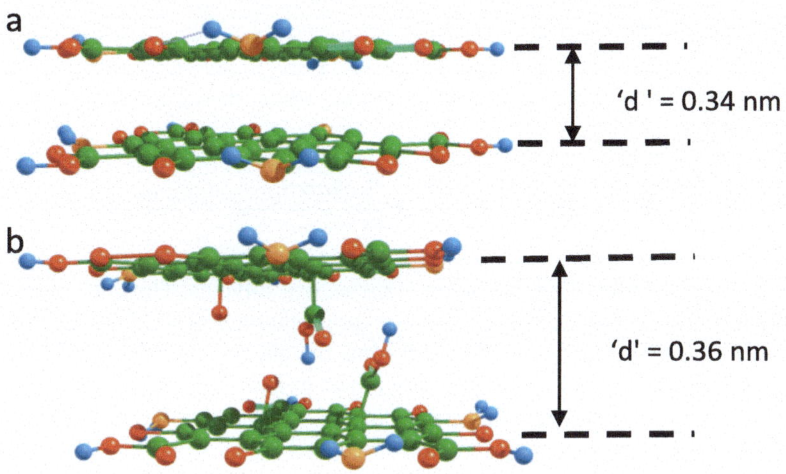

Fig. 7.20 Schematic representation of graphene layers **a** without and **b** with functional groups in between the graphene layers

7.11 Optical Properties of GQDs and GQD Solid Sheets

To obtain the optical properties of the raw samples (containing GQDs and composite sheets), absorption and emission studies were conducted. Figure 7.21a and b show the absorption and emission spectra from the raw samples (Z0–Z9). The absorption spectra of the raw samples (Fig. 7.21a) have shown an absorption band at 234 nm due to the π-π* transition of the C = C bond present in the graphene. Likewise, an absorption band at 334 nm was also noticed due to the n-π* transition of the C = O/

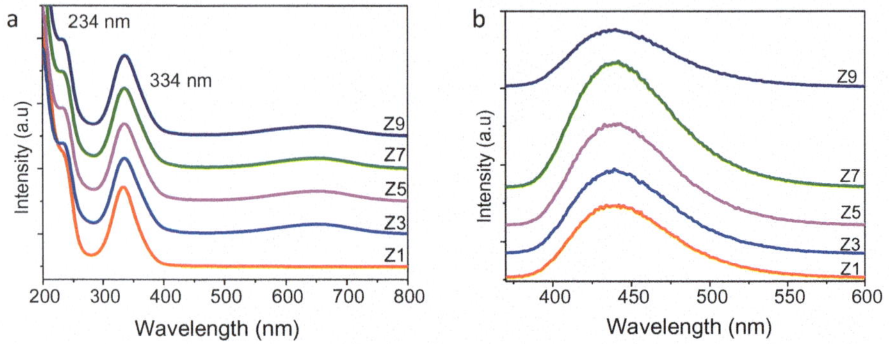

Fig. 7.21 **a** UV–Visible absorption spectra of GQD solid sheet structures showing two absorption bands at 234 and 334 nm correspond to the π-π* (C = C) and n-π* (C = O/C = N) transitions, and **b** Photoluminescence emission spectra of GQD solid sheet structures, showing an increase in emission intensity up to Z7 and further decrease for Z9 sample

C = N bonds [9]. The photoluminescence property of the raw samples was analyzed using a spectrofluorimeter instrument. The quantum yield values were calculated in comparison with the emission from the quinine sulfate dye as a reference by using the following relation [35],

$$QY_{QD} = QY_{Dye} X \frac{Abs_{Dye}}{Abs_{QD}} X \frac{Peakarea_{QD}}{Peakarea_{Dye}} X \frac{\eta_{QDsolvent^2}}{\eta_{Dyesolvent^2}}$$

where η represents the refractive index of respective solvents. The QY values of the as-prepared samples and reference dye are represented in Fig. 7.22a, b, in which the integrated emission intensity is plotted against the absorption intensity. The results have shown an increasing trend in the emission yield with the increase in zinc concentration. The GQD (Z_0) sample has given out QY of 18% at 440 nm. Other samples Z1, Z3, Z5, and Z7 have shown an increasing trend in QY from 20 to 36% with the increase of the average sheet size. On the contrary, the Z9 sample has shown a QY value of 18%.

In GQDs, domain size, edge, and nitrogen dopant effects [9, 25] are the major contributors to blue emission. The QY trend in other samples indicated that there is a correlation between the increasing trends of the average size and emission yield, viz, the larger the sheet size, the stronger the emission yield.

The strong blue emission from the sheet sample is an unusual result in graphene research. The scientific reason for this new emission behavior has to be understood. To answer this question, it was subjected the as-prepared raw samples to a filtering process so that the possible presence of graphene QDs and their contribution to the emission yield could be excluded. In this way, two sets of samples from each raw sample and measured emission yield values were obtained.

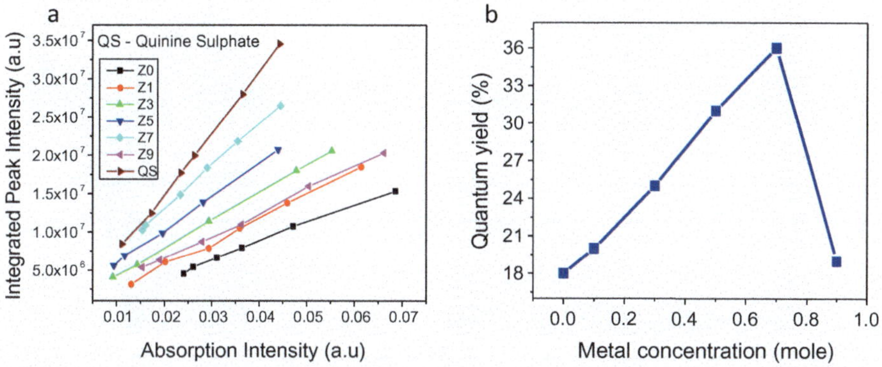

Fig. 7.22 **a** Integrated emission intensities of GQDs and GQD solid sheets compared to that of quinine sulfate dye and **b** Quantum yield values as a function of metal concentration. It indicates the increase of PL emission yield with an increase of metal concentration up to the Z7 sample and further a decrease for the Z9 sample

7.12 Filtration Process

A dialysis membrane was used for the filtration process. The double-distilled (DDW) served as the filtering medium. After 30 min of immersion in DDW at 90 °C, the dialysis membrane was activated. Following activation, the raw material was placed into a dialysis tube and placed in a beaker with the solvent. While the solvent was passing through the dialysis membrane, the quantum dots—which are smaller than the membrane's pores—moved from the solution to the solvent. The solvent was found to have changed hue, suggesting that the quantum dots had been removed by filtering. For six hours, the solvent was changed every four hours. The samples that had been filtered out and those that had not were gathered individually and kept for later analysis.

7.13 Morphology and Optical Properties of Filtered Samples

From the divided samples, the optical excitation and emission spectra were captured. After being filtered out of the membrane, the Z7 sample's GQDs reveal 18% QY at 440 nm. At 440 nm, the sample without filtering exhibits a lower QY of 3%. Further, 36% of QY is shown in the raw sample (Z7). Consequently, the obtained size and shape of samples QY are the cause of the discrepancy. 2–5 nm is present in the filtered-out GQD data. Nonetheless, the unfiltered sample has a micron-scale sheet shape. Figure 7.23a, b shows the TEM images of the filtered-out GQDs and the unfiltered composite sheet samples. Figure 7.24a, b shows the absorption and emission spectra of the corresponding samples.

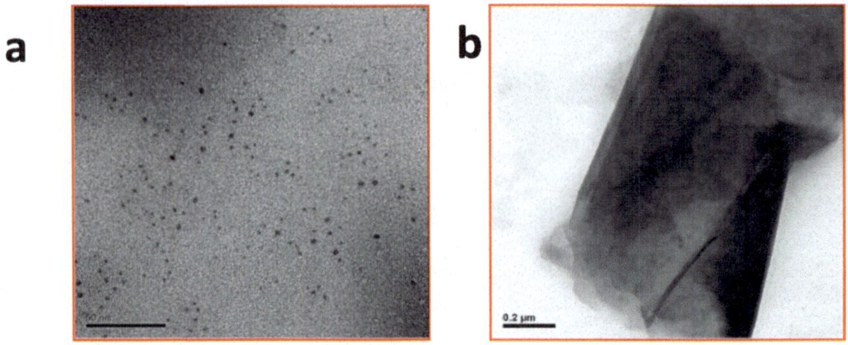

Fig. 7.23 TEM images **a** GQD, **b** graphene composite sheets separated from the Z7 sample. The scale bar represents the measure of 50 nm in **a** and 0.2 μm in **b**

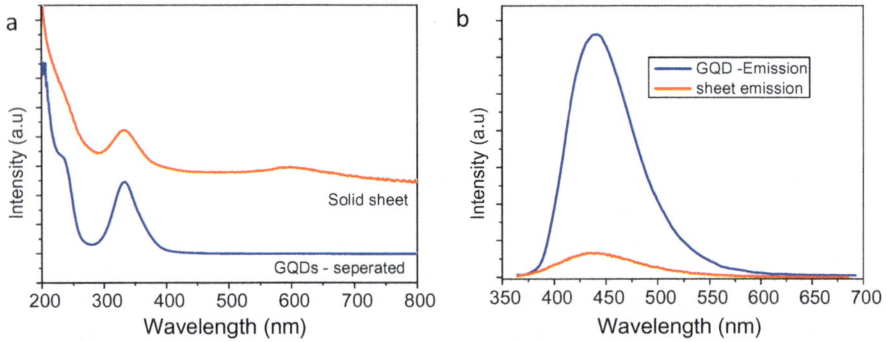

Fig. 7.24 **a** UV–visible absorption spectra, **b** PL emission spectra of filtered-out GQDs and filtered-in solid sheet samples separated from the Z7 raw sample

The existence of sp^2 domains with varying sizes in the composite-type sheet sample is indicated by a red shift in the emission maxima. For separated composite sheets with varied excitation wavelengths, a weak emission spectrum was observed (Fig. 7.25). Both the raw GQD sample (Z0) and the filtered-out GQD sample show a QY of 18%. Remarkably, compared to the raw sample, the composite of graphene composite sheets and filtered-out GQD yielded an emission yield that was 36% greater (Z7). The high QY of 18% for raw GQDs was generated by the interaction between GQDs and sheets. There may be an interaction of the Förster Resonance Energy Transfer (FRET) type, in which energy is transferred from the excited donor to the acceptor non-radiatively.

The excitation spectra of unfiltered and filtered-out GQD composite sheet samples were examined to validate the FRET mechanism. The excitation peak of the composite sheet sample is located at 410 nm, which is redshifted from the 350 nm excitation peak of GQDs. Because of the strong excitation band of GQDs, the specific excitation peak at 410 nm does not arise for the raw sample. Because of the higher

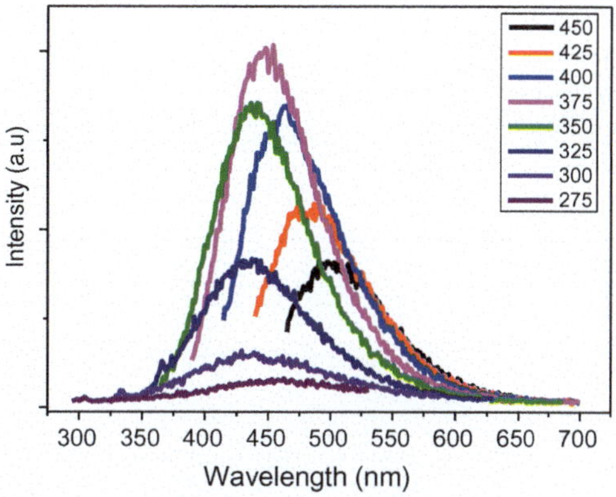

Fig. 7.25 PL emission spectra of filtered-in solid sheet samples separated from the Z7 raw sample recorded as a function of different excitation wavelengths

metal content in the Z3 sample, the peak just marginally emerged (Fig. 7.26). Moreover, the GQD emission band and the excitation peak at 410 nm overlap (Fig. 7.27). By increasing the sheet size, the percentage of overlap between the emission spectra of the GQD sample and the excitation spectra of the composite sheet sample was found. Therefore, it proves that GQD and graphene composite sheet samples have a higher energy transfer interaction. One of the requirements for the energy transfer interaction is encountered as the GQD acts as a donor and the composite sheet acts as an acceptor.

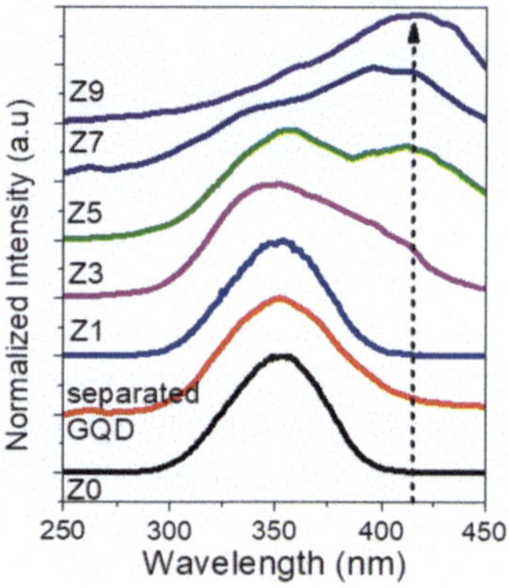

Fig. 7.26 Excitation spectra of GQDs and solid sheets separated from the raw samples

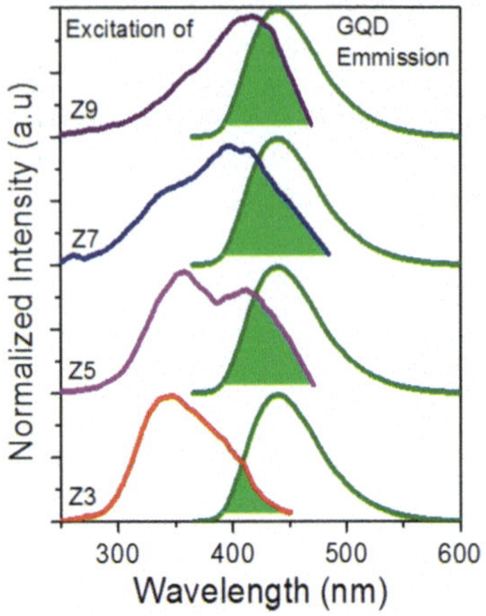

Fig. 7.27 Spectral match between GQD emission spectra and solid sheet excitation spectra

7.14 pH-Dependent Photoluminescence (PL) Measurements

The intrinsic nature of the emission from GQD solid sheets is confirmed by the pH-sensitive photoluminescence studies (Fig. 7.28a–d). The emission of GQDs from their carbene states and other surface states is known to be suppressed at low pH levels (pH = 1) [20]. A tiny quantity of emission at a low pH level is left behind in GQDs where the emission is slightly quenched. This minimal unaffected emission from the GQDs is owing to the quantum size effect. But for sheets, the emission is not affected by pH, since the size effect alone is responsible for the solid sheet sample's emission.

Because GQD is a pH-sensitive donor and shows high energy-transfer-assisted emission from the raw samples, the raw samples also displayed a sensitivity to pH. Therefore, the pH-independent emission-producing ZnS QDs system has been used as the donor to cause robust pH-insensitive emission from the sheet sample (Fig. 7.29a, b). The energy transfer method caused pH-insensitive strong blue emission at 440 nm from the solid sheet sample with an emission yield of 20%, with the defect-related emission from ZnS QDs acting as the donor. Thus, the pH-independent emission behavior of solid sheet samples was confirmed. In summary, the main cause of the emission maximum and related energy transfer interaction is determined by the excitation peak at 410 nm from the graphene composite-type sheet samples. This new excitation band at 410 nm is strongly related to the sp^2 domains' quantum size impact. The surface-to-volume ratio of graphene composite sheets is small, indicating a small surface effect. Consequently, the quantum-size-effect-induced excitation band begins to emerge. It was therefore unable to bring GQDs close together and join them into sheets in a methodical manner. Without sacrificing optical transparency, metal atoms facilitated the interconnection of GQDs into solid sheets into a composite. Shortly,

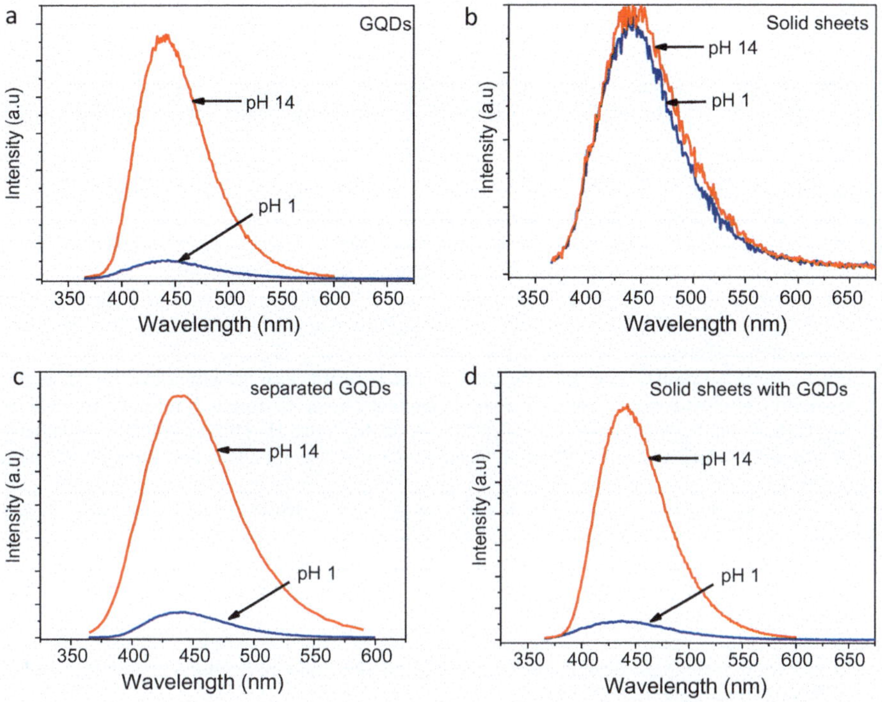

Fig. 7.28 pH-dependent emission spectra of **a** GQDs, **b** solid sheets separated from the Z7 sample **c** GQDs separated from the Z7 sample, and **d** Z7 sample containing solid sheets with GQDs

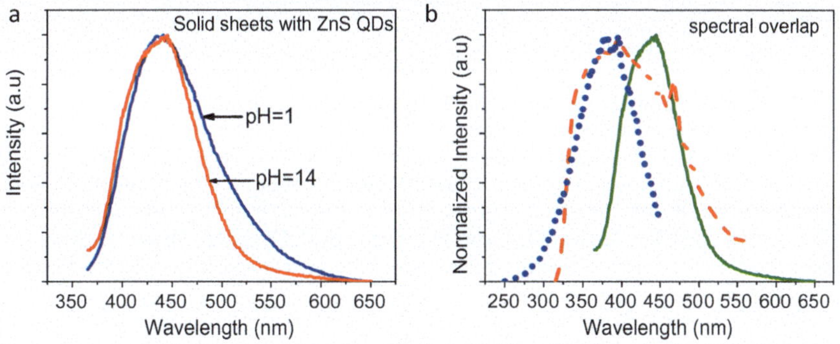

Fig. 7.29 **a** Emission spectra of solid sheets with ZnS QDs measured at pH 1 and 14, **b** spectral overlap between ZnS QD emission (red dashed line) and solid sheet excitation (blue dotted line) along with solid sheet emission when mixed with ZnS QDs (green solid line)

these energy band gap-induced sheets can be employed as an active device in the production of optoelectronic devices.

References

1. Geim AK, Novoselov KS (2007) The rise of graphene. Nat Mater 6:183–191
2. Kim TY, Park CH, Marzari N (2016) The electronic thermal conductivity of graphene. Nano Lett 16:2439–2443
3. Griep MH, Sandoz-Rosado E, Tumlin TM, Wetzel E (2016) Enhanced graphene mechanical properties through ultrasmooth copper growth substrates. Nano Lett 16:1657–1662
4. Novoselov KS (2009) Graphene: the magic of flat carbon. ECS Trans 19:3
5. Meija R, Signetti S, Schuchardt A, Meurisch K, Smazna D, Mecklenburg M, Pugno NM (2017) Nanomechanics of individual aero graphite tetrapods. Nat Commun 8:1–9
6. Mecklenburg M, Schuchardt A, Mishra YK, Kaps S, Adelung R, Lotnyk A, Schulte K (2012) Aerographite: ultra-lightweight, flexible nanowall, carbon microtube material with outstanding mechanical performance. Adv Mater 24:3486–3490
7. Wu T, Zhang X, Yuan Q, Xue J, Lu G, Liu Z, Jiang M (2016) Fast growth of inch-sized single-crystalline graphene from a controlled single nucleus on Cu–Ni alloys. Nat Mater 15:43–47
8. Narita A, Feng X, Hernandez Y, Jensen SA, Bonn M, Yang H, Müllen K (2014) Synthesis of structurally well-defined and liquid-phase-process able graphene nanoribbons. Nat Chem 6:126–132
9. Qu D, Zheng M, Zhang L, Zhao H, Xie Z, Jing X, Sun Z (2014) Formation mechanism and optimization of highly luminescent N-doped graphene quantum dots. Sci Rep 4:1–11
10. Chernov AI, Fedotov PV, Talyzin AV, Suarez Lopez I, Anoshkin IV, Nasibulin AG, Obraztsova ED (2013) Optical properties of graphene nanoribbons encapsulated in single-walled carbon nanotubes. ACS Nano 7:6346–6353
11. Kumawat MK, Thakur M, Gurung RB, Srivastava R (2017) Graphene quantum dots from Mangifera Indica: application in near-infrared bioimaging and intracellular nanothermometry. ACS Sust Chem Eng 5:1382–1391
12. Zhang JH, Niu A, Li J, Fu JW, Xu Q, Pei DS (2016) In vivo characterization of hair and skin derived carbon quantum dots with high quantum yield as long-term bioprobes in zebrafish. Sci Rep 6:1–12
13. Tang L, Ji R, Li X, Bai G, Liu CP, Hao J, Lau SP (2014) Deep ultraviolet to near-infrared emission and photoresponse in layered N-doped graphene quantum dots. ACS Nano 8:6312–6320
14. Hummers WS Jr, Offeman RE (1958) Preparation of graphitic oxide. J Am Chem Soc 80:1339–1339
15. Yu H, Zhang B, Bulin C, Li R, Xing R (2016) High-efficient synthesis of graphene oxide based on improved hummers method. Sci Rep 6:1–7
16. Paton KR, Varrla E, Backes C, Smith RJ, Khan U, O'Neill A, Coleman JN (2014) Scalable production of large quantities of defect-free few-layer graphene by shear exfoliation in liquids. Nat Mater 13:624–630
17. Chang MM, Gaidukov L, Jung G, Tseng WA, Scarcelli JJ, Cornell R, Weiss R (2019) Small-molecule control of antibody N-glycosylation in engineered mammalian cells. Nat Chem Biol 15:730–736
18. Ambrosi A, Pumera M (2016) Electrochemically exfoliated graphene and graphene oxide for energy storage and electrochemistry applications. Chem–A Europ J 22:153–159
19. Quintana M, Grzelczak M, Spyrou K, Kooi B, Bals S, Van Tendeloo G, Prato M (2012) Production of large graphene sheets by exfoliation of graphite under high power ultrasound in the presence of tiopronin. Chem Commun 48:12159–12161
20. Pan D, Zhang J, Li Z, Wu M (2010) Hydrothermal route for cutting graphene sheets into blue-luminescent graphene quantum dots. Adv Mater 22:734–738
21. Zhang Y, Tang TT, Girit C, Hao Z, Martin MC, Zettl A, Wang F (2009) Direct observation of a widely tunable bandgap in bilayer graphene. Nature 459:820–823
22. Novoselov KS, Geim AK, Morozov SV, Jiang D, Katsnelson MI, Grigorieva I, Firsov A (2005) Two-dimensional gas of massless Dirac fermions in graphene. Nature 438:197–200

23. Peng L, Xu Z, Liu Z, Wei Y, Sun H, Li Z, Gao C (2015) An iron-based green approach to 1-h production of single-layer graphene oxide. Nat Commun 6:1–9
24. Stankovich S, Dikin DA, Dommett GH, Kohlhaas KM, Zimney EJ, Stach EA, Ruoff RS (2006) Graphene-based composite materials. Nature 442:282–286
25. Yoon H, Chang YH, Song SH, Lee ES, Jin SH, Park C, Jeon S (2016) Intrinsic photoluminescence emission from subdomained graphene quantum dots. Adv Mater 28:5255–5261
26. Nourbakhsh A, Cantoro M, Vosch T, Pourtois G, Clemente F, Van Der Veen MH, Sels BF (2010) Bandgap opening in oxygen plasma-treated graphene. Nanotechnology 21:435203
27. Gokus T, Nair RR, Bonetti A, Bohmler M, Lombardo A, Novoselov KS, Hartschuh A (2009) Making graphene luminescent by oxygen plasma treatment. ACS Nano 3:3963–3968
28. Shang J, Ma L, Li J, Ai W, Yu T, Gurzadyan GG (2012) The origin of fluorescence from graphene oxide. Sci Rep 2:1–8
29. Tripathi KM, Singh A, Bhati A, Sarkar S, Sonkar SK (2016) Sustainable feasibility of the environmental pollutant soot to few-layer photoluminescent graphene nanosheets for multifunctional applications. ACS Sust Chem Eng 4:6399–6408
30. Hwang E, Hwang HM, Shin Y, Yoon Y, Lee H, Yang J, Lee H (2016) Chemically modulated graphene quantum dot for tuning the photoluminescence as novel sensory probe. Sci Rep 6:1–10
31. Ke CC, Yang YC, Tseng WL (2016) Synthesis of blue-, green-, yellow-, and red-emitting graphene-quantum-dot-based nanomaterials with excitation-independent emission. Part Part Sys Charact 33:132–139
32. Yang S, Sun J, Li X, Zhou W, Wang Z, He P, Jiang M (2014) Large-scale fabrication of heavy-doped carbon quantum dots with tunable photoluminescence and sensitive fluorescence detection. J Mater Chem A 2:8660–8667
33. Ganguly A, Sharma S, Papakonstantinou P, Hamilton J (2011) Probing the thermal deoxygenation of graphene oxide using high-resolution in situ X-ray-based spectroscopies. J Phys Chem C 115:17009–17019
34. Zhao J, Tang L, Xiang J, Ji R, Yuan J, Zhao J, Song L (2014) Chlorine doped graphene quantum dots: Preparation, properties, and photovoltaic detectors. Appl Phys Lett 105:111116
35. Shi W, Fan H, Ai S, Zhu L (2015) Preparation of fluorescent graphene quantum dots from humic acid for bioimaging application. New J Chem 39:7054–7059
36. Bharathi G, Nataraj D, Premkumar S, Sowmiya M, Senthilkumar K, Daniel Thangadurai T, Khyzhun OY, Gupta M, Phase D, Patra N, Jha SN, Bhattacharyya D (2017) Graphene quantum dot solid sheets: Strong blue-light-emitting and photocurrent-producing bandgap opened nanostructures. Sci Rep 7:10850

Chapter 8
Physicochemical Properties of Graphene Quantum Dots

8.1 Introduction

Zero-dimensional (OD) nanomaterials known as GQDs possess intriguing features that render them very desirable for a range of applications, including biosensing [1, 2]. Among GQD's physicochemical properties are: (i) size effect—GQDs typically have a lateral size below 100 nm, (ii) structure effect—They consist of one or a few layers of graphene [3], (iii) stability—GQDs are chemically and physically stable [4], (iv) surface area—large surface-to-mass ratio makes them ideal for various applications [5], (v) water dispersibility—functional groups at the edges enable easy dispersion in water [6], (vi) oxidation control—It is suggested by the findings that oxidation in GQDs has a significant role in the photoluminescence (PL) emission of GQDs [7]. The inherent PL features of GQDs derive from the tiny sp^2 domains bound by oxygen functional groups in their basal plane [8], (vii) surface functionalization- The luminescence properties of GQDs can be tuned based on the electronegativity or electron-withdrawing/donating characteristics of the atom or functional groups when they are functionalized or doped with heteroatoms [9], (viii) Tuning of photoluminescence's- By modifying the degree of oxidation, one may tune the amount of extrinsic emission and therefore the PL. Quantum yield increases when surface passivation (polyethylene glycol (PEG)) is used to block traps or flaws [10].

GQD possesses a high surface-to-volume ratio, chemical stability, edge effect, quantum confinement, and biocompatibility. By adding dopants inside the graphene structure and many chemical functional groups at the edges, one can alter the structural and morphological characteristics of GQDs. GQDs have been investigated in the development of nanoscaled optoelectronic and electronic devices because of their strong quantum-confined electronic state, distinct edge structure effect, and open band gap structure, which provide them with unique optoelectronic features. The edge states whose energy is contained within the finite-size quantization pseudo gap give GQD its magnetism. The cluster's geometric shape has a significant influence on

N. Manjubaashini et al., *Graphene Quantum Dots*, Materials Horizons: From Nature to Nanomaterials, https://doi.org/10.1007/978-981-97-5722-0_8

the magnetic response and edge state spectrum structure of GQDs. It has been hypothesized that the edge crystallographic orientation of GQDs plays a crucial role in determining their electrical structure. Broad photon energies, optical spin blockage, and gate tunable optical properties were determined by controlled magnetic moment through carrier density manipulation.

8.2 Size and Structural Effects of GQDs

Temperature, reactant concentration, and reaction time can all affect GQD size. GQD syntheses, especially those carried out top-down, typically yield GQDs with a broad spectrum of sizes [11]. The photoluminescence spectrum is excitation-dependent because of this inhomogeneity. Gel electrophoresis or ultrafiltration via membranes with specified pore diameters can be used to separate GQDs of various sizes [12]. Longer fluorescence emission wavelengths result from lower bandgaps caused by larger sizes [13]. GQDs have the same crystal structure as graphene but are quasi-zero dimensional (Fig. 8.1).

Two sets of nested sublattices (lattices A and B) with a distance of 0.142 nm between adjacent cells make up the ideal infinite graphene crystal [13]. Together they create a periodic six-membered ring structure in the graphene plane, with carbon atom C occupying lattice points and three x-bonds inside each lattice [15]. They also belong to various sublattices. The electronic structure of this material differs from other materials in that the C-atoms are closely stacked; each carbon atom is connected to three other carbon atoms by sp^2. Three of the four valence electrons are paired with the valence electrons of the surrounding three carbon atoms to form δ-bonds, and the unpaired π-electrons work together to form large π-bonds and this unique electronic structure confers an allotropic nature that is different from that of other carbons [16]. Although the lower and upper constraints on size fluctuation are highly dependent on the preparation procedure, GQDs can range in size from approximately 1.5 to 60 nm [17]. It is well known that the size of GQDs is connected with their shape, height, and edge states [18]. Since most techniques for creating GQDs include several

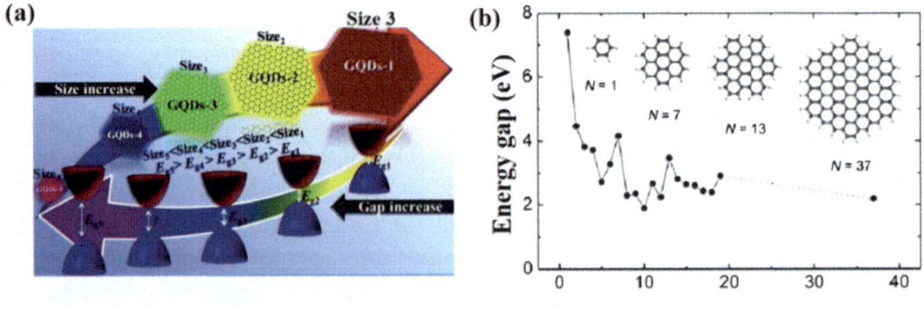

Fig. 8.1 **a** Schematics of size effect on GQDs, **b** DFT values for hexagonal GQDs [39]

chemical reactions, it stands to reason that GQDs should have oxygen-containing groups including carbonyl, carboxyl, hydroxyl, and epoxy groups [19].

To elucidate the impact of the oxygen groups on the optical characteristics of graphene oxide quantum dots (GOQDs), a series of GOQDs with varying oxygen contents were created by oxidizing graphite nanoparticles to varying degrees. Subsequently, the GOQDs' oxygen content was progressively decreased through chemical reduction in stages, resulting in a series of reduced GOQDs (rGOQDs) with varying oxygen contents [20]. Numerous researchers have demonstrated that well-defined GQDs have a high degree of crystalline structure [21, 52]. It has been reported that the inter-layer gap varies between 0.343 and 0.481 mm, however, it appears that this may be altered by doping distinct groups on the GQD edges. Diffraction planes play a major role in determining the 0.18–0.24 mm graphite in-plane spacing [22].

8.3 Oxidation Process

By adding intermediate n orbitals, the formation of oxygenated groups on GQDs redshifts photoluminescence (PL), but they also reduce the quantum yield of PL by serving as nonradiative centers [23]. Conversely, the use of reductive agents to remove the oxygenated groups results in a blueshift in PL and an increase in quantum yield (QY), but at the expense of GQD dispersibility [24]. The composition of oxygenated species is determined by the particular oxidants or reductants used during synthesis, whereas the number of oxygenated groups can be regulated by adjusting the oxidative cutting time, oxidant strength, or post-synthesis reduction [25]. The –C = O and –COOH groups on GQD work together as peroxidase mimics, while the –COOH and –SO_3H groups can operate as acid catalysts for a variety of processes [26]. Diverse oxygenated species provide varied functions for these groups.

8.4 Heteroatom Doping

One useful method for modifying the electrical characteristics of GQDs is heteroatom doping [27]. GQDS can be applied in a variety of multidisciplinary disciplines due to its sensitivity or reactivity to a particular environment depending on different heteroatom doping (such as N, F, S, Se, B, etc.) [28]. The alteration of photoluminescence intensity and emission wavelength that occurs with doping is one of the fascinating benefits [29]. However, because nitrogen and carbon have similar sizes, nitrogen doping is the doping mechanism that has been investigated the most [30]. Doped heteroatoms provide GQDs with different catalytic properties and electronic bandgap architectures [31]. Post-synthesis procedures can be used to achieve heteroatom doping. For example, N-doping can be achieved by heating GQDs in ammonia gas or hydrothermally treating GQDs in ammonia solution [32]. N-doping increases the PL lifespan and narrows the bandgap [33].

8.5 Electronic Properties

Tunneling spectroscopy analysis has revealed that the electrical structure of GQDs ranging in lateral dimensions from 2 to 20 nm is dependent on the graphene edge lattice symmetry [34, 41]. Because of the existence of zigzag-edge states, predominantly zigzag-edge GQDs with average diameters of 7–8 nm were demonstrated to be metallic. For armchair edge GQDs, the computed energy gap falls off as 1/L in the range of 0–3 eV, where L is the hexagonal GQD's length on each edge. This energy gap is compatible with the QCE [35]. In contrast to a typical semiconductor quantum dot, the bandgap of GQDs is highly dependent on their size [36]. According to Suk-Ho [37], the absorption spectrum of GQDs was predicted to involve three distinct photon energy scales: transitions within the ZB (terahertz energies), transitions from the VB to the zero-energy degenerate band (ZB), and from the ZB to the CB ($\approx$2.8 eV),and transitions from the VB to the conduction band (CB) ($\approx$1.4 eV).

One of the primary benefits of using quantum dots in a material is the ability to create a regulated energy gap [38]. A qualitative illustration of how shrinking the size of the GQDs causes the energy gap to widen [39]. Regarding the quantitative energy gap values, Eda et al. indicated that, in theory, the energy gap of a single aromatic ring can reach up to around 7 eV, and that it decreases as the number of aromatic rings increases. The electrical characteristics of graphene with nanoscale dimensions are greatly influenced by the crystallographic orientation of its edges [34, 41]. The experimentally estimated energy gap (Eg)–size (L) connection for GQDs with 2–20 nm lateral dimensions, was made possible by the invention of an atomically clean, in situ deposition process [42]. Due to the presence of metallic zigzag-edge states, which spatially decay into the graphene interior with a decay length of 1.0–1.2 nm, mostly zigzag-edge GQDs with average lengths of 7–8 nm are metallic and deviate from the Eg–L scaling rule [43]. Using different functional groups, edge passivation can also change the electronic characteristics. Through theoretical modeling and simulations, the impact of chemical functionalization on the energy bandgap was investigated. GQD conformation was shown to have a greater influence on the energy bandgap (3.269 eV in AM and 0.273 eV in ZZ), with functionalization just slightly altering the bandgap [44].

8.6 Optical Properties (Absorption and Photoluminescence)

A strong peak located deep in the UV range of approximately 4.6 to 6.2 eV and a tail extending into the visible range (down to ~2.1 eV) is the typical resolution of the broad absorption spectrum of GQDs. These tails originate from the n $\rightarrow$ π^* transition of –C = O bonds and the $\pi \rightarrow \pi^*$ transition of –C = C– bonds, respectively, and are similarly observed in GO sheets [45, 46]. Following the experimental findings, it was also anticipated that the interband transitions of a graphene sheet would generate strong absorption peaks at 4.55 and 5.15 eV with/without exciton effects, respectively [13]. The luminescent behaviors of GQDs are a subject of great controversy at the moment because their explanations have been based on a variety of origins, including size, shape, excitation wavelength, solvent, pH, functional groups, and/or combinations of these [47, 48]. PL is defined as a GQD's capacity to absorb photons, excite electrons to a higher energy level, and then emit photons when the electrons return to their stable energy level [49]. To date, several GQDs have been synthesized with a range of photoluminescence colors from ultraviolet to visible light and even near-infrared. Despite this, opinions in the literature differ about the source of the photoluminescence in GQDs. Their photoluminescence is thought to be influenced by their size, heteroatom dopants, and surface functional groups [50].

Based on these parameters, writers are presently characterizing the photoluminescence behavior of GQDs via the extrinsic and intrinsic emission channels. The amount and size of aromatic rings in GQDs' structure govern bandgap opening and quantum confinement, which are at the core of the intrinsic emission pathway [51]. GQDs and C-dots exhibit the same PL characteristics [21, 52]. Regardless of how they were created, every known GQD is photoluminescent, although it is still unclear where their radiative characteristic came from [53]. With a QY ranging from 2 to 22.4%, GQDs made using various techniques can emit deep UV, blue [54], green, yellow [20], red, or white [55]. After being excited by UV light, the most frequently reported GQDs have emission peaks in the blue to green portion of the spectrum [56]. Moreover, the PL of GQDs depends on solvent, pH, and excitation. Under UV light, the PL is quite stable and shows just slight photo-bleaching [57]. Carbon-based nanomaterials known as GQDs exhibit a range of properties, including tunable photoluminescence, strong chemical inertness, strong photostability, high biocompatibility, low toxicity, and high solubility in both aqueous and non-aqueous solvents. R. Matshitse et al. examined the impact of the charge on the physicochemical characteristics of ZnPc derivatives in the presence of GQDs,the corresponding Förster resonance energy transfer (FRET) efficiencies were determined to be 0.81, 0.80, and 0.28 [58]. Through non-covalent surface contact, GQDs-ZnPc, GQDs-ZnTPPcQ, and GQDs-ZnTSPc conjugates were surface created. It was demonstrated that GQDs increased the Pcs' triplet quantum yields. In contrast to ZnPcs alone, as-synthesised nanoconjugate systems produced less singlet oxygen in water. Additionally, we were able to demonstrate that the physicochemical properties of as-synthesized nanoconjugates are influenced by the type of phthalocyanine, loading, and solvent. The highest triplet quantum yield was seen in DMSO for both positively and neutrally charged conjugates, indicating the drawback of combining negatively charged Pcs with GQDs.

Denis M. Basko et al. uses precise ab initio many-body GW and Bethe–Salpeter simulations to analyze the electrical and optical characteristics of GQD. Dark low-energy singlet excitations that quench fluorescence characterize the pure GQD in this, including formations with irregular forms. Thus, by making use of the chiral symmetry of the low-energy electronic states in graphene, the low-energy excitation phenomenon may be rationalized. It is demonstrated that by warping the sp^2 backbone planar symmetry, edge sp^3 functionalization effectively brightens these low-lying excitations [59]. The suspension of GQDs in water is studied at 355, 532, and 1064 nm in the picosecond regime, as demonstrated by H. Wang et al. The NL absorption coefficients and the third-order nonlinear (NL) refractive index are found. Furthermore, the NL only exhibits a greater reaction when exposed to UV light. Furthermore, using a straightforward model, the NL refractive index and the saturable absorption are calculated for a single nanoparticle that makes up the GQDs. The obtained Kerr coefficient value is three orders of magnitude lower with the opposite sign as that discovered for monolayer graphene, and it is in the order of magnitude of that found in bulk materials [60].

8.7 Optoelectronic Properties

Using the tight-binding approach, the electrical and interband optical characteristics of vertically connected stacked graphene quantum dots are studied (Fig. 8.3a–d). Configurations with armchair edges and zigzags are both considered [61]. Specifically, the main focus is on the effect of the geometric shape and the twisting angle between the layers. The computed imaginary component of the dielectric function is used to examine the optical response [62]. It is discovered that the dot size and geometry have a significant influence on the interband absorption threshold. While zigzag triangular bilayer graphene dots are small gap systems in which the twisting causes the appearance of zero-gap states associated with the variation of HOMO and LUMO states resulting from the breaking of zero-energy degeneracy, armchair triangular bilayer graphene dots exhibit a moderate increase in optical gap for smaller angles of twisting and the structure behaves as an intermediate to a wide gap semiconductor [63]. In the latter instance, it is demonstrated that the primary optical response of the structures is caused by the low-energy transitions between those states, suggesting potential uses in THz optoelectronics [64]. According to Jelinek et al. [65], circular dots exhibit stronger low-energy absorption thresholds and are selected in commensurate arrangements. For sufficiently high dot radii, a unique feature that emerges in this instance is the existence of Bravais–Moir′e patterns in the 2D probability density distributions [66].

By adding amino ligands to the edges, Ding et al. were able to modify the photoluminescence of hBN-QDs to include full-color emission (420–610 nm) [67]. When particular amine ligands are attached to hBN-QDs in various solvents, full-color photoluminescence emission, and UV–visible spectra can be obtained, as shown in Fig. 8.2a–c. Specifically, urea and N-methyl pyrrolidone are attached to produce blue hBN-QDs (B-QDs), while p-phenylenediamine and sulfuric acid are attached to produce green hBN-QDs (GQDs). By passivating with thiourea ligand in acetone, p-phenylenediamine in dimethylformamide, and p-phenylenediamine in ethanol, respectively, green-yellow, yellow, and red hBN-QDs are produced [68]. Through doping processes, the electrical characteristics of GQDs can be adjusted. GQDs' semiconductive characteristic was discovered by Q. Jiajia et al., who also converted hydrothermally produced n-type GQDs into p-type GQDs by altering $P = O$ groups

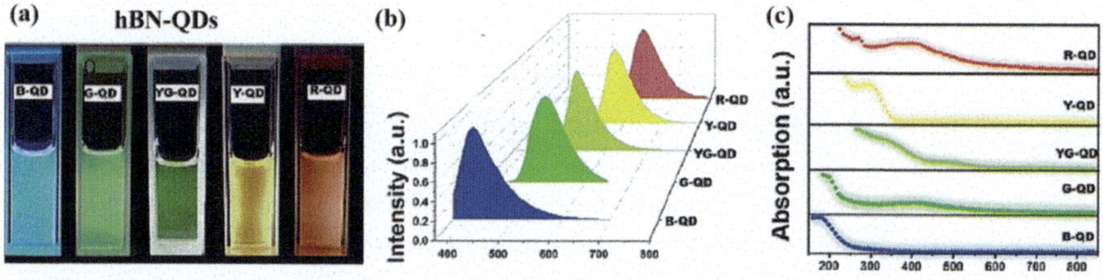

Fig. 8.2 Images of functionalized hBN-QDs under UV light (360 nm), **b** and **c** Emission and excitation spectra of hBN-QDs [38]

that could remove electrons. The interface between P-GQDs and n-type carbon nitride formed a p–n junction, which led to a considerable increase in the separation rate of photogenerated electron–hole pairs as compared to the undoped GQDs-based composites. According to [69], doping enhances the stability and reusability of different photo-oxidant reactions as well as photoreduction reactions (Fig. 8.4a, b).

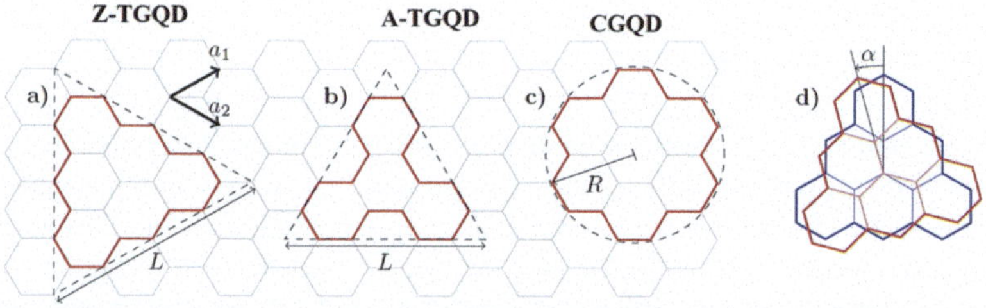

Fig. 8.3 Illustration of triangular and circular graphene quantum dots; **a** zigzag, **b** armchair, **c** circular shaped GQDs, and **d** center of symmetry of GQDs [61]

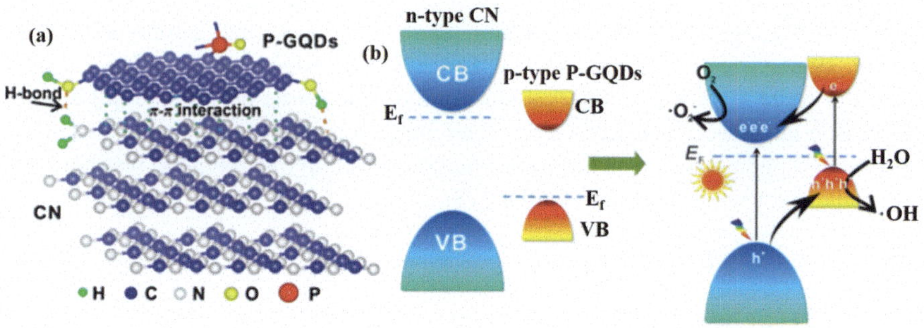

Fig. 8.4 **a**, **b** Structural illustrations and band alignment of P-doped GQDs [69]

8.8 Magnetic Properties

Because of the limited spin–orbit coupling between the C-atoms in the planar structure, graphene has a lengthy spin relaxation time and diffusion length [70]. Because of this characteristic, graphene is a viable option for spintronics. Graphene's delocalized distribution of π-bonds on the periodic lattice, however, makes it non-magnetic, a significant disadvantage that prevents its use in spintronic devices [71]. For example, a spin-valve based on graphene [72] exhibits a relatively minimal magnetoresistance of 10%. The magnetic state of graphene can be effectively controlled by cutting it into zigzag-edged nanoribbons or quantum dots, which greatly increases the magnetoresistance of the valves [73, 74]. By applying a suitably crafted non-uniform magnetic field and temperature cycle that can trap different GQD in different points of space, one can naturally separate graphene clusters based on their size and shape, thanks to the strong dependence of GQDs' magnetic properties on their geometry, size, and temperature [75]. Studying the carefully sliced nanoclusters of highly ordered pyrolytic graphite would also be intriguing because they may contain discrete graphene sheets with a Dirac-like spectrum with magnetic behavior.

Sultan et al. investigated the magnetic characteristics of as-synthesized N-GQDs by measuring the magnetization versus temperature (M–T) and the magnetization versus magnetic field (M–H). N-GQDs with a high doping level (N/C) of 1.4 and an average diameter of 3.45 nm were created using the pulsed laser ablation (PLA) technique. The presence of defect states and graphitic nitrogen bonding arrangement was suggested to be the source of superparamagnetism (Fig. 8.7a–d). The superparamagnetic behavior of N-GQDs was revealed by M–H plots measured in the 2–300 K temperature range [76]. For a variety of applications, the combination of magnetic nanoparticles with nanomaterial GQDs offers a special mix of optical and magnetic capabilities. The room temperature ferromagnetism caused by pure Fe_3O_4 nanoparticles and the significant spin-polarized edges of GQD nanoparticles has been demonstrated by the magnetic characteristics of the GQDs-Fe_3O_4 nanocomposites (NC) samples (Fig. 8.6a–c). Using a pulsed laser discharge approach, Sajjad M et al. synthesize Fe_3O_4 NCs and GQDs [77]. The edge states that have energy inside the finite-size energy interval surrounding the Dirac point—the edge states energy domain—are what give GQDs their magnetism (Figs. 8.5a, b and 8.8a–h). The cluster's geometric shape has a significant impact on the magnetic response and edge-state spectrum structure of GQDs [78].

8.9 Electrochemical Luminescence (ECL) Properties

According to Cho et al. [79], electrochemical processes in solution cause the luminous phenomenon known as ECL. ECL is an analytical technique that combines the benefits of reaction electrochemical analysis using easily controllable electrode potentials with the advantages of high sensitivity, strong selectivity, and lack of acoustic noise,

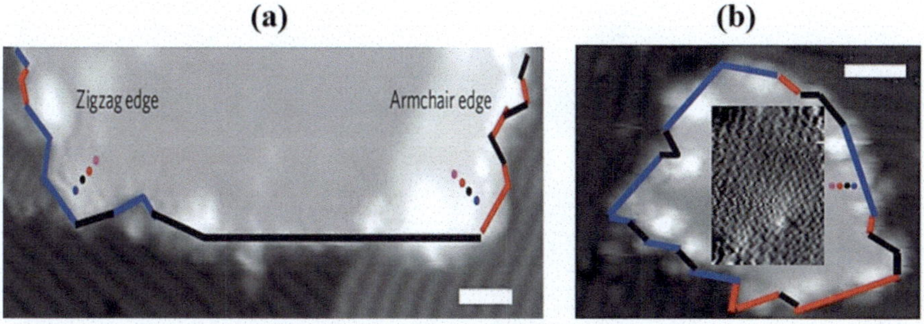

Fig. 8.5 **a** GQDs STM topography, **b** spectrum of dI/dv [78]

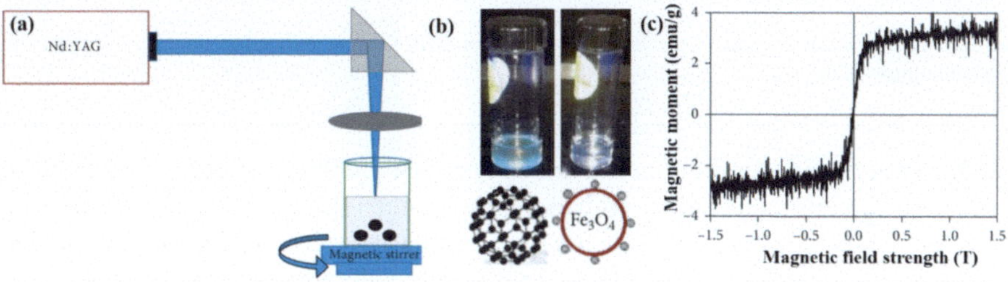

Fig. 8.6 **a** Synthesis procedure for GQDs-Fe_3O_4 NC, **b** Optical images of light emitting GQDs-Fe_3O_4 NC, and **c** Magnetization curves of GQD-Fe_3O_4 NC [77]

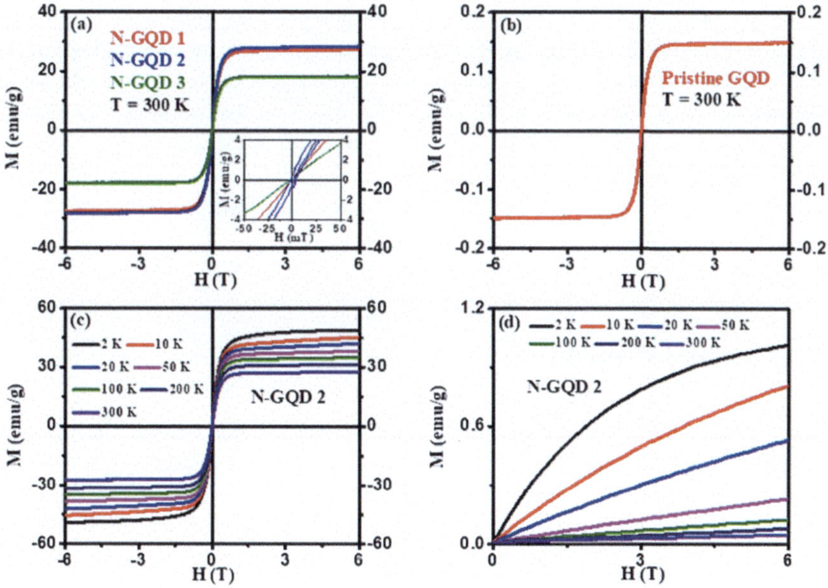

Fig. 8.7 M–H hysteresis curves of **a** Pristine GQD and N-GQD-1, 2, 3, **b** Pristine GQD, **c** N-GQD-2, and **d** Field dependence of magnetization of N-GQDs-2 [76]

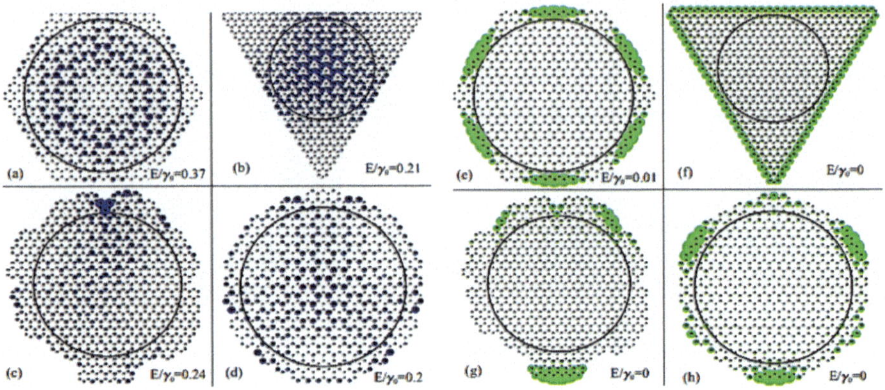

Fig. 8.8 **a–d** The amplitude of the wave function for bulk GQDs, **e–h** edge states in GQDs with different shapes [78]

with the benefits of chemiluminescence analysis without background light signals. It must have exceptional benefits, such as a broad range [80]. Researchers have shown that while GQDs have a higher specific surface area and richer surface states, making them more promising as ECL luminophores, semiconductor quantum dots have ECL characteristics [81]. GQDs may have richer surface states because GQD nanosheets, particularly monolayer GQDs, often have a greater specific surface area than CQD nanoparticles. Furthermore, because oxidized GQDs have more oxygen-containing functional groups, they have more emission sites and produce higher electrochemiluminescence effects [82]. Recent years have seen a significant amount of research on the ECL of GQDs. The high ECL green-emission of GQDs has been seen when using $K_2S_2O_8$ as a co-reactant,however, this does not affect the emission of GO [83, 84]. ECL investigations indicate that at a positive potential, GQDs exhibit an emission that is identical to that of C-dots. During a continuous scan, the electroluminescence of GQDs in a buffer medium demonstrates outstanding stability with only a 1.3% standard deviation [52].

8.10 Thermal Properties

Because of the nanomaterials' quantum confinement, GQDs exhibit the transition between field and heat conduction [85]. Electrons are trapped inside GQDs with their edges acting as limits when energy is incident on them. The margins of the GQDs become charge aggregation zones that facilitate the transfer of charge in GQD hybrid material due to the localized electrons and high electron mobility of the GQDs [86].

8.11 Electrochemical Properties

Different functional groups, heteroatom doping, graphene core, and GQDs all have a major impact on the electrochemical characteristics. In certain molecules, the efficient transfer of electrons is facilitated by the high specific surface area and excellent edge sites; the transfer of charges or electrons preferentially occurs at the edge surface. GQDs produced from biowaste have also been shown to have electrochemical characteristics [87]. The GQDs produced by chemical oxidation on charcoal were measured electrochemically. The three electrodes made from the combination of acetylene black, polyvinylidene fluoride (PVDF), and graphene quantum dots in ethanol were used for the electrochemical tests that were published [88]. Additionally, these electrodes were set up in a KOH electrolyte solution to investigate the super capacitive characteristics using a variety of tests, including galvanostatic charge–discharge (GCD), electrical impedance spectroscopy (EIS), and cyclic voltammetry (CV) [89]. A discernible shift in the scan rate attested to the electrode's high-rate capabilities and capacitive performance, which may pave the way for future developments in energy storage applications [90].

References

1. Li Y, Shu H, Wang S, Wang J (2015) Electronic and optical properties of graphene quantum dots: the role of many-body effects. J Phys Chem C 119:4983–4989
2. Yan X, Li B, Cui X, Wei Q, Tajima K, Li L (2011) Independent tuning of the band gap and redox potential of graphene quantum dots. J Phys Chem Lett 2:1119–1124
3. Pan D, Zhang J, Li Z, Wu M (2010) Hydrothermal route for cutting graphene sheets into blue-luminescent graphene quantum dots. Adv Mater 22:734–738
4. Jin SH, Kim DH, Jun GH, Hong SH, Jeon S (2012) Tuning the photoluminescence of graphene quantum dots through the charge transfer effect of functional groups. ACS Nano 7:1239–1245
5. Shi H, Barnard AS, Snook IK (2012) Quantum mechanical properties of graphene nano-flakes and quantum dots. Nanoscale 4:6761–6767
6. Ma WL, Li SS (2012) Electrically controllable energy gaps in graphene quantum dots. Appl Phys Lett 100:163109
7. Yoon H, Park M, Kim J, Novak TG, Lee S, Jeon S (2021) Toward highly efficient luminescence in graphene quantum dots for optoelectronic applications. Chem Phys Rev 2:031303
8. Zhu S, Zhang J, Liu X, Li B, Wang X, Tang S, Meng Q, Li Y, Shi C, Hu R (2012) Graphene quantum dots with controllable surface oxidation, tunable fluorescence and up-conversion emission. RSC Adv 2:2717
9. Mei Q, Zhang K, Guan G, Liu B, Wang S, Zhang Z (2010) Highly efficient photoluminescent graphene oxide with tunable surface properties. Chem Commun 46:7319
10. Yan Y, Gong J, Chen J, Zeng Z, Huang W, Pu K, Liu J, Chen P (2019) Recent advances on graphene quantum dots: from chemistry and physics to applications. Adv Mater 31:1808283
11. Gan Z, Xu H, Hao Y (2016) Mechanism for excitation-dependent photoluminescence from graphene quantum dots and other graphene oxide derivates: consensus, debates and challenges. Nanoscale 8:7794
12. Zhang F, Liu F, Wang C, Xin X, Liu J, Guo S, Zhang J (2016) Effect of lateral size of graphene quantum dots on their properties and application. ACS Appl Mater Interf 8:2104

13. Kwon W, Kim YH, Lee CL, Lee M, Choi HC, Lee TW, Rhee SW (2014) Electroluminescence from graphene quantum dots prepared by amidative cutting of tattered graphite. Nano Lett 14:1306
14. Coluci VR, Galvao DS, Baughman RH (2004) Theoretical investigation of electromechanical effects for graphyne carbon nanotubes. J Chem Phys 121:3228–3237
15. Geim AK, MacDonald AH, Graphene: exploring carbon flatland. Phys Today 60:35–41
16. Cui Y, Liu L, Shi M, Wang Y, Meng X, Chen Y, Huang Q, Liu C (2024) A review of advances in graphene quantum dots: from preparation and modification methods to application. J Carbon Res 10:7
17. Li L, Wu G, Yang G, Peng JM, Zhao J, Zhu JJ (2013) Focusing on luminescent graphene quantum dots: current status and future perspectives. Nanoscale 5:4015–4039
18. Liu R, Wu D, Feng X, Muellen K (2011) Bottom-up fabrication of photoluminescent graphene quantum dots with uniform morphology. J Am Chem Soc 133:15221–15223
19. Li Y, Hu Y, Zhao Y, Shi G, Deng L, Hou Y, Qu L (2011) An electrochemical avenue to green-luminescent graphene quantum dots as potential electron acceptors for photovoltaics. Adv Mater 23:776–780
20. Peng J, Gao W, Gupta BK, Liu Z, Romero-Aburto R, Ge L, Song L, Alemany LB, Zhan X, Gao G, Vithayathil SA, Kaipparettu BA, Marti AA, Hayashi T, Zhu JJ, Ajayan PM (2012) Graphene quantum dots derived from carbon fibers. Nano Lett 12:844
21. Tang L, Ji R, Cao X, Lin J, Jiang H, Li X, Teng KS, Luk CM, Zeng S, Hao J, Lau SP (2012a) Deep ultraviolet photoluminescence of water-soluble self-passivated graphene quantum dots. ACS Nano 6:5102–5110
22. Zhang X, Wei C, Li Y, Yu D (2019) Shining luminescent graphene quantum dots: synthesis, physicochemical properties, and biomedical applications. Trends Anal Chem 116:109–121
23. Yoon H, Chang YH, Song SH, Lee ES, Jin SH, Park C, Lee J, Kim BH, Kang HJ, Kim YH, Jeon S (2016) Intrinsic photoluminescence emission from subdomained graphene quantum dots. Adv Mater 28:5255
24. Tetsuka H, Asahi R, Nagoya A, Okamoto K, Tajima I, Ohta R, Okamoto A (2012) Optically tunable amino-functionalized graphene quantum dots. Adv Mater 24:5333
25. Chen S, Ullah N, Wang T, Zhang R (2018) Tuning the optical properties of graphene quantum dots by selective oxidation: a theoretical perspective. J Mater Chem C 6:6875
26. Sun H, Zhao A, Gao N, Li K, Ren J, Qu X (2015) Deciphering a nanocarbon-based artificial peroxidase: chemical identification of the catalytically active and substrate-binding sites on graphene quantum dots. Angew Chem Int Ed 54:7176
27. Kundu S, Yadav RM, Narayanan TN, Shelke MV, Vajtai R, Ajayan PM, Pillai VK (2015) Synthesis of N, F and S co-doped graphene quantum dots. Nanoscale 7:11515–11519
28. Bang JH, Suslick KS (2010) Applications of ultrasound to the synthesis of nanostructured materials. Adv Mater 22:1039–1059
29. Ozhukil VM, Pillai VK, Subbiah A (2017) Spotlighting graphene quantum dots and beyond: synthesis, properties and sensing applications. Appl Mater Today 9:350–371
30. Srivastava R, Thakur M, Kumawat MK, Bahadur R (2021) Physicochemical properties and toxicity analysis. In: Next generation graphene nanomaterials for cancer theranostic applications. Springer, Singapore
31. Feng J, Dong H, Pang B, Shao F, Zhang C, Yu L, Dong L (2018) Theoretical study on the optical and electronic properties of graphene quantum dots doped with heteroatoms. Phys Chem Chem Phys 20:15244
32. Zhang L, Peng D, Liang RP, Qiu JD (2015) Nitrogen-doped graphene quantum dots as a new catalyst accelerating the coordination reaction between cadmium (II) and 5,10,15,20-Tetrakis(1-methyl-4-pyridinio) porphyrin for Cadmium (II) Sensing. Anal Chem 87:10894
33. Sun L, Luo Y, Li M, Hu G, Xu Y, Tang T, Wen J, Li X, Wang L (2017) Role of Pyridinic-N for Nitrogen-doped graphene quantum dots in oxygen reaction reduction. J Colloid Interface Sci 508:154
34. Ritter KA, Lyding JW (2009a) The influence of edge structure on the electronic properties of graphene quantum dots and nanoribbons. Nat Mater 8:235–242

35. Zhang ZZ, Chang K, Peeters FM (2008) Tuning of energy levels and optical properties of graphene quantum dots. Phys Rev B 77:235411
36. Guclu AD, Potasz P, Hawrylak P (2010) Excitonic absorption in gate-controlled graphene quantum dots. Phys Rev B 82:155445
37. Suk-Ho A (2017) Unique properties of graphene quantum dots and their applications in photonic/electronic devices. J Phys D 50:103002
38. Hazem A, Fang ZQ (2022) Properties and applications of quantum dots derived from two-dimensional materials. Adv Phys X 7:2048966
39. Tian P, Tang L, Teng KS, Lau SP (2018) Graphene quantum dots from chemistry to applications. Mater Today Chem 10:221–258
40. Eda G, Lin YY, Mattevi C et al (2022) Blue photoluminescence from chemically derived graphene oxide. Adv Mater 10:505
41. Ritter KA, Lyding JW (2009b) The influence of edge structure on the electronic properties of graphene quantum dots and nanoribbons. Nat Mater 8:235–242
42. Gunlycke D, Areshkin DA, White CT (2007) Semiconducting graphene nanostrips with edge disorder. Appl Phys Lett 90:142104
43. Ritter KA, Lyding JW (2008) Characterization of nanometer-sized, mechanically exfoliated graphene on the H-passivated Si(100) surface using scanning tunneling microscopy. Nanotechnology 19:015704
44. Seokhwan C, Revia RA, Miqin Z (2021) Graphene quantum dots and their applications in bioimaging, biosensing, and therapy. Adv Mater 33:e1904362
45. Yang L, Deslippe J, Park CH, Cohen ML, Louie SG (2009) Excitonic effects on the optical response of graphene and bilayer graphene. Phys Rev Lett 103:186802
46. Luo Z, Lu Y, Somers LA, Johnson ATC (2009) High yield preparation of macroscopic graphene oxide membranes. J Am Chem Soc 131:898–899
47. Hu C, Liu Y, Yang Y, Cui J, Huang Z, Wang Y, Yang L, Wang H, Xiao Y, Rong J (2013) One-step preparation of nitrogen-doped graphene quantum dots from oxidized debris of graphene oxide. J Mater Chem B 1:39–42
48. Liu F, Jang MH, Ha HD, Kim JH, Cho YH, Seo TS (2013) Facile synthetic method for pristine graphene quantum dots and graphene oxide quantum dots: origin of blue and green luminescence. Adv Mater 25:3657–3662
49. Sheikh Mohd Ghazali SAI, Fatimah I, Zamil ZN, Zulkifli NN, Adam N (2023) Graphene quantum dots: a comprehensive overview. Open Chem 21:20220285
50. Deng J, Lu Q, Li H, Zhang Y, Yao S (2015) Large scale preparation of graphene quantum dots from graphite oxide in pure water via one-step electrochemical tailoring. RSC Adv 5:29704
51. Liang L, Kong Z, Kang Z, Wang H, Zhang L, Shen JW (2016) Theoretical evaluation on potential cytotoxicity of graphene quantum dots. ACS Biomater Sci Eng 2:1983
52. Tang L, Ji R, Cao X, Lin J, Jiang H, Li X, Teng KS, Luk CM, Zeng S, Hao J, Lau SP (2012) Deep ultraviolet photoluminescence of water-soluble self-passivated graphene quantum dots. ACS Nano 6:5102
53. Bourlinos AB, Stassinopoulos A, Anglos D, Zboril R, Georgakilas V, Giannelis EP (2008) Photoluminescent carbogenic dots. Chem Mater 20:4539
54. Pan D, Guo L, Zhang J, Xi C, Xue Q, Huang H, Li J, Zhang Z, Yu W, Chen Z, Li Z, Wu M (2012) Cutting sp^2clusters in graphene sheets into colloidal graphene quantum dots with strong green fluorescence. J Mater Chem 22:3314
55. Mueller ML, Yan X, McGuire JA, Li LS (2010) Triplet states and electronic relaxation in photoexcited graphene quantum dots. Nano Lett 10:2679
56. Shoujun Z, Yubin S, Joy W, Hao W, Yuan Z, Yang N, Bai Y (2017) Photoluminescence mechanism in graphene quantum dots: quantum confinement effect and surface/edge state. Nano Today 13:10–14
57. Priya R, Ranjeet D, Sunita S (2023) Effect of surface modification on optical and electronic properties of graphene quantum dots. Appl Surf Sci 609:155379
58. Matshitse R, Kutloano ES, Ojodomo JA, Nyokong T (2017) Characterization and physicochemical studies of the conjugates of graphene quantum dots with differently charged zinc phthalocyanines. J Coord Chem 70:3308–3324

59. Basko M, Duchemin I, Blasé X (2020) Optical properties of graphene quantum dots: the role of chiral symmetry. 2D Materials 7: 025041
60. Wang H, Ciret C, Cassagne C, Boudebs G (2019) Measurement of the third order optical nonlinearities of graphene quantum dots in water at 355 nm, 532 nm, and 1064 nm. Optical Express 9:339
61. Tiutiunnyk A, Duque CA, Caro-Lopera FJ, Mora-Ramos ME, Correa JD (2019) Opto-electronic properties of twisted bilayer graphene quantum dots. Phys E 112:36–48
62. Orlof A, Shylau AA, Zozoulenko IV (2015) Electron-electron interactions in graphene field-induced quantum dots in a high magnetic field. Phys Rev B 92:075431
63. Mirzakhani M, Zarenia M, Vasilopoulos P, Peeters FM (2017) Electrostatically confined trilayer graphene quantum dots. Phys Rev B 95:155434
64. Guclu AD, Potasz P, Hawrylak P (2013) Zero-energy states of graphene triangular quantum dots in a magnetic field. Phys Rev B 88:155429
65. Jelinek R (2016) Carbon quantum dots: synthesis, properties and applications (carbon nanostructures). Springer
66. Reich S, Thomsen C, Maultzsch J (2004) Carbon nanotubes: basic concepts and physical properties. Wiley-VCH
67. Bertel R, Mora-Ramos ME, Correa JD (2019) Electronic properties and optical response of triangular and hexagonal MoS_2 quantum dots. A DFT Approach Phys E 109:201
68. Ding Y, He P, Li S et al (2021) Efficient full-color boron nitride quantum dots for thermostable flexible displays. ACS Nano 15:14610
69. Jiajia Q, Chao S, Jing Y, Fengna X, Xiaoping D, Jiyang L (2018) Tailoring the electronic properties of graphene quantum dots by p doping and their enhanced performance in metal-free composite photocatalyst. J Phys Chem C 122:349–358
70. Han W, Kawakami RK, Gmitra M et al (2014) Graphene spintronics. Nat Nanotechnol 9:794
71. Sepioni M, Nair RR, Rablen S et al (2010) Limits on intrinsic magnetism in graphene. Phys Rev Lett 105:207205
72. Hill W, Geim A, Novoselov K et al (2006) Graphene spin valve devices. IEEE Trans Magn 42:2694
73. Kim WY, Kim KS (2008) Prediction of very large values of magnetoresistance in a graphene nanoribbon device. Nat Nanotechnol 3:408
74. Silva PV, Saraiva-Souza A, Maia DW et al (2018) High efficiency spin-valve and spin-filter in a doped rhombic graphene quantum dot device. J Magn Magn Mater 451:532
75. Lukyanchuk I, Kopelevich Y, El Marssi M (2009) Dirac fermions in graphite: the state of art. Phys B Cond Matt 404:404
76. Shehzad SM, Makarov VI, Frank M, Muhammad S, Barrionuevo Diestra DG, Inbanathan FPN, Eli S, Premadasa UI, Cimatu KLA, Khaled H, Jadwisienczak WM, Weiner BR, Gerardo M (2024) Superparamagnetic properties of metal-free nitrogen-doped graphene quantum dots. J Appl Phys 135:083904
77. Sajjad M, Makarov V, Sultan MS, Jadwisienczak WM, Weiner BR, Morell G (2018) Synthesis, optical, and magnetic properties of graphene quantum dots and iron oxide nanocomposites. Adv Mater Sci Eng Article ID 3254081, 8
78. Espinosa-Ortega T, Lukyanchuk LA, Rubo YG (2013) Magnetic properties of graphene quantum dots. Phys Rev B 87:205434
79. Cho H, Yang Y, Kang DJ, Kim BJ (2015) Surface engineering of graphene quantum dots and their applications as efficient surfactants. ACS Appl Mater Interf 7:8615
80. Dong Y, Tian W, Ren S, Dai R, Chi Y, Chen G (2014) Graphene quantum dots/L-cysteine co-reactant electrochemiluminescence system and its application in sensing lead(ii) ions. ACS Appl Mater Interf 6:1646–1651
81. Gao Y, Tian E, Zhang Y, Mo J (2022) Utilizing electrostatic effect in fibrous filters for efficient airborne particles removal: principles, fabrication, and material properties. Appl Mater Today 26:101369
82. Shang YD, Chen XH, Ma WH, Li SY, Wang YC, Xiang FW (2017) Preparation and optical properties research on graphene quantum dots. Key Engin Mater 727:303–308

83. Li LL, Ji J, Fei R, Wang CZ, Lu Q, Zhang JR, Jiang LP, Zhu JJ (2012) Fabrication of highly fluorescent graphene quantum dots using L-glutamic acid for in vitro/in vivo imaging and sensing. Adv Funct Mater 22:2971
84. Wang X, Cao L, Yang ST, Lu F, Meziani MJ, Tian L, Sun KW, Bloodgood MA, Sun YP (2010) Bandgap-like strong fluorescence in functionalized carbon nanoparticles. Angew Chem Int Ed 49:5310
85. Yu Y, Mei L, Shi Y, Zhang X, Cheng K, Cao F, Zhang L, Xu J, Li X, Xu Z (2020) Ag-conjugated graphene quantum dots with blue light-enhanced singlet oxygen generation for ternary-mode highly-efficient antimicrobial therapy. J Mater Chem B 8:1371–1382
86. Pin T, Libin T, Kar-Seng Teng T, Shu-Ping L (2024) Graphene quantum dots: preparations, properties, functionalizations and application. Mater Futures 3:022301
87. Singh A, Kumar S, Ojha AK (2020) Charcoal derived graphene quantum dots for flexible supercapacitor oriented applications. New J Chem 44:11085–11091
88. Zheng XT, Ananthanarayanan A, Luo KQ, Chen P (2015) Glowing graphene quantum dots and carbon dots: properties, syntheses, and biological applications. Small 11:1620–1636
89. Shinde DB, Pillai VK (2013) Electrochemical resolution of multiple redox events for graphene quantum dots. Angew Chemie Int Ed 52:2482–2485
90. Neeraj S, Sonali S, Malode SJ, Soumen B, Banibrata M, Shetti NP (2023) Bioresource-based graphene quantum dots and their applications: a review. ACS Appl Nano Mater 6:10925–10943

Chapter 9
Modeling and Simulation of Graphene Quantum Dots

9.1 Introduction

GQD simulations enable intelligent material design, environmental applications, and scientific advances by bridging theory and experiment. GQDs are graphene particles at the nanoscale with special optical and electrical characteristics. Several methods can be used to model and simulate GQDs, depending on the particular properties or phenomena you are interested in. Among the modulation techniques are (i) time-dependent density functional theory (TDDFT), which reveals the optical properties and excitations in GQDs; (ii) density functional theory (DFT), which is performed to calculate the electronic structure of GQDs; (iii) Monte Carlo simulation, which is derived for studying statistical thermodynamics and exploring phase transitions in GQDs; (vi) computational software, such as Gaussian, VASP, and Quantum ESPRESSO, for quantum mechanical calculations and simulations.

9.2 Analysis of Enhancement of GQDs Properties Using DFT

Majid G et al. looked into the electrical architectures and optical characteristics of bilayer GQDs. For both in-plane and out-of-plane polarization, the bilayer GQDs' dielectric tensors, absorption spectra, and refractive indices were determined [1]. Additionally, the TDDFT absorption spectra in the linear response domain were computed. Out-of-plane polarization light absorption peaks in bilayer GQDs arise in the near-infrared range of 500–1600 nm, comparable to those in bilayer graphene sheets [2–4] (Fig. 9.1a, b, c). The absorption of light for out-of-plane polarization occurs in the ultraviolet wavelength range of 85–250 nm [5]. Bilayer GQDs are appropriate for integrated optics and optical communication applications because of

N. Manjubaashini et al., *Graphene Quantum Dots*, Materials Horizons: From Nature to Nanomaterials, https://doi.org/10.1007/978-981-97-5722-0_9

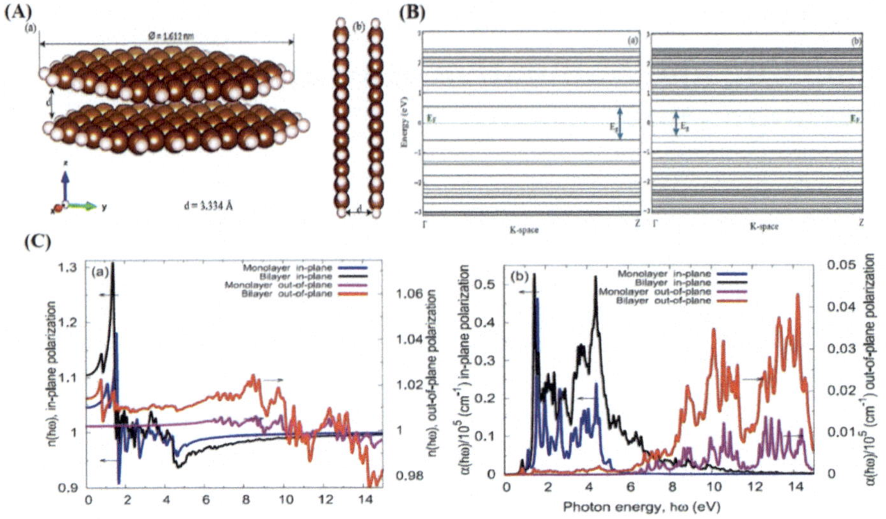

Fig. 9.1 **a** AB-stacked bilayer $C_{132}H_{40}$ GQDs with a hexagonal shape and zigzag edge: (**a**) perspective view and (**b**) XY view. **b** Energy levels of $C_{132}H_{40}$ GQD's DFT calculations: (**a**) monolayer and (**b**) bilayer system. **c** Optical properties: (**a**) refractive index and (**b**) absorption coefficient spectra [1]

their out-of-plane polarization light absorption maxima in the near-infrared range [6].

According to Keles O et al., molecular dynamics simulations were used to determine how the mechanical stability of GQDs affected the characteristics of epoxy [7]. All of the MD simulations were carried out using the Large-scale Atomic/Molecular Massively Parallel Simulator (LAMMPS) software program (Fig. 9.2). A single-layer graphene quantum dot containing six reactive amine groups increased the epoxy yield strength by 17% and stiffness by 6%, according to mechanical simulations; in contrast, pristine GQDs decreased the strength by 6% and stiffness by 10% [8].

To improve the properties of composites, Prathamesh P Deshpande et al. tried to fill the role of GQDs as nanofillers. With LAMMPS, a virtual curing simulation was run at 499 K for 1.5 ns. The tool used for this was REACTER (Fig. 9.3a–b). An internal Python script and LAMMPS were used to model the functionalized GQD [9]. The principal amine (NH_2) was the preferred functional group, and it was dispersed at random along the GQD's edge. Different numbers of amine groups were used to model two f-GQDs [10]. The overall number of functional groups for the first subsystem (4N-GQD-epoxy) was 5.2% of the GQD's total mass. 128 f-GQD and 5616 epoxy atoms made up the model. The model underwent cross-linking after annealing [11].

GQD photonics properties (empirical and virtual worlds) were explored by Mikhail F Budyka et al. The primary factor controlling the peculiarity of the empirical spectral features of GQDs is this size-content-shape-fractal inhomogeneity [12]. The empirical characteristics pertain to the dispersions of stacked nanoscale framed

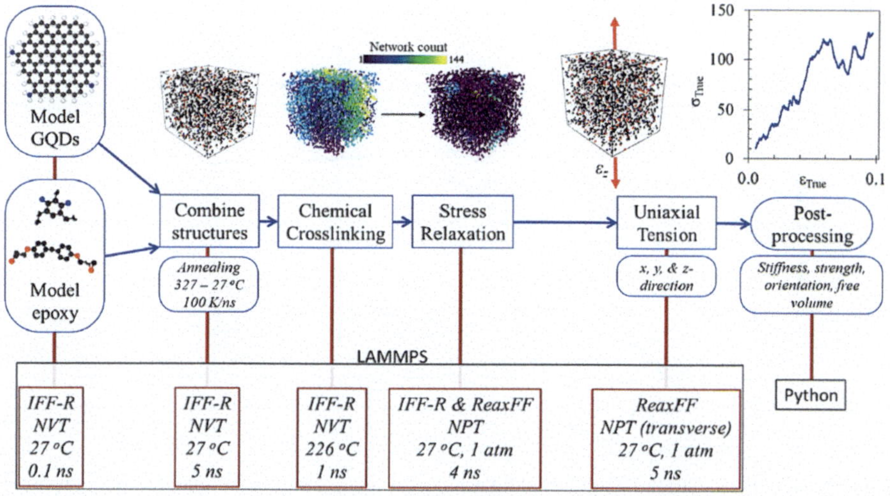

Fig. 9.2 MD framework of epoxy resins and corresponding GQD composites [7]

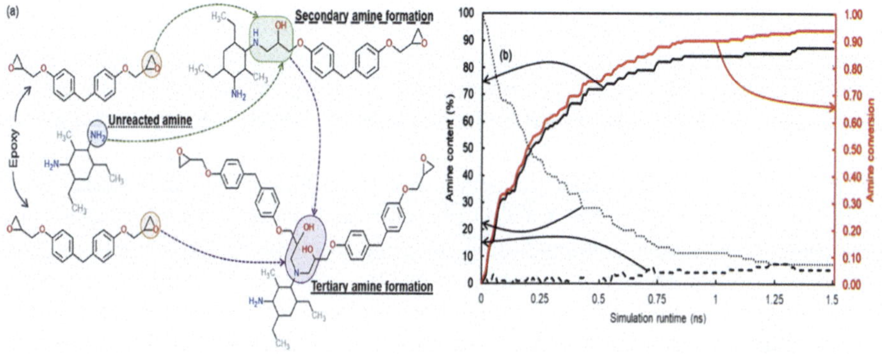

Fig. 9.3 **a** Epoxy-amine reaction pathway of bisphenol-F epoxy (DGEBF) and diethyl toluene diamine (DETDA) and **b** amine content and conversion during the chemical cross-linking simulation in a representative model with 90% cross-link density [9]

graphene sheets or molecules that are liquid or solidified (Fig. 9.4). These dispersions are created via circumferential polyderivatization, which preserves the sp^2 electron configuration of carbon atoms in the basal plane [13]. GQD photonics is virtually ascribed to single-framed molecules, ignoring fractal inhomogeneity and concentrating solely on chemical content. For the computational absorption process, the new open-shell electrical properties of GQDs are highly relevant [14].

A collection of model molecules, such as "bulk" sp^3 and sp^2 graphene molecules, are included in the computational model. The molecules' ZINDO/S excited states and UHF ground states were examined. When it comes to "bulk" closed-shell molecules, the UHF-ZINDO/S combination makes sense because the ground state outcomes

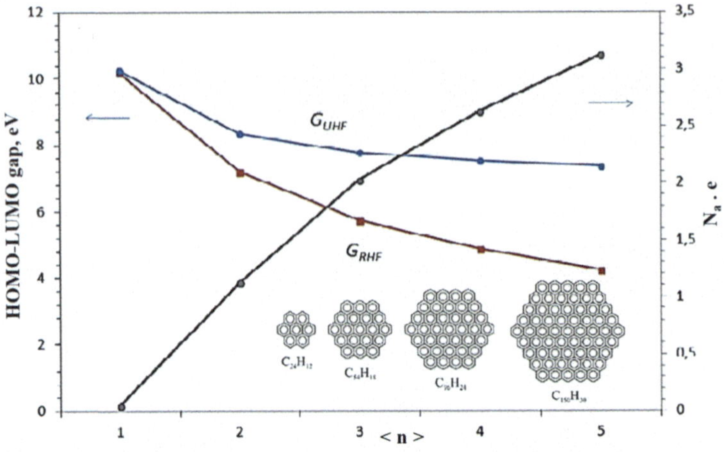

Fig. 9.4 *G*UHF and *G*RHF HOMO–LUMO gaps (left axis) and the number of effectively unpaired electrons *N*D (right axis) as a function of *n*. UHF AM1 calculations. (Inset Equilibrium structure of coronene-based molecules) [12]

for both RHF and UHF are the same [15]. The restricted nature of the ZINDO/S formalism reveals the incoherence of the UHF and ZINDO/S approaches in the case of open-shell molecules, as evidenced by a significant decrease in the energy gap between the highest occupied molecular orbital (HOMO) and lowest unoccupied molecular orbital (LUMO). The characteristic causes an extreme, implausible red shift in the absorption spectra, which directly conflicts with important experimental reality for GQDs. Since GQD photonics is a largely extending topic, the development of feasible computational methods to characterize excited states of large open-shell molecules is greatly sought [16]. Because of their very spatially confined extent, changeable size, shape, and doping, GQDs are showing promise as a molecular engineering tool for controlling light. GQDs are excellent materials for photonic technologies because of their collective excitation, which exhibits high energy plasmon frequency and frequencies in the terahertz (THz) range [17].

The strengths of semiempirical and first-principles methods were used to conduct a thorough analysis of the linear and nonlinear optical properties of vast variations of GQDs (about 400 systems) in terms of size and topology [18]. The GQDs with inequivalent sublattice atoms always have reduced HOMO—LUMO gap, broadband absorption, and high nonlinear optical coefficients among the triangular, stripe, and random-shaped GQDs [19]. Additionally, most GQDs exhibit intriguing zigzag-edge linear and nonlinear optical properties; however, this isn't always the case (Fig. 9.5a, b). Therefore, it is quite likely that the results can serve as a reference for designing GQDs in optical modulators, higher harmonic generators, and parametric oscillators [20]. DFT and thermal DFT (thDFT) computations were used to assess the energy band structure, bandgap, and total energy of several GQDs. The exchange–correlation functional and norm-conserving pseudopotentials were calculated using the

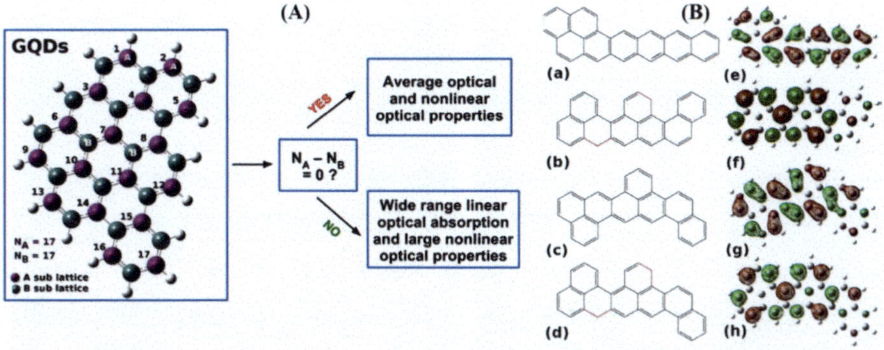

Fig. 9.5 **a** Systematic study of the linear and nonlinear optical properties of large varieties of GQDs; **b** Representation of four structural isomers: (**a**–**d**) $C_{32}H_{18}$ GQD and (**e**–**h**) HOMO isosurfaces of $C_{32}H_{18}$ [18]

local density approximation in DFT computations [21]. For examination, the 1–3 nm-diameter, multi-hundred-atom GQDs with zigzag and armchair edges were taken into consideration. Degenerate zero-energy shells were revealed by analysis of GQDs in which the A and B sublattice symmetries were disrupted [22]. Additionally, the temperature dependency of the GQD bandgaps and total energies were evaluated via the electron's kinetic energy and the entropy term using the thDFT calculations, which were conducted at temperatures as high as 1400 K. The electrical characteristics of GQDs can be ascertained using ground state DFT computations up to room temperature [5].

The optoelectronic properties of small penta-graphene quantum dots (PGQDs) with different edge-functionalized groups, such as hydrogen, halogen (fluorine, chlorine, and bromine), and hydroxyl functional groups, are studied by Dang MT et al. using first-principles calculations applied by DFT calculations [23]. The CASTEP package's DFT approach was used to optimize the structures. Utilized was the Perdew–Burke–Ernzerhof (PBE) exchange–correlation functional generalized gradient approximation (GGA). These quantum dots are thermally stable in a vacuum, especially the ones passivated by hydrogen atoms [24]. Furthermore, the formation energy on stability may become more negative as quantum dots grow in size, resulting in the development of thermodynamically more stable quantum dots (Fig. 9.6a, b). Every PGQD under investigation has semiconductor characteristics [25]. Due to the hybridization of sp^2 and sp^3 carbon atoms as well as charge depletion or accumulation between the passivated atoms and the major components during interactions, their bandgaps drop as the size of the quantum dots increases [26].

Consequently, it has been suggested that by passivating these quantum dots with different edge-functionalized groups, the DFT computations actively modify the electronic bandgaps and the absorption spectra of PGQDs [27]. The hydroxyl groups passivated the PGQDs, which have the smallest bandgaps. This is because the strong hybridization effect of the carbon and oxygen p-orbitals of these quantum dots causes the LUMO states to downshift toward the Fermi level, while the HOMO energies

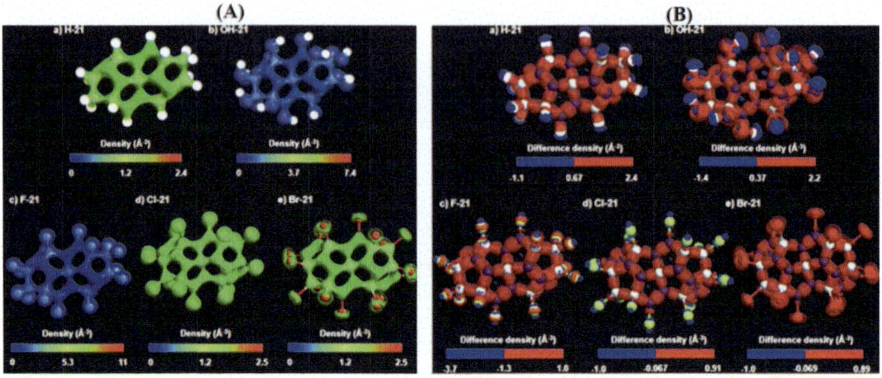

Fig. 9.6 **a** Electron density of edge-functionalized PGQDs with an odd number of carbon atoms (**a**–**e**). **b** Electron difference density of edge-functionalized PGQDs with an odd number of carbon atoms (**a**–**e**) [23]

remain nearly unchanged [28]. Moreover, PGQDs with an odd number of carbon atoms exhibit optical anisotropy due to orbital interactions and spatially directed bond orientations [29]. Additionally, edge-functionalized groups are used in penta-GQD development to create next-generation optoelectronic devices [30].

TDDFT analysis of 63 distinct GQDs with systematically varied sizes, dopant types, and dopant percentages was published by Ozonder et al. [31]. The systematics and trends in the UV–vis absorption spectra of GQDs—doped with elements B, N, O, S, and P with the dopant percentages of 1.5, 3, 5, and 7%—were used to investigate the matter (Fig. 9.7a, b). Using the hybrid functional B3LYP and the basis set 6-31G(d), the geometry optimization and excited-state computations of GQDs were carried out using Gaussian16 [32, 33].

Shi H et al. investigate whether structural alterations can be helpful for wavefunction engineering and whether the wavefunction, which is quantified by the distribution of the highest occupied molecular orbital, has a direct and trustworthy relationship

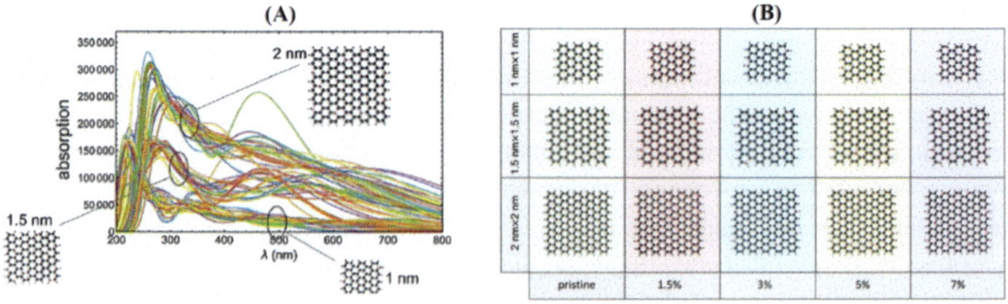

Fig. 9.7 **a** The UV–visible absorption spectrum of GQDs (63) was calculated by TDDFT, **b** GQDs excited states, and UV—visible spectra calculated topography [31]

to the physical structure using quantum mechanical simulations [34], since the techniques for altering and regulating graphene quantum dot electrical and quantum mechanical properties are very important. Small molecule graphene structures can have wavefunctions that differ from their larger nanoscale counterparts, and the geometric shape has a significant impact on the distribution of the highest occupied molecular orbital. An excellent technique for examining the electrical characteristics of graphene has been the density functional tight-binding method with self-consistent charges (SCC-DFTB), a two-center approach to DFT. Using the generalized gradient approximation (GGA) and self-consistent density functional computations of weakly confined neutral atoms, the Kohn–Sham density functional is enlarged to second order around the reference electron density. According to Barnard and Snook [35], the confinement potential is tuned to predict the charge density and effective potential in molecules and solids. In addition to accounting for double counting terms in the internuclear repulsion and the Coulomb and exchange–correlation contributions, a universal short-range repulsive potential is also established, and one- and two-center tight-binding matrix elements are explicitly calculated within DFT.

Since most academics have not yet conducted much study on the PL characteristics of GQDs, the DFT and TDDFT for photoluminescence (PL) property enhancement were discovered by Alam et al. [36]. The PL sensitivity of GQDs can be improved by defects, heteroatom doping, edge configuration, shape, and associated chemical functions using DFT and TDDFT calculations. Furthermore, it is found that the embedded tiny sp^2 clusters segregated by sp^3 carbons essentially influence the PL of a large GQD consisting of a heterogeneously hybridized carbon network [37]. The computation was made to determine the emission wavelength of pristine zigzag-edged GQDs with varying diameters to unveil the size-dependent PL of GQD. It is found that by changing the diameter from 0.89 to 1.80 nm, the emission of pure GQDs with zigzag edges can span the whole visible light spectrum. The GQD PL emission is blue-shifted by the armchair edge and pyrrolic N-doping, and red-shifted by chemical functionalities and defects. A heterogeneously hybridized GQD's isolated tiny sp^2 domains control its PL emission. One may customize GQDs to emit a broad variety of wavelengths. Therefore, the study validates the intriguing potential of GQDs to function as universal fluorophores for a range of imaging applications. It will also offer crucial insights and direction for the development of techniques to synthesize GQDs with precise and targeted properties in a controlled manner.

Shujun W et al. discovered the dual nature of the GQD functional group toward the emission spectrum from PL to enrich the PL property of GQDs. The DFT calculations were performed using Version 7.4.3 of the ABINIT programs. For the exchange–correlation function, the localized density approximation (LDA) was used [38]. To systematically probe the effects of the size of the aromatic core, functional groups, and disordered structures on the electronic structure of GQDs, a DFT simulation was used to confirm the PL mechanism from the experimental work [39]. The electronic structures of oxidized GQDs (oGQDs, or GQDs with four different types of oxygen functional groups of varying sizes) and pristine GQDs were solved by the DFT calculation. Furthermore, by changing one carbon atom in the 7-ring pGQD model into a methyl group, a defective GQD (dGQD) was produced that had both a defect and a sp^3 structure. Based on theoretical simulation studies, it was found that the functional groups play two roles in the phase separation of GQDs: they enable the $\pi^* \rightarrow$ n and $\sigma^* \rightarrow$ n transitions, which result in molecular-type PL components that are spectrally invariant with changes in particle size or excitation energy, just like defects and sp3 carbon. Additionally, they induce structural deformation to the aromatic core, which leads to irregular hybridization of the π system, providing mid-gap states and letting $\pi^* \rightarrow$ ESiD $\rightarrow \pi$ transitions that cause bathochromic broadening of the integrated PL spectra [40].

The effects of five different types of oxygen-containing functional groups (-COOH, -COC-, -OH, -CHO, and -OCH_3) on the optical and electronic properties of GQDs were projected and evaluated by F Jianguang et al. using DFT and TDDFT. First, using the Becke three-parameter Lee–Yang–Parr hybrid functional (B3LYP) and 6-31G* basis set (B3LYP/6-31G*), geometry optimizations for the ground states of GQDs are computed by DFT [41]. Following the optimizations, vertical excitations of GQDs in the improved ground state shape are obtained using the same level of theory using TD-DFT calculations [42]. In all calculations, at least the first 20 lowest excited states are taken into account. According to an analysis of the edge-functionalization effect, oxygen-containing groups on graphene quantum dots (GQDs) have a minor impact on optical and electrical properties, but decorating the basal planes of GQDs results in a major effect [43]. The surface- and edge-functionalized GQDs with five oxygen-containing groups have computed HOMO–LUMO gap trends that are nearly in the opposite order. For GQDs with various oxygen-containing groups, the electron–hole coherence and exciton delocalization of charge transfer at electronic transitions are studied using the three-dimensional (3D) real space representations (NTOs and EDDs).

Using DFT simulations, Deb J et al. examined the nonlinear optical characteristics of a series of T-GQD with varying sizes and shapes. These systems' first- and second-order hyperpolarizability, as well as their electronic polarizability, are examined to provide insight into their stability and electrical characteristics [44]. All of the T-GQDs undergo structural relaxation without any symmetry constraints. The Lee–Yang–Parr hybrid functional (B3LYP) with three parameters is calculated using the triple-ζ valence basis set with the diffuse and polarize function joined together, or 6–311 + g(d,p) [45]. Frequency calculation is used to characterize the saddle point's nature. Since all of these structures have genuine frequencies, they have effectively reached true local minima in the potential energy surface [46]. Because T-GQDs have a negative cohesive energy, they are energetically stable. For the majority of QDs, strong nonlinear optical (NLO) responses have been seen, and these responses sharply increase when an external field is applied, particularly when a frequency of 0.0428 au is applied [47]. Since the incident light frequency has a significant impact on the third-order NLO responses, in particular, it can be adjusted by varying the frequency. Based on calculations of aromaticity, the NLO characteristics are intimately associated with electron delocalization [48].

Using DFT and time-dependent DFT, the structural stabilities and optical characteristics of GQDs and holey GQDs with varying sizes, edge terminations, and pore passivation are examined [49]. The size of GQDs has a major influence on their optical spectra. The spectra are slightly affected by edge termination, which shifts the absorption peaks by about 0.2 eV and becomes less significant for larger GQDs [50]. New peaks appear in the absorption spectra of GQDs as a result of pore creation. The absorption spectra of holey GQDs appear to be most affected by pore passivation, and their termination causes minor peak shifts [51].

The energy gap was adjusted by Hazem A et al. using edge passivation by various elements or groups. DFT is used to study the energy gap and dipole moment of chemically functionalized graphene quantum dots [52]. The results of the computations indicate that all the structures—AHEX, ATRI, and ZHEX—are stable, with a big energy gap detected in AHEX and a minor energy gap in ZTRI. The passivation of clusters with oxygen can drastically reduce this enormous energy gap, that is, when oxygen is used instead of hydrogen for passivation, the energy gap reduces to less than half its value in AHEX. High electronegativity elements, like fluorine, can also be used to passivate ZTRI and modify its distinctive HOMO/LUMO orbitals and tiny energy gap.

According to Ladd [53], the DFT calculations were carried out in Gaussian 09 by reducing the total energy without the use of symmetry constraints. DFT and TD-DFT calculations were made using the Gaussian09 software package; the HOMO–LUMO gaps of N-doped GQDs with graphitic N, pyrrolic N, and pyridinic N are, respectively, 0.77, 0.25, and 2.69 eV [54]. Three N-doping combinations are used in the co-doped GQDs described by Jianguang F et al. to expose the underlying mechanisms of optical characteristics. Using TD-DFT, heteroatoms (B, P, and S) and three N-doping patterns were examined [55]. For some N-doping patterns, the absorption spectral change trend for co-doped GQDs may be in the opposite direction. For instance, the primary absorption peak of GQD-NpBC2O clearly shows a blue shift in comparison to GQD-Np, but those of GQD-NpPC2O and GQD-NpSC2O show a red shift [56]. However, in contrast to GQD-Nh, GQD-NhBC2O's absorption spectrum shows a red shift, while GQD-NhPC2O's and GQD-NhSC2O's show a blue shift [45]. It was discovered by the electron transition analysis that Np has a greater influence than B/P/S-doping on fine-tuning the electronic structure of co-doped GQDs. In the process of fine-tuning the HOMO–LUMO gap of co-doped GQDs with the Nh doping pattern, B/P/S doping is more noticeable [18]. This work is anticipated to yield important insights and guidance for the development of methods to controllably synthesize GQDs with well-defined and desirable properties for specific purposes. It is also anticipated to yield valuable knowledge for understanding and interpreting the electronic and optical properties of GQDs co-doped with different atoms [57].

9.3 DFT Calculation of GQDs upon Surface Functionalization

Using MD simulations and DFT calculations, Morteza and Zahra examined the contribution of various N-functionalities to the drug delivery efficacy of N-GQDs [58]. One medicine of interest is gemcitabine (GC). GC, a deoxycytidine analog, has been used either alone or in combination with other medications as a standard treatment for a variety of solid tumors. Using the Gaussian 09 computer package, full optimization and property calculations for the adsorption of GC on the nanocarrier surfaces were carried out at the M06-2X/6-31G(d) level of theory. The self-consistent reaction field theory, which is based on Tomasi's polarized continuum model (PCM), was used in all calculations. The simulation's findings showed that all nanocarriers had the GC drug loaded on their surface and that in the acidic conditions of cancer tissues, the only place where favorable drug release could happen was from the center N-GQDs. Lastly, steered molecular dynamics (SMD) was used to investigate and describe the process by which the drug-loaded nanocarriers penetrated the cell membrane [59]. According to the findings, center N-GQDs performed better in drug delivery than edge and immaculate N-GQDs.

To ascertain the sensing capacity of a magnesium-decorated graphene quantum dot ($C_{24}H_{12}$-Mg) surface toward XH_3 gases, where X = As, N, and P, in four distinct phases, gas, benzene solvent, ethanol solvent, and water, Agwamba EC et al. developed the surface using DFT computations at the ωB97XD/6–311 + + G(2p,2d) level of theory [60]. Selectivity, sensitivity, conductivity, and efficacy have all been studied and analyzed using the DFT approach using the WB97XD functional and 6–311 + + G(2p,2d) basis set [61]. The energy gap in the various phases is found to follow the order $NH_3@C_{24}H_{12}$-Mg < $PH_3@C_{24}H_{12}$-Mg < $AsH_3@C_{24}H_{12}$-Mg (Fig. 9.8a, b), according to an analysis of the electronic characteristics [62]. The adsorption investigations' outcomes demonstrate that all computed adsorption energies are negative, demonstrating the chemisorption character of the adsorption [63]. According to Singsen [64], the adsorption energies can be placed in the following increasing order: $NH_3@C_{24}H_{12}$-Mg < $PH_3@C_{24}H_{12}$-Mg.

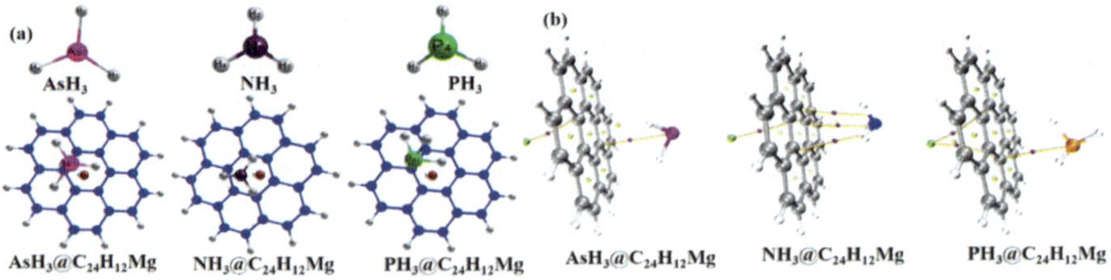

Fig. 9.8 **a** Formation of AsH_3, NH_3, and PH_3 and gas-adsorbed surfaces of $AsH_3@C_{24}H_{12}$-Mg, $NH_3@C_{24}H_{12}$-Mg, and $PH_3@C_{24}H_{12}$-Mg. **b** Representation of the quantum theory of atoms-in-molecules (QTAIM) analysis [60]

GQDs and a pyrene-functionalized azobenzene (AZO) composite were studied by Magdalena Kazmierczak et al. to understand the kinetics of photoinduced energy and charge transfer. The creation of photo-switchable nanoelectronics provides radiation-induced structural modifications that are non-invasive [65]. A multiscale computational technique combined with electronic structure computations is used to describe the impact of conformation, isomerization, and temperature effects on photophysics [50]. Considering AZO molecules that are either trans or cis physisorbed on graphene, the GDQ–AZO interface is constructed by running MD simulations to evaluate the stability and strength of the interactions occurring at the interface [66]. CHARMM36 force field and GROMACS package are used for the simulations. When azobenzene molecules are in their trans configuration, photoinduced reactions for them are always faster than those involving the cis isomer (Fig. 9.9a, b, c). This is because the trans isomer, which is thought to be the active state, interacts with the GQDs more strongly and has a favorable overlap [66]. Conversely, the inactive species can be thought of as the cis configuration, which has slower photoinduced energy and charge transfer activities. However, its contribution to photophysics overall is still significant.

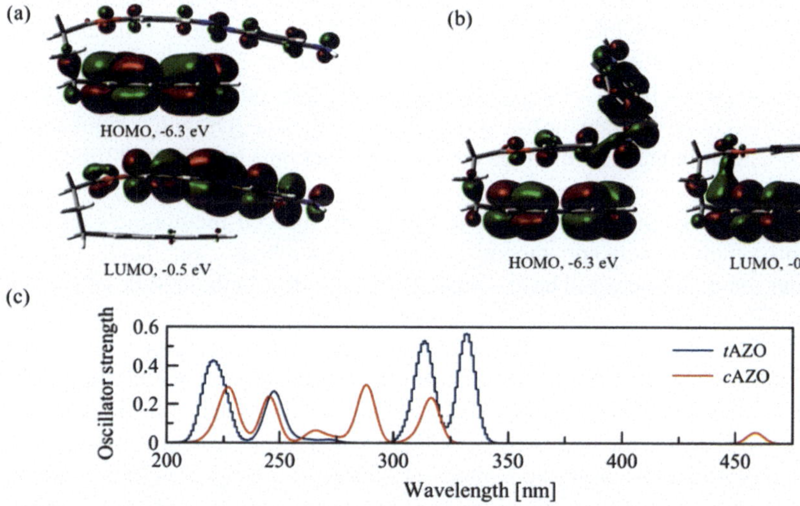

Fig. 9.9 **a** Trans isomer, **b** Cis isomer of isolated AZO molecules, and **c** Computed UV–Vis spectra for the isolated AZO molecules in trans (tAZO) and cis (cAZO) configurations [65]

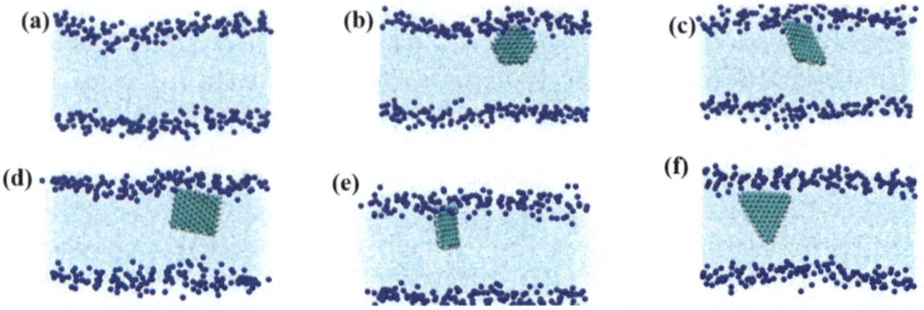

Fig. 9.10 **a** Protocell membrane, **b** Cir-GQD37, **c** Par-GQD49, **d** Rec-GQD50, **e** Squ-GQD46, and **f** Tri-GQD45 of GQDs after 100 ns MD simulation [67]

9.4 Simulation and Modeling of GQDs for Biological Claims

With the aid of MD, the permeation process of the variously shaped GQDs via a 1-palmitoyl-2-oleoylphosphatidylcholine (POPC) membrane was investigated (Fig. 9.10a–f). Kong Z et al. investigate the GQD transport mechanism on the cell membrane using DFT [67]. Consequently, Fang et al. [68] and Chang et al. [69] used MD simulations to investigate the transport mechanisms of GQDs with various forms on the POPC membrane. The findings demonstrate that on a nanosecond timescale, all small-sized GQDs with various geometries are translocated through the lipid membrane [70]. The GQDs initially tend to stay on the cell membrane's surface, subsequently, their corners spontaneously enter the membrane, and, lastly, the full GQDs enter the membrane and have a tendency to stabilize in the middle of the cell membrane [71].

P Zhang et al. revealed that VMD software and MD simulations were used to assess the GQDs' penetration process through various lipid membranes. The study conducted by Zhang et al. [72] employed molecular dynamics simulations to investigate the process of GQD penetration through various cell membranes, such as 1-palmitoyl-2-oleoyl-sn-glycero-3-phosphocholine (POPC), 1,2-dioleoyl-sn-glycero-3 phosphoethanolamine (DOPE), and 1-palmitoyl-2-oleoyl phosphatidylethanolamine (POPE). According to Yuqi L et al., a thorough study of the toxicity profile of GQDs is necessary to ensure their biosafety. The possible toxicity of GQDs on the CYP3A4 enzyme was examined using MD (Fig. 9.11). A pair of exemplar simulation systems are built, with a CYP3A4 encircled by 10 GQDs or GOQDs in each. Direct contact between GQDs and the bottleneck residues of Channels 2a and 2b suggests that GQDs may have a negative impact on these two channels. The possible toxicity of GQDs to the CYP3A4 enzyme offers important information for the responsible and efficient use of such nanomaterials in biomedicine [73].

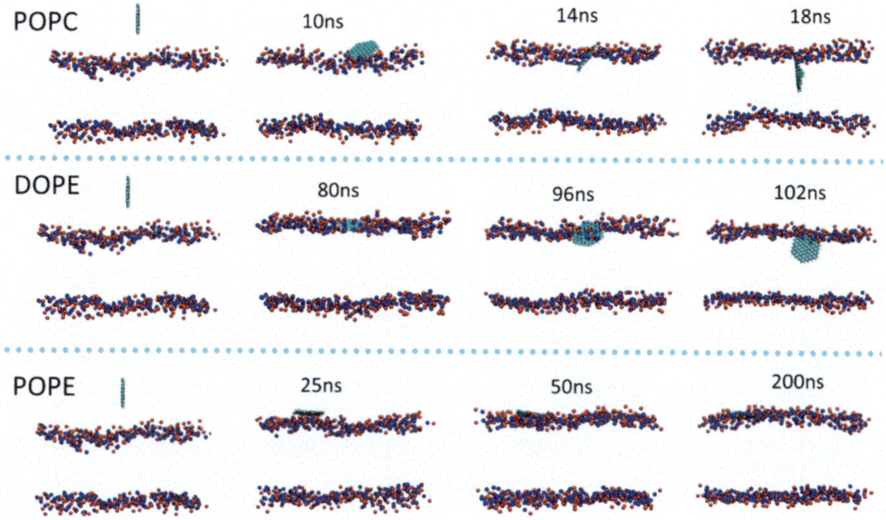

Fig. 9.11 Permeation of GQDs through different cell membranes [72]

References

1. Majid G, Ghafar D, Mohammad Kazem MF (2020) Properties of bilayer graphene quantum dots for integrated optics: an ab initio study. Photonics 7:78
2. Sun L, Wang W, Lv D, Gao ZY, Li Q, Li BC (2021) Magnetic and thermodynamic behaviors of the graphene-like quantum dots: a Monte Carlo study. J Magnetism Magnet Mater 528:167820
3. Priya R, Ranjeet D, Sunita S (2023) Effect of surface modification on optical and electronic properties of graphene quantum dots. Appl Surf Sci 609:155379
4. Napiorkowska E, Milcarz K, Szeleszczuk L (2023) Review of applications of density functional theory (DFT) quantum mechanical calculations to study the high-pressure polymorphs of organic crystalline materials. Int J Mol Sci 15:14155
5. Giannozzi P, Baroni S, Bonini N, Calandra M, Car R, Cavazzoni C, Ceresoli D, Chiarotti GL, Cococcioni M, Dabo I et al (2009) Quantum ESPRESSO: a modular and open-source software project for quantum simulations of materials. J Phys Condens Matter 21:395502–395514
6. Giannozzi P, Andreussi O, Brumme T, Bunau O, Nardelli MB, Calandra M, Car R, Cavazzoni C, Ceresoli D, Cococcioni M et al (2017) Advanced capabilities for materials modeling with quantum ESPRESSO. J Phys Condens Matter 29:465901–465917
7. Keles O, Deshpande PP (2024) Mechanical behavior of graphene quantum dot epoxy nanocomposites: a molecular dynamics study. Mater Lett 362:136206
8. Thompson AP, Aktulga HM, Berger R, Bolintineanu DS, Brown WM, Crozier PS, Veld PJ, Kohlmeyer A, Moore SG, Nguyen TD, Shan R, Stevens MJ, Tranchida J, Trott C, Plimpton SJ (2022) LAMMPS - a flexible simulation tool for particle-based materials modeling at the atomic, meso, and continuum scales. Comput Phys Commun 271:108171
9. Deshpande PP, Keles O (2024) Simulation data for engineering graphene quantum dot epoxy nanocomposites using molecular dynamics. Data Brief 53:110169
10. Karimi B, Ramezanzadeh B (2017) A comparative study on the effects of ultrathin luminescent graphene oxide quantum dot (GQD) and graphene oxide (GO) nanosheets on the interfacial interactions and mechanical properties of an epoxy composite. J Colloid Interface Sci 493:62–76

11. Gissinger JR, Jensen BD, Wise KE (2020) REACTER: a heuristic method for reactive molecular dynamics. Macromolecules 53:9953–9961
12. Budyka MF, Sheka EF, Popova NA (2017) Graphene quantum dots: theory and experiment. Rev Adv Mater Sci 51:35–49
13. Starcke LH, Wormit M, Dreuw A (2009) Unrestricted algebraic diagrammatic construction scheme of second order for the calculation of excited states of medium-sized and large molecules. J Chem Phys 130:024104
14. Wenzel J, Wormit M, Dreuw A (2014) Calculating X-ray absorption spectra of open-shell molecules with the unrestricted algebraic-diagrammatic construction scheme for the polarization propagator. J Chem Theory Comput 10:4583
15. Lopata K, Reslan R, Kowalska M, Neuhauser D, Govind N, Kowalski K (2011) Excited-state studies of polyacenes: a comparative picture using EOMCCSD, CR-EOMCCSD(T), range-separated (LR/RT)-TDDFT, TD-PM3, and TD-ZINDO. J Chem Theory Comput 7:3686
16. Sheka EF (2013) Topological modelling of nanostructures and extended systems, in carbon materials: chemistry and physics. (Eds.) Ashrafi AR, Cataldo F, Iranmanesh A, Ori O, vol 7. Springer: Berlin, p 137
17. Kuc A, Heine T, Seifert G (2018) Structural and electronic properties of graphene nanoflakes. Phys Rev B 81:085430
18. Sharma Yamijala SRKC, Mukhopadhyay MM, Pati SK (2015) Linear and nonlinear optical properties of graphene quantum dots: a computational study. J Phys Chem C 119:12079–12087
19. Rieger R, Muullen K (2010) Forever young: polycyclic aromatic hydrocarbons as model cases for structural and optical studies. J Phys Org Chem 23:315–325
20. Cocchi C, Prezzi D, Ruini A (2014) Anisotropy and size effects on the optical spectra of polycyclic aromatic hydrocarbons. J Phys Chem A 118:6507–6513
21. Sholl DS, Steckel JA (2009) Density functional theory: a practical introduction. Wiley, New Jersey
22. Guclu AD, Potasz P, Hawrylak P (2013) Zero-energy states of graphene triangular quantum dots in a magnetic field. Phys Rev B 88:155429–155443
23. Dang MT, Bich Thao PT, Ngoc Thao TT, Thanh Tien NT (2022) First principles study of electronic and optical properties of small edge-functionalized penta-graphene quantum dots. AIP Adv 12:065008
24. Clark SJ, Segall MD, Pickard CJ, Hasnip PJ, Probert MIJ, Refson K, Payne MC (2005) First principles methods using CASTEP. Z Kristallogr Cryst Mater 220:567–570
25. Jippo H, Ohfuchi M (2013) First-principles study of edge-modified armchair graphene nanoribbons. J Appl Phys 113:183715
26. Li Y, Shu H, Wang S, Wang J (2015a) Electronic and optical properties of graphene quantum dots: the role of many-body effects. J Phys Chem C 119:4983–4989
27. Lee SH, Kim DY, Lee J, Lee SB, Han H, Kim YY, Mun SC, Im SH, Kim TH, Park OO (2019) Synthesis of single-crystalline hexagonal graphene quantum dots from solution chemistry. Nano Lett 19:5437–5442
28. Hu W, Lin L, Yang C, Dai J, Yang J (2016) Edge-modified phosphorene nanoflake heterojunctions as highly efficient solar cells. Nano Lett 16:1675–1682
29. Derbenyova NV, Konakov AA, Shvetsov AE, Burdov VA (2017) Electronic structure and absorption spectra of silicon nanocrystals with a halogen (Br, Cl) coating. JETP Lett 106:247–251
30. Cao L, X. Li X, Y. Li Y, G. Zhou G, (2020) Electrical properties and spintronic application of carbon phosphide nanoribbons with edge functionalization. J Mater Chem C 8:9313–9321
31. Ozonder S, Unlu C, Guleryuz C, Trabzon L (2023) Doped graphene quantum dots UV−vis absorption spectrum: a high-throughput TDDFT study. ACS Omega 8:2112–2118
32. Escudero D, Laurent AD, Jacquemin D (2017) In Handbook of Computational Chemistry, (Eds) Leszczynski J, Kaczmarek-Kedziera A, Puzyn TG, Papadopoulos M, Reis HK, Shukla M. Springer International Publishing, Cham, pp 927−961
33. Ajeel FN, Mohammed MH, Khudhair AM (2019) Effects of lithium impurities on electronic and optical properties of graphene nanoflakes: a DFT-TDDFT study. Chin J Phys 58:109–116

34. Shi H, Barnard AS, Snook IK (2012) Quantum mechanical properties of graphene nano-flakes and quantum dots. Nanoscale 4:6761
35. Barnard S, Snook IK (2010) Size- and shape-dependence of the graphene to graphane transformation in the absence of hydrogen. J Mater Chem 20:10459
36. Alam SKM, Arundithi A, Lin H, Kok Hwa L, Peng C (2014) Revealing the tunable photoluminescence properties of graphene quantum dots. J Mater Chem C 2:6954
37. Li Y, Hu Y, Zhao Y, Shi G, Deng L, Hou Y, Qu L (2011) An electrochemical avenue to green-luminescent graphene quantum dots as potential electron-acceptors for photovoltaics. Adv Mater 23:776–780
38. Shujun W, Cole IS, Zhao D, Li Q (2015) The duo roles of functional groups in the photoluminescence of graphene quantum dots. Nanoscale 8:7449–7458
39. Wang WL, Meng S, Kaxiras E (2008) Graphene nanoflakes with large spin. Nano Lett 8:241–245
40. Li M, Cushing SK, Zhou X, Guo S, Wu N (2012) Fingerprinting photoluminescence of functional groups in graphene oxide. J Mater Chem 22:23374
41. Jianguang J, Hongzhou D, Liyan Y, Lifeng D (2017) Optical and electronic properties of graphene quantum dots with oxygen-containing groups: a density functional theory study. J Mater Chem C 5:5984–5993
42. Li L, Wu G, Yang G, Peng J, Zhao J, Zhu JJ (2013) Focusing on luminescent graphene quantum dots: current status and future perspectives. Nanoscale 5:4015–4039
43. Peng J, Gao W, Gupta BK, Liu Z, Romeroaburto R, Ge L, Song L, Alemany LB, Zhan X, Gao G (2012) Graphene quantum dots derived from carbon fibers. Nano Lett 12:844–849
44. Deb J, Paul D, Sarkar U (2020) Density functional theory investigation of nonlinear optical properties of T-graphene quantum dots. J Phys Chem A 124:1312–1320
45. Frisch MJ, Trucks GW, Schlegel HB, Scuseria GE, Robb MA, Cheeseman JR, Scalmani G, Barone V, Mennucci B, Petersson GA et al (2009) Gaussian 09, revision D.01; Gaussian, Inc.: Wallingford, CT
46. Zhang TG, Zhao Y, Asselberghs I, Persoons A, Clays K, Therien MT (2005) Design, synthesis, linear, and nonlinear optical properties of conjugated (Porphinato)zinc(II)-based donor−acceptor chromophores featuring Nitrothiophenyl and Nitrooligothiophenyl Electron-accepting moieties. J Am Chem Soc 127:9710–9720
47. Zhong RL, Xu HL, Muhammad S, Zhang J, Su M (2012) The stability and nonlinear optical properties: encapsulation of an excess electron compound LiCN···Li within boron nitride nanotubes. J Mater Chem 22:2196–2202
48. Xu X, Hu CL, Kong F, Zhang JH, Mao JG, Sun JL (2013) Cs2GeB4O9: a new second-order nonlinear-optical crystal. Inorg Chem 52:5831–5837
49. Abdelati MA, Fadlallah MM, Gamal YEED, Maarouf AA (2021) Pristine and holey graphene quantum dots: Optical properties using time independent and dependent density functional theory. Phys E 128:114602
50. Li Y, Shu H, Wang S, Wang J (2015b) Electronic and optical properties of graphene quantum dots: the role of many body effects. J Phys Chem C 119:4983–4989
51. Yin H, Zhang H (2012) Plasmons in graphene nanostructures. J Appl Phys 111:103502
52. Hazem A, Hanan E, Ibrahim MA (2018) Tuning electronic properties in graphene quantum dots by chemical functionalization: density functional theory calculations. Chem Phys Lett 695:138–148
53. Ladd TD, Jelezko F, Laflamme R, Nakamura Y, Monroe C, O'Brien JL (2010) Quantum computers. Nature 464:45
54. Jianguang F, Guo Q, Song N, Liu H, Dong H, Chen Y, Yu L, Dong L (2021a) Density functional theory study on optical and electronic properties of co-doped graphene quantum dots based on different nitrogen doping patterns. Diam Relat Mater 113:108264
55. Ayers PW (2007) The physical basis of the hard/soft acid/base principle. Faraday Discuss 135:161–190
56. Zhang S, Sui L, Dong H, He W, Dong L, Yu L (2018) High-performance supercapacitor of graphene quantum dots with uniform sizes. ACS Appl Mater Interfaces 10:12983–12991

57. Jauja-Ccana VR, Cordova-Huaman AV, Feliciano GT, Rosa-Toro Gómez AL (2021) Experimental and molecular dynamics study of graphene oxide quantum dots interaction with solvents and its aggregation mechanism. J Mol Liquids 335:116136
58. Morteza V, Zahra S (2019) Revealing the role of different nitrogen functionalities in the drug delivery performance of graphene quantum dots: a combined density functional theory and molecular dynamics approach. J Mater Chem B 7:6156
59. Van Khai T, Na HG, Kwak DS, Kwon YJ, Ham H, Shim KB, Kim HW (2012) Significant enhancement of blue emission and electrical conductivity of N-doped graphene. J Mater Chem 22:17992–18003
60. Agwamba EC, Louis H, Olagoke PO, Gber TE, Okon GA, Fidelisa CF, Adeyinka AS (2023) Modeling of magnesium-decorated graphene quantum dot nanostructure for trapping AsH_3, PH_3 and NH_3 gases. RSC Adv 13:13624
61. Henna TK, Pramod K (2020) Graphene quantum dots redefine nanobiomedicine. Mater Sci Eng C 110:110651
62. Gber TE, Louis H, Owen AE, Etinwa BE, Benjamin I, Asogwa FC, Eno EA (2022) Heteroatoms (Si, B, N, and P) doped 2D monolayer MoS_2 for NH_3 gas detection. RSC Adv 12:25992–26010
63. Nematollahzadeh A, Basharnavaz H, Habibi-Yangjeh A, Kamali SH (2021) Enhancing AsH_3 gas adsorption potentials of graphitic carbon nitride by codoping Cr/P, Mo/P, and W/P atoms: a DFT investigation. J Nanostructures 11:638–646
64. Singsen S, Thasami N, Tangpakonsab P, Bae H, Lee H, Hussain T, Kaewmaraya T (2022) Transition-metals Graphdiyne monolayer as efficient sensor toward Phosphide (PH_3) and Arsine (AsH_3). Phys Chem Chem Phys 24:26622–26630
65. Magdalena K, Samuele G, Silvio O (2024) Photoinduced energy and electron transfer at graphene quantum dot/azobenzene interfaces. J Mater Chem C 12:143
66. Min M, Seo S, Lee SM, Lee H (2013) Voltage-controlled nonvolatile molecular memory of an azobenzene monolayer through solution-processed reduced graphene oxide contacts. Adv Mater 25:7045–7050
67. Kong Z, Zhang P, Chen J, Zhou H, Ma X, Wang H, Shen JW, Liang LJ (2021) Effect of shape on the entering of graphene quantum dots into a membrane: a molecular dynamics simulation. ACS Omega 6:10936–10943
68. Fang J, Liu Y, Chen Y, Ouyang D, Yang G, Yu T (2018) Graphene quantum dots-gated hollow mesoporous carbon nano platform for targeting drug delivery and synergistic chemo-photothermal therapy. Int J Nanomed 13:5991–6007
69. Chang Y, Yang ST, Liu JH, Dong E, Wang Y, Cao A, Liu Y, Wang H (2011) In vitro toxicity evaluation of graphene oxide on A549 cells. Toxicol Lett 200:201–210
70. Kong Z, Wang H, Liang L, Zhang Z, Ying S, Hu Q, Shen JW (2017) Investigation of the morphological transition of a phospholipid bilayer membrane in an external electric field via molecular dynamics simulation. J Mol Model 23:113
71. Wu L, Zhang P, Zhou H, Li J, Shen X, Li T, Kong Z, Hu W, Zhang Y (2022) Molecular dynamics simulation of the interaction between graphene oxide quantum dots and DNA fragment. Materials 15:8506
72. Zhang P, Jiao F, Wu L, Kong Z, Hu W, Liang L, Zhang Y (2022) Molecular dynamics simulation of transport mechanism of graphene quantum dots through different cell membranes. Membranes 12:753
73. Yuqi L, Jinjun L, Zonglin G, Yaoxing H (2023) Graphene quantum dots blocking the channel egresses of cytochrome P450 enzyme (CYP3A4) reveals potential toxicity. Sci Rep 13:21091
74. Jianguang F, Qian G, Na S, Haiying L, Hongzhou D, Yingjie C, Liyan Y, Lifeng D (2021b) Density functional theory study on optical and electronic properties of co-doped graphene quantum dots based on different nitrogen doping patterns. Diam Relat Mater 113:108264

Chapter 10
Graphene Quantum Dots in Nanoelectronics Application

10.1 Introduction

GQDs' special qualities have attracted applications in nanoelectronics. These fragments of nanoscale graphene, usually smaller than 10 nm, show size-dependent optical and electrical properties, which makes them very desirable for use in next-generation electronic devices [1].

Important GQD features in nanoelectronics are,

(i) **Quantum Confinement Effects**: Because of their small size, GQDs exhibit quantum confinement phenomena, which result in discrete electronic energy levels. By using this characteristic, their electronic band structure can be modified to suit particular applications, giving them more control over their electronic characteristics.

(ii) **Tunable Bandgap**: Unlike bulk graphene, which has zero bandgaps and behaves like a semimetal, GQDs exhibit tunable bandgaps depending on their size and edge morphology. This tunability makes them suitable for applications requiring semiconducting behavior, such as transistors and light-emitting diodes (LEDs).

(iii) **High Charge Carrier Mobility**: High charge carrier mobility is a characteristic of GQDs that is usually on par with or even higher than that of graphene. To create high-speed electronic gadgets with low power consumption, this great mobility is essential.

(iv) **Size and Surface Effects**: Compared to bulk graphene, GQDs have a high surface-to-volume ratio, which may result in improved surface interactions and special electrical characteristics. GQDs are appealing for applications like sensors, where interactions with analytes at the surface are crucial, because of their size and surface effects [2].

(v) **Solution Processability**: GQDs can be integrated into flexible and large-area electronic devices by methods including spin coating, drop casting, or inkjet printing. They can also be produced in solution. Their compatibility with

N. Manjubaashini et al., *Graphene Quantum Dots*, Materials Horizons: From Nature to Nanomaterials, https://doi.org/10.1007/978-981-97-5722-0_10

scalable production processes is improved by this solution's processability [3].

(vi) **Chemical and Mechanical Stability**: The mechanical and chemical stability of GQDs is beneficial for long-term performance and device reliability. Because of their resilience, GQD-incorporated nanoelectronics devices can function in a variety of operational and environmental environments [4].

(vii) **Optoelectronic Applications**: Additionally, GQDs have superb optoelectronic qualities, such as broad absorption, robust photoluminescence, and high quantum yield. These properties render them appropriate for application in solar cells, photodetectors, and other light-emitting devices [5].

(viii) **Compatibility with Existing Fabrication Techniques**: GQDs can be made in several ways, including top-down techniques like chemical exfoliation of graphene and bottom-up techniques like chemical vapor deposition. Furthermore, their integration into pre-existing nanoelectronics device fabrication techniques facilitates their acceptance in real-world applications [6].

GQDs are employed in charge sensors composed of single electron transistors (SET), as opposed to graphene, which is used in field-effect transistors (FETs). Here is a list of the few uses for GQDs:

(a) **Quantum Dot Transistors**: GQDs can be used in transistors as active parts. They are appropriate for high-performance transistors because of their compact size, which enables fine control over electronic properties.

(b) **Photodetectors**: Because of their effective charge transfer and tunable bandgap, GQDs have outstanding photoresponse. For sensitive light detection, they can be included in photodetectors [7].

(c) **Memory Devices**: GQDs can be used in memory devices as components for charge storage. Their potential for use in non-volatile memory applications stems from their capacity to capture and release charge carriers.

(d) **Sensors**: GQDs that have been functionalized are responsive to environmental changes. They can be employed as sensors to find other analytes, gasses, and biomolecules.

(e) **Flexible Electronics**: Because of their electrical and mechanical qualities, GQDs can be integrated into flexible electronic devices. Wearable electronics and flexible displays are two examples of applications.

(f) **Energy Harvesting**: GQDs can transform energy from external sources, such as ambient light, into electrical energy. This characteristic holds significance for autonomous gadgets.

(g) **Optoelectronics**: GQDs can emit both visible and near-infrared light. Light-emitting diodes (LEDs) and other optoelectronic devices employ them [8].

10.2 Functioning of GQD-Based Nanoelectronic Devices

GQDs are quickly rising to the top of the list of cutting-edge functional materials, due to the bandgap opening caused by the quantum confinement effect, which exhibits special optical and electrical properties. Chlorine (Cl)-GQDs, or chlorine-doped GQDs, were created by chemically exfoliating HCl-treated carbon fibers (CFs), which were first made from degreasing cotton and then annealed for 30 min at 1000 °C (Fig. 10.1a–d). Using a 405 nm laser irradiation under the reverse bias voltage, Cl-GQDs-based photovoltaic detectors have been demonstrated. Both the light-absorbing and electron-accepting roles for Cl-GQDs in photodetection have been found, leading to an exceptionally big ratio of photocurrent to dark current, was higher at room temperature [9]. The doping impact of GQDs in a nematic liquid crystal (NLC) device was described by M J Cho. The application of GQD results in enhanced electro-optical (EO) characteristics, including a lower threshold voltage and quicker response times. ZnO-doped NLCs are utilized as white light-emitting diodes [10]. By trapping impurity ions, the GQDs added to the NLCs lowered the threshold voltage and sped up reaction times while retaining other characteristics of traditional twisted nematic (TN)-LCDs.

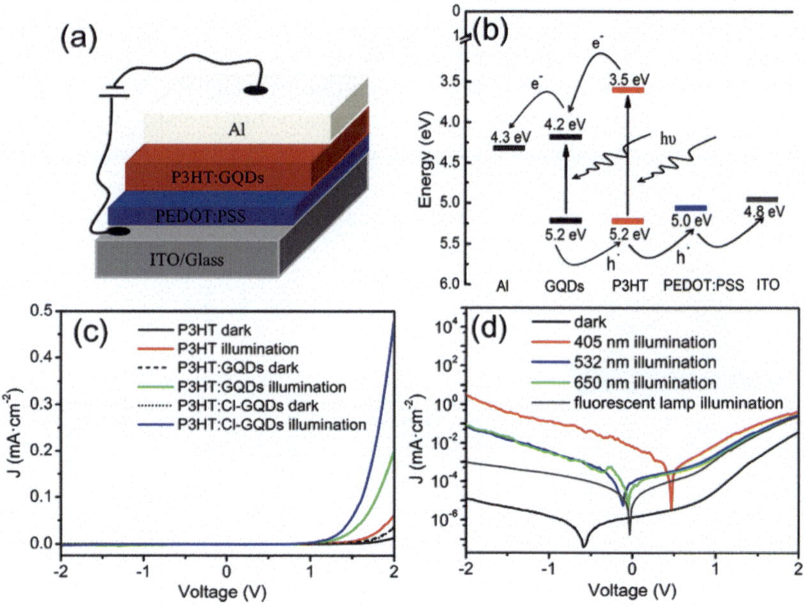

Fig. 10.1 ITO/PEDOT: PSS/P3HT: GQDs/Al photovoltaic detector: **a** preparation schematics, **b** energy level diagram, **c** J-V curves of photovoltaic detector using P3HT or P3HT:GQDs or P3HT:Cl-GQDs, and **d** log current density as functions of voltage for the Cl-GQDs detector under different light illuminations. Reprinted with permission from @ Appl. Phys. Lett. 105, 111,116 (2014)

Two-dimensional (2D) graphene films and graphene derivatives have garnered a lot of attention due to their enormous potential in optoelectronic applications. It is still very difficult to improve the performance of photodetectors based on graphene films and graphene derivatives. For near-infrared light detection, a functional vertically oriented graphene (VOG) is constructed on the germanium (Ge) heterojunction (GQDs/VOG/Ge) by substituting VOG for graphene films. VOG is then functionalized by GQDs. The synergistic effects of VOG and GQDs improve the photodetector's light absorption and electron transport capabilities. Enhancing the intrinsic potential of the Schottky junctions, controlling the Fermi level of VOG, and facilitating the separation of photoinduced electron and hole pairs can all be accomplished effectively through functional modification of VOG [11]. At a wavelength of 1,550 nm, the as-fabricated photodetector exhibits high responsivity (1.06×10^6 AW^{-1}) and detectivity (2.11×10^{14} cm $Hz^{1/2}W^{-1}$). Examining the photoresponse showed great reproducibility and long-term stability along with response rates with microsecond rise/fall times (Fig. 10.2a–f).

The production of GQDs that are unoxidized at the sp^2 surfaces and regulated in size and dimensions is sought for the generation of optoelectronic devices. Bulky Frechet's dendritic wedges functionalized at the periphery of GQDs were synthesized. Atomic force microscopy revealed the single-layered, size-regulated architectures of the dendronized GQDs [12]. An unusual feature resulted from the edge-functionalization of the GQDs: white-light emission. Created edge-functionalized graphene quantum dots (GQD-2 and GQD-3a–c) that emit white light by reacting with 4-propynyloxybenzylamine (4) after edge-oxidized GQDs (GQD-1) were first produced (Fig. 10.3a–f). To create optoelectronically functionalized carbon-based materials, a variety of chemical changes in organic solvents are made possible by

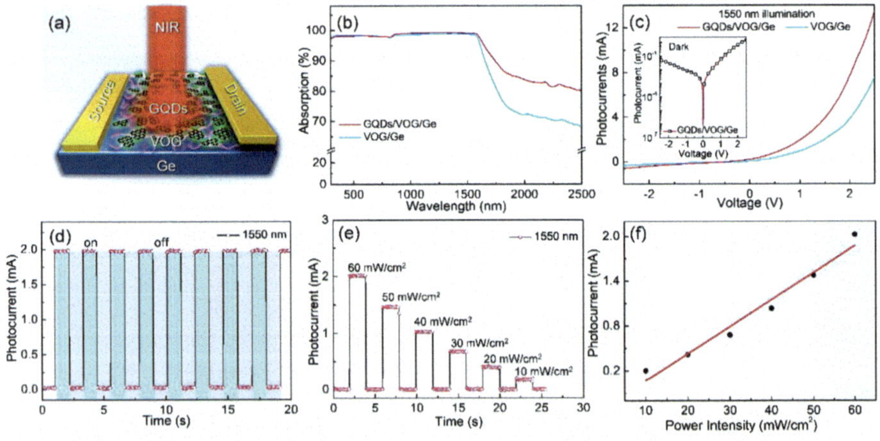

Fig. 10.2 Photoresponse of NIR photodetector based on the hybrid architecture of VOG/Ge heterojunction: **a** Schematic diagram of the hybrid photodetector, **b** UV-NIR absorption spectra of VOG/Ge and GQDs/VOG/Ge, **c–d** I-V curves of the photodetector, and **e–f** Photo-switching curve of the photodetector under pulsed light irradiation of 1,550 nm [11]

GQD-2's lipophilicity. In fact, by functionalizing the edges with dendritic wedges, the optical characteristics of GQD-2 were adjusted, and the resultant GQDs showed white-light emission. These lipophilic GQDs have the potential to be used in carbon-based materials, which could lead to the development of novel optoelectronic devices because they are chemically accessible [13]. For the first time, GQDs were created via simple hydrothermal processing and raw plant leaf extracts of fenugreek (*Trigonella foenum-graecum*) and neem (*Azadirachta indica*). N-GQDs and F-GQDs have high QYs of 41.2 and 38.9%, respectively, and a very vivid green, pH-independent PL (Fig. 10.4a–d). Additionally, N-GQDs were employed as a light converter to create white light by combining them in the proper ratios according to the RGB principle with QS and CPY, which were also extracted from neem leaf extracts [14]. With a photocurrent density of just 0.09 $\mu A/cm^2$, the photocurrent response of the thin layer of citric acid (CA)-GQDs is hardly perceptible. On the other hand, the photocurrent density of the N- and O-doped GQD (E-GQD) film is two times higher than that of the CA-GQD film, indicating that the E-GQD film is significantly more effective in producing photocurrent [15].

Stretchable device production is anticipated to heavily rely on 2D materials. A very stretchable and sensitive photodetector based on hybrid graphene and GQD is demonstrated by Chiang CW et al. The graphene layer is supported by a special poly(dimethylsiloxane) ripple structure that allows it to expand in response to external strain. Because of the many photon reflections between the ripples, the device's ripple can boost carrier generation in GQDs and surpass graphene's innate stretchability limit (Fig. 10.4a–f). Under various strain circumstances, the ripple-structured photodetector demonstrated remarkably stable conductance properties and photoresponsivity [16]. This undulating framework offers a great chance to create elastic textiles with stretchability. Stretchability from the rippling structure can be used in many different substance systems, including 2D materials, in addition to the graphene and GQD composite.

A high-performance composite-based triboelectric nanogenerator (CTENG) was created by Farah et al. using GQDs as conductive media, polydimethylsiloxane (PDMS) as a polymeric matrix, and barium titanite (BTO) nanopowders as dielectric fillers. The higher concentration of 30 wt% GQD, which is twice as high as nanocomposite films without GQDs (PDMS/BTO alone), improves the output performance of TENGs. PDMS/BTO/G30 achieves the greatest output performance with a voltage, current, and power density of around 310.0 V, 23.0 μA, and 1.6 W/m^2, respectively [17]. The TENGs' consistent and high-quality electrical output power indicates that the technology may be useful for nanoenergy applications including energy harvesting (Fig. 10.5a–d). Hemen K et al. reported on the field effect properties of lithographically produced FET with as-prepared GQDs and GQDs treated with hydrazine as channel material. With the ability to control doping and the strength of interdot coupling in the quantum dot array, the graphene quantum dot surface can be modified by external adsorbates (Fig. 10.6a–c). This presents intriguing opportunities for the development of size-tunable solid-state quantum dot electronics with yield hole mobility of 0.01 $cm^2\ V^{-1}\ s^{-1}$ and an I_{on}/I_{off} ratio is 45 [18].

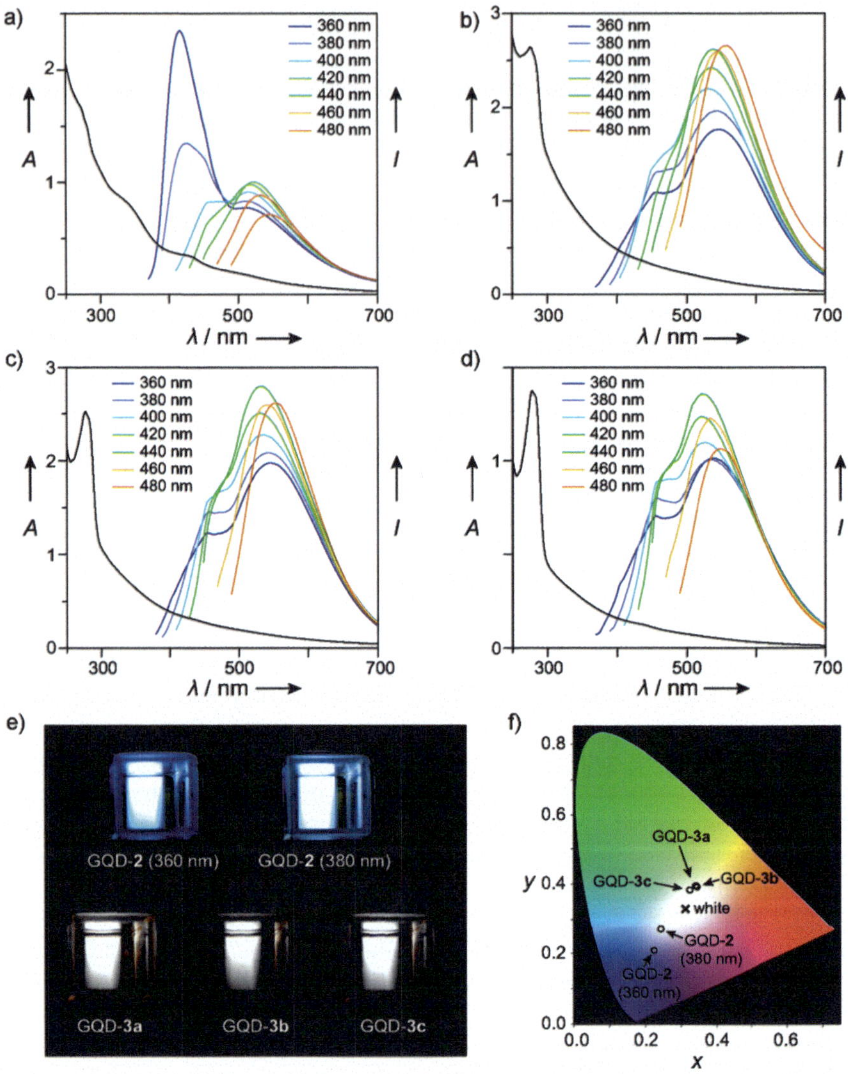

Fig. 10.3 UV–visible absorption and fluorescence spectra: **a** GQD-2, **b** GQD-3a, **c** GQD-3b, **d** GQD-3c in CH_2Cl_2, **e** Optical images of GQD-2 (360 and 380 nm), GQD-3a (360 nm), GQD-3b (360 nm), and GQD-3c (360 nm), and **f** the CIE 1931 chromaticity diagram of GQD-2 at 360 nm (x = 0.223, y = 0.214), GQD-2 at 380 nm (x = 0.240,y = 0.265), GQD-3a (x = 0.336, y = 0.394), GQD-3b (x = 0.340,y = 0.392), and GQD-3c (x = 0.324, y = 0.383) [12]

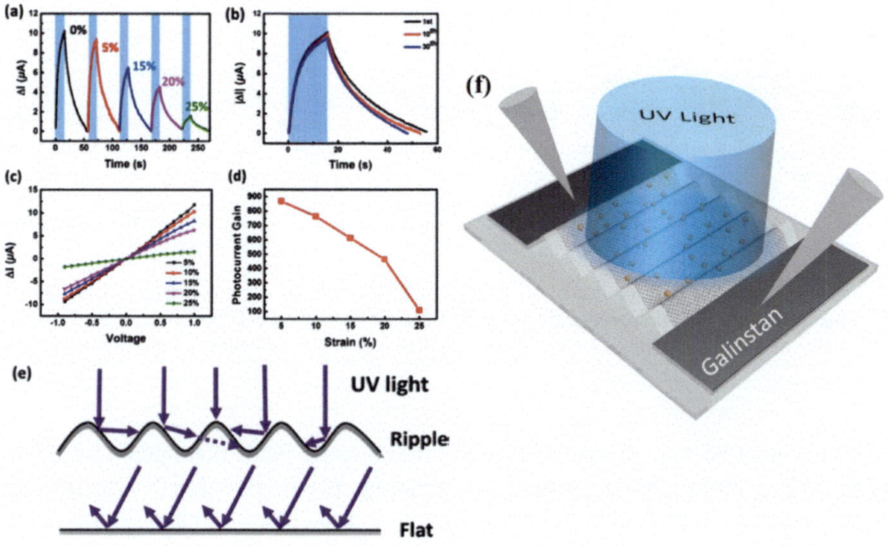

Fig. 10.4 **a** Dynamic photoresponse under different strains, **b** Fatigue test, **c** Photocurrent gain under different strain conditions, **d** Dependence of the photocurrent gain on the external strain, **e** Schematic of multiple reflections of incident light arising from the rippled structure, and **f** setup of hybrid photodetector device measurement [16]

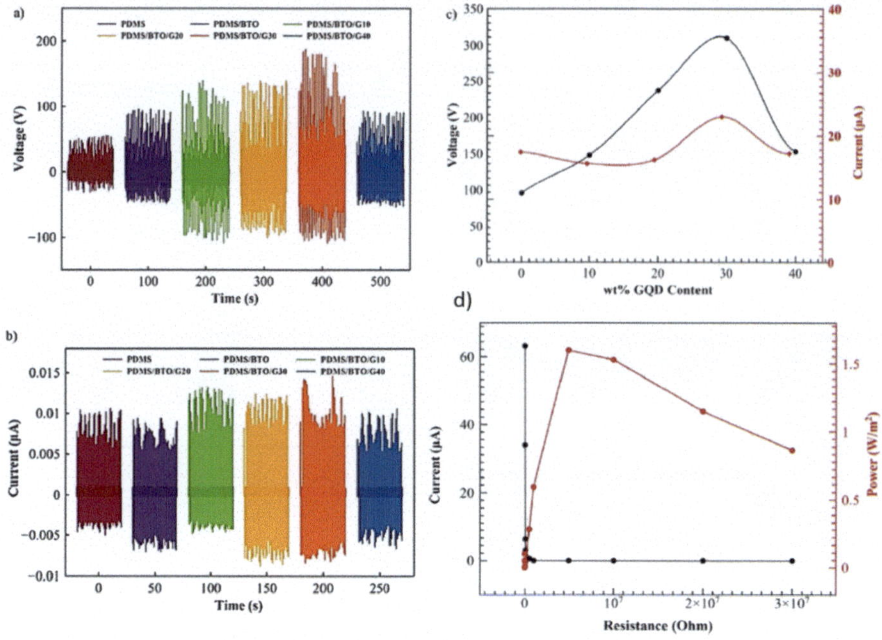

Fig. 10.5 **a** *V*OC, **b** *I*SC, **c** peak-to-peak value of the PDMS/BTO-based TENGs as a function of GQD, and **d** output current and power as a function of the external load resistance [17]

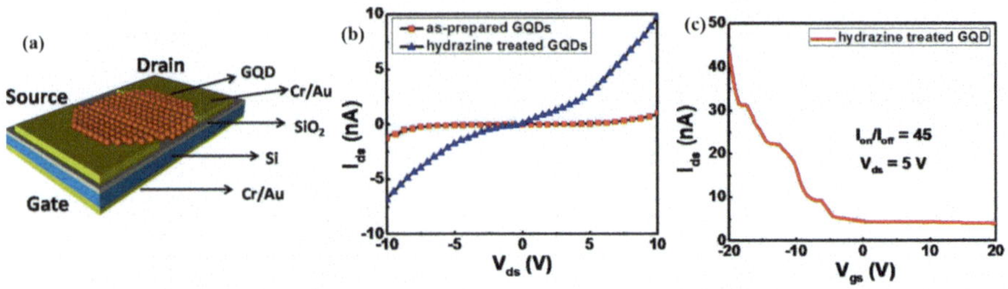

Fig. 10.6 **a** Pictorial representation of device prepared for the electrical measurements, **b** *Ids* versus *Vds* characteristics of as-prepared and hydrazine-treated GQDs, and **c** Transfer characteristics *Ids* versus *Vgs* of hydrazine-treated GQDs at fixed bias at $Vds = 5$ V showing p-type characteristics [18]

With tunnel barriers determined by external electric fields that open a bandgap, Allen MT demonstrated a method for creating quantum-confined structures in suspended bilayer graphene that eliminates both substrate and edge disorder. We demonstrate pure quantum dot production in two regimes: B > 0 utilizing the quantum Hall v = 0 gap for confinement, and at zero magnetic field B using the energy gap created by a perpendicular electric field. Periodicity corresponding with electrostatic calculations based on local top-gate geometry is demonstrated directly via Coulomb blockade oscillations, indicating local control over graphene's band structure [19]. With the integration of single electron transit, excellent device quality, and vibrational mode access, this technology opens up a wide range of applications, including quantum bits and electromechanical sensors (Fig. 10.7a–c).

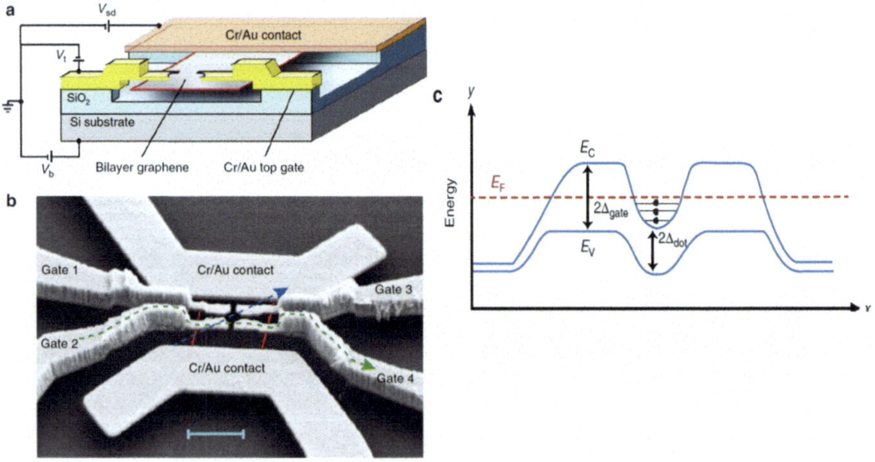

Fig. 10.7 **a** Schematic cross-section of a suspended gate-defined bilayer graphene quantum dot, **b** Scanning electron micrograph of quantum dot device, and **c** Quantum dot formation at $B = 0$, illustrated in a cross-sectional cut of energy versus position. Reprinted with permission from @ Nature Communication 3, 934 (2012)

Nitrogen-doped graphene QD (NGQD) heterostructures comprising photoresponsive and photodetective optoelectronic devices are placed over a monolayer (ML) of tungsten diselenide (ML WSe_2). The enhanced photoluminescence of ML WSe_2 is ascribed to the n-doping effect's dominant neutral exciton emission. A pure ML WSe_2 photodetector's photoresponsivity was only 48% of that of N-GQD-covered ML WSe_2, due to the substantial light absorption and charge transfer from N-GQDs to ML WSe_2. With 46% of its performance retained after 30 days in ambient settings, the hybrid photodetector demonstrates good environmental stability (Fig. 10.8a–c). The enhancement of hybrid photodetector performance is also significantly influenced by the photogating effect. The hybrid photodetector has a substantially higher responsivity of 2578 A W^{-1} when applying back-gate voltage modulation than the ML WSe_2-based device [20]. K Gerasimos et al. developed a hybrid photodetector that is made of monolayer or bilayer graphene coated in a thin layer of colloidal quantum dots (Fig. 10.9a–c). Due to the lengthy trapped-charge lifetimes in the quantum dot layer and the high charge mobility of graphene, strong and controllable light absorption generates electric charges that are transferred to the graphene and recirculated numerous times [21]. Benefits of the device include compatibility with current circuit technology, spectral selectivity from short-wavelength infrared to visible, and gate-tunable sensitivity and speed (731,013 Jones).

A straightforward solution method for creating emissive hybrid quantum dots with a ZnO core encased in a single-layer graphene shell was presented by Son DI. Next, quantum dots were incorporated into a 798 cd m^{-2} brilliant white LED. Two more

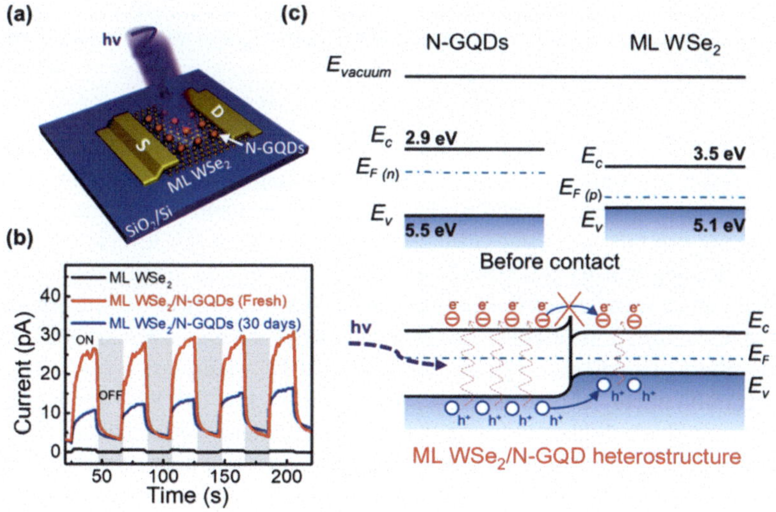

Fig. 10.8 **a** Schematic illustration of an ML WSe_2/N-GQD Photodetector, **b** Time-dependent photocurrent of ML WSe_2 (black line) and ML WSe_2/N-GQD devices after fabrication (red line) and after 30 days under ambient (blue line) at VDS = 1 V and VG = 0 V, and **c** Energy band diagram of N-GQDs and ML WSe_2 before contact and after formation of heterostructure under 405 nm light illumination [20]

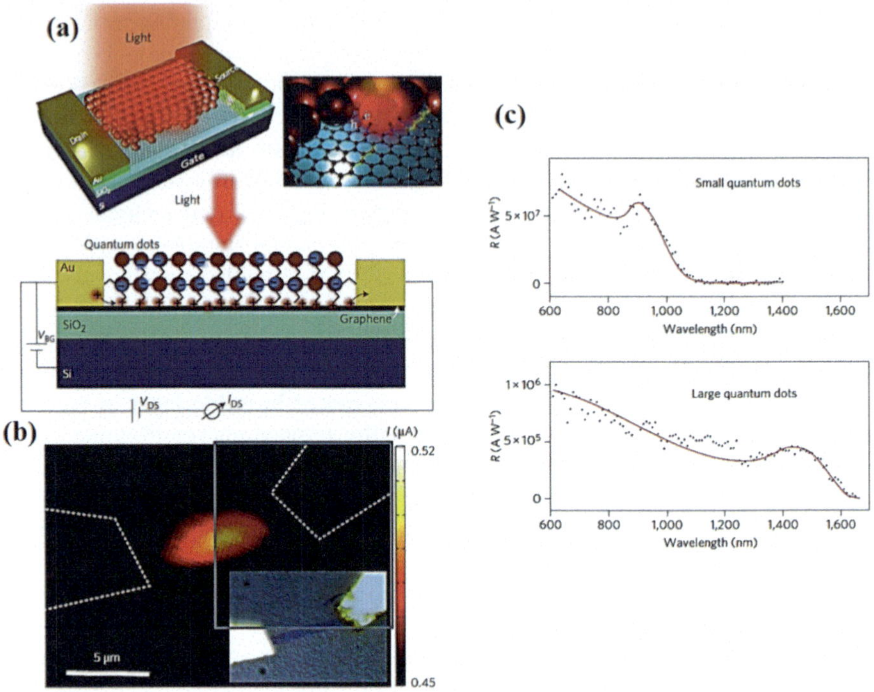

Fig. 10.9 **a** Schematic of the GQD hybrid phototransistor, **b** Spatial photocurrent profile using a laser beam focused at 532 nm with a power of 1.7 pW, and **c** Spectral responsivity of two phototransistors of single- and bilayer graphene. Reprinted with permission from @ Nature Nanotechnology 7, 363–368 (2012)

blue emission peaks are seen in the quantum dot's luminescent spectrum, and the strain caused by curvature opens an electronic bandgap of 250 meV in the graphene (Fig. 10.11a–d). These additional peaks are caused by a splitting of the graphene's lowest empty orbitals into three orbitals with different energy levels, according to calculations using density functional theory [22]. By mixing the quantum dots with additional emissive materials in a multilayer light-emitting diode, white emission is produced. Quantum dots of epitaxial graphene on SiC display a remarkable temperature variation of resistance (greater than 430 M\K^{-1} below 6 K) as a result of quantum confinement. This results in responsivities of 1×10^{10}VW^{-1}, which is five orders of magnitude higher than other types of graphene hot-electron bolometers (Fig. 10.10a–c). With its exceptional electrical noise-equivalent power (about 2×10^{-16}WHz$^{-1/2}$ at 2.5 K) and great responsiveness, the bolometers surpass those of commercially available cooled bolometers [23]. E Abu Bakar et al. created a flexible piezoelectric nanogenerator based on $BaTiO_3$ and GQDs. The initial, bending, and release states—all of which may be similarly illustrated—determine the device's generated output voltage for this flexible device (Fig. 10.12a–d). The finger tension mechanism was used in this investigation. The distinct output power with positive

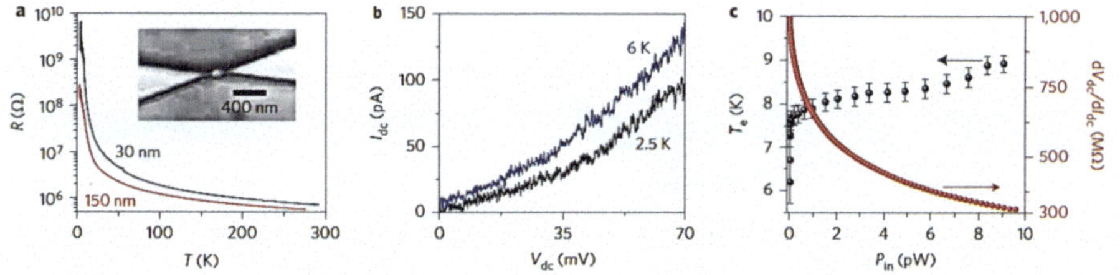

Fig. 10.10 Temperature dependence and electrical power characterization of GQD: **a** Resistance versus temperature for two quantum dots with different diameters at Vdc = 5 mV, **b** IV characteristic of a 30 nm dot at 2.5 and 6 K, **c** Differential resistance versus Joule power and electron temperature as a function of joule power at 6 K. Reprinted with permission from @ Nature Nanotechnology 11, 335–338 (2016)

and negative pulse sinusoidal waveform shape was traced by each bending release step [24].

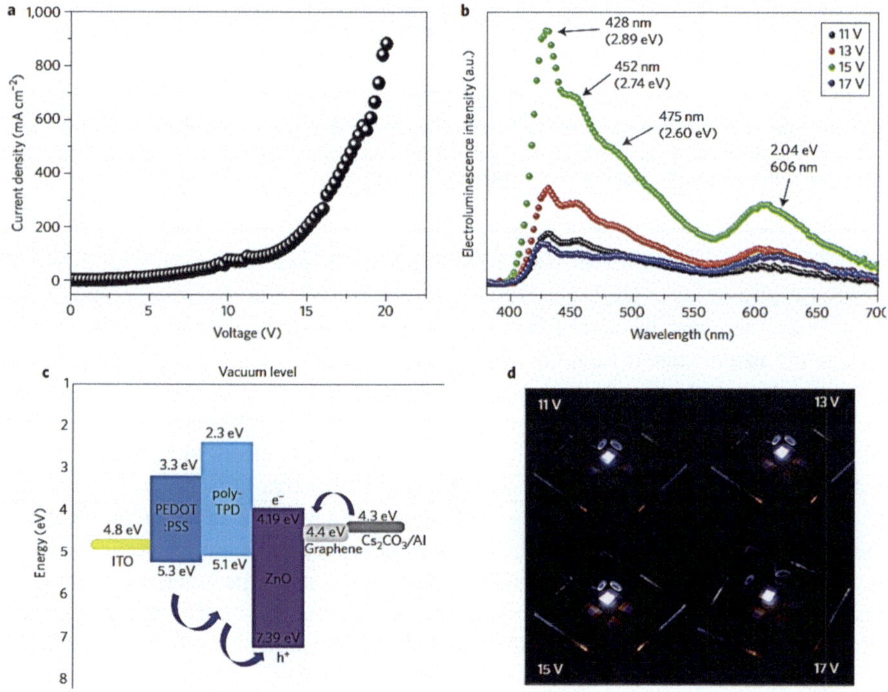

Fig. 10.11 Characterization of ZnO–GQD LEDs: **a** Current density–voltage (J-V) characteristics for the fabricated LED device, **b** Electroluminescence spectra of the fabricated ZnO–graphene quasi-quantum dot LED device with applied voltages from 11 to 17 V, **c** Band diagram of the fabricated LED device, and **d** Photograph of light emission at 11, 13, 15, and 17 V applied bias. Reprinted with permission from @ Nature Nanotechnology 7, 465–71 (2012)

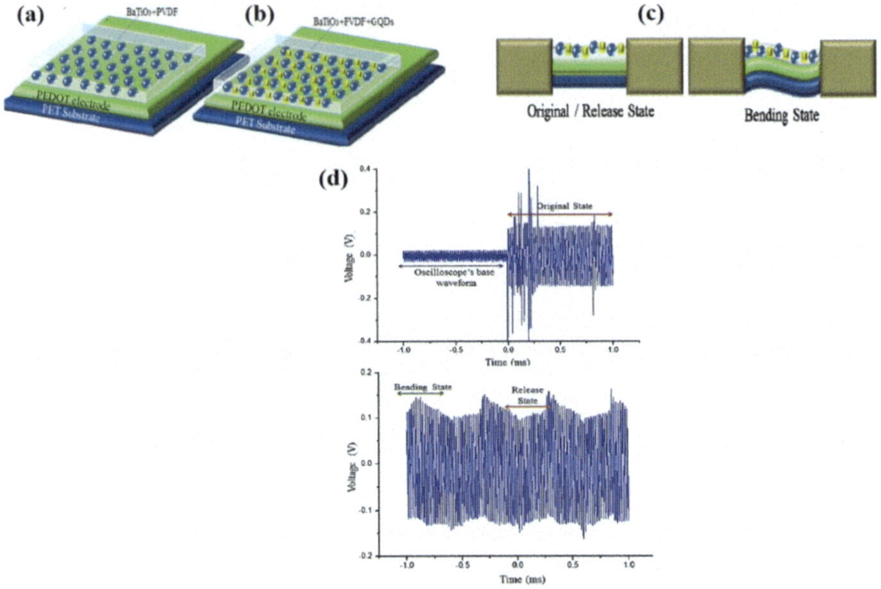

Fig. 10.12 Schematic structures: **a** PEDOT-PSS/$BaTiO_3$ + PVDF, **b** PEDOTPSS/$BaTiO_3$ + PVDF + GQDs nanocomposite thin film on PET substrate, and **c–d** Original/release and bending state; Waveform determination of original state and Waveform determination of release and bending state. Reprinted with permission from @ IEEE RSM 54–57 (2017)

High-efficient photocurrent (PC) characteristics of graphene sheets sandwiched between multiple-layer GQDs are reported by Kim CO et al. for photovoltaic cells (PDs). Since the gain in GQD PD devices is determined by the ratio of the carrier lifetime (τ) to the transit time (τ_0) across the electrodes, G~τ/τ_0, and the lifetime of the electron–hole pair is projected to be many tens of μs (τ~($\tau_1 + \tau_2$)) from the transient PC, extremely high gain is expected. After a light pulse, electrons and holes are generated in GQDs and move to the graphene sheets in a matter of tens of nanoseconds (~τ_0) (Fig. 10.13a–f). Therefore, effective QEs in the PD devices are well over 80% when the current gain in the PD reaches an order of 10^3 [25]. Chen CC has demonstrated highly responsive photodetectors based on the blend of GQDs and MoS_2. When GQDs are added to MoS_2, an n–n type heterostructure with much higher carrier mobility is formed. This is because the GQDs inject electrons into MoS_2. The enhanced light–matter interactions brought about by the transfer of holes from the GQDs to MoS_2, the re-absorption of photons emitted from the GQDs by MoS_2, and the tunneling of photo-excited carriers from the GQDs to MoS_2 are responsible for the improved photodetection performance (Fig. 10.14a–h) [26].

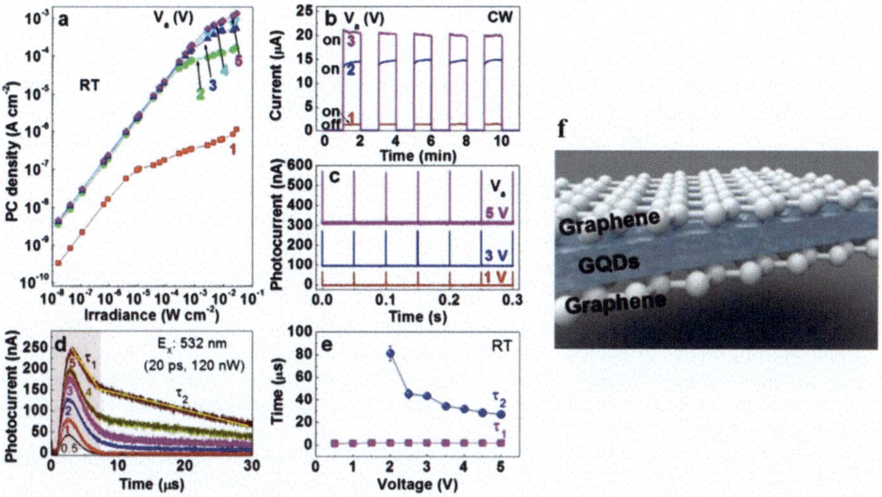

Fig. 10.13 Linearity and time response: **a** Irradiance-dependent photocurrent densities at different forward-bias voltages from 1 to 5 V. Repeated **b** long- and **c** short-term photocurrent switching at various bias voltages by using chopped-continuous-wave-and pulsed-laser lines at 532 nm, **d** Transient photocurrents at different forward biases from 0.5 to 5 V, **e** the characteristic times t1 and t2 as functions of forward bias, and **f** diagrams of typical GQD PD device and its band diagrams [25]

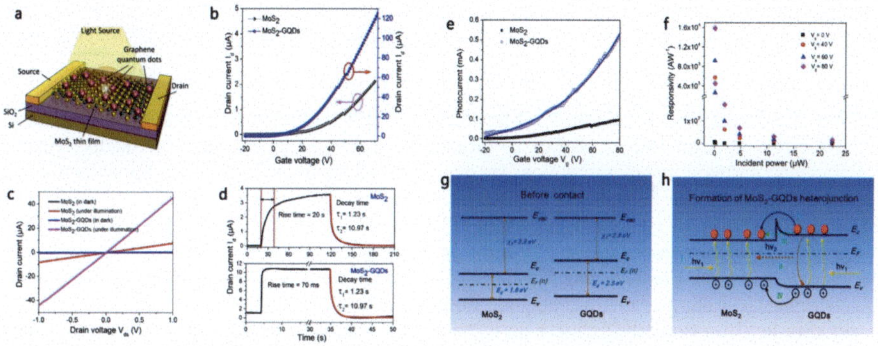

Fig. 10.14 Optoelectronic performance of a MoS_2-GQDs phototransistor: **a** Schematic of a MoS_2-GQDs heterostructure phototransistor, **b** Typical transfer curves of MoS_2 and MoS_2-GQD transistor devices, **c** Drain current (*I*d) as a function of source-drain voltage (*V*ds) without and with light illumination for MoS_2 and MoS_2-GQDs phototransistors, **d** Time-dependent photoresponse of MoS_2 (top) and MoS_2-GQDs (bottom) devices, **e** Photocurrent as a function of back-gate voltage for MoS_2 and MoS_2-GQDs devices—Incident light power: 30.1 μ W, **f** the dependence of photoresponsivity on incident light power under back gate modulation, **g** energy diagram of MoS_2 and GQDs before contact, and **h** energy diagram of the interface between MoS_2 and GQDs after the formation of a heterojunction [26]

References

1. Kushagra A, Himanshu R, Sandip M (2023) Quantum dots: an overview of synthesis, properties, and applications. Mater Res Express 10:062001
2. Yan Y, Chen J, Li N, Tian J, Li K, Jiang J, Liu J, Tian Q, Chen P (2018) Systematic bandgap engineering of graphene quantum dots and applications for photocatalytic water splitting and CO_2 reduction. ACS Nano 12:3523–3532
3. Ansari SA (2022) Graphene quantum dots: novel properties and their applications for energy storage devices. Nanomaterials (basel) 12:3814
4. Francesco B, Antonio L, Tawfique H, Zhipei S, Luigi C, Ferrari AC (2012) Production and processing of graphene and 2d crystals. Mater Today 15:564–589
5. Lin TN, Santiago SRM, Yuan CT et al (2017) Enhanced performance of GaN-based ultraviolet light emitting diodes by photon recycling using graphene quantum dots. Sci Rep 7:7108
6. Deepshikha S, Akanksha R, Naina A, Sonu G (2020) Chapter 8: graphene based sensors, in analytical applications of graphene for comprehensive analytical chemistry. In: Hussain CM (ed). Elsevier, pp 175–199
7. Asim K, Muhammad I, Rehman A, Mustansar AS (2021) Chapter 16-application of quantum dots in electronics. In: Inamuddin, Rajender B, Asiri AM, Muzibur RM (eds) Green sustainable process for chemical and environmental engineering and science. Elsevier, pp 319–336
8. Subramani MS, Soumyo C, Kallol M (2023) Nanoelectronic applications of carbon quantum dots. In: Kumar BS, Basudev P, Kallol M, Ranjan BR, Amit B (eds) Woodhead publishing series in electronic and optical materials, carbon quantum dots for sustainable energy and optoelectronics. Woodhead Publishing, pp 183–203
9. Jianhong Z, Libin T, Jinzhong X, Rongbin J, Jun Y, Jun Z, Ruiyun Y, Yunjian T, Song SL (2014) Chlorine doped graphene quantum dots: Preparation, properties, and photovoltaic detectors. Appl Phys Lett 105:111116
10. Cho MJ, Park HG, Jeong HC, Lee JW, Jung YH, Kim DH, Kim JH, Lee JW, Seo DS (2014) Superior fast switching of liquid crystal devices using graphene quantum dots. Liq Cryst 41:761–767
11. Wei Z, Zhongying X, Gang W, Menghan Z, Da C, Qinglei G, Zhiduo L, Xiaoqiang F, Guqiao D, Chu PK, Di Z (2020) Graphene quantum dot-decorated vertically-oriented graphene/germanium heterojunctions for near-infrared photodetectors. ACS Appl Nano Mater 3:6915–6924
12. Ryo S, Yuichiro U, Hideki M, Takeharu H (2014) White-light-emitting edge-functionalized graphene quantum dots. Angew Chem Int Ed 53:1–6
13. Zhipan Z, Jing Z, Nan C, Liangti Q (2012) Graphene quantum dots: an emerging material for energy-related applications and beyond. Energy Environ Sci 5:8869
14. Prathik R, Arun Prakash P, Chiashain C, Yi-Rou L, Yang-Fang C, Joseph J, Chi-Te L, Huan-Tsung C (2014) Plant leaf-derived graphene quantum dots and applications for white LEDs. New J Chem 38:4946–4951
15. Ma CB, Zhu ZT, Wang HX, Huang X, Zhang X, Qi X, Zhang HL, Zhu Y, Deng X, Peng Y, Yu H, Zhang H (2015) A General solid-state synthesis of chemically-doped fluorescent graphene quantum dots for bioimaging and optoelectronic applications. Nanoscale 7:10162–10169
16. Chiang CW, Haider G, Tan WC, Liou YR, Lai YC, Rini R, Huan-Tsung C, Yang-Fang C (2016) Highly stretchable and sensitive photodetectors based on hybrid graphene and graphene quantum dots. ACS Appl Mater Interf 8:466–471
17. Farah HF, Aniq Shazni SMHM, Ambri MM (2024) Enhanced-performance triboelectric nanogenerator based on polydimethylsiloxane/barium titanate/graphene quantum dot nanocomposites for energy harvesting. ACS Omega 9:5608−5615
18. Hemen K, Harikrishnan V, Aslam M (2014) Field effect transport properties of chemically treated graphene quantum dots. Int J Nanotechnol 11:75–84
19. Allen MT, Martin J, Yacoby A (2012) Gate-defined quantum confinement in suspended bilayer graphene. Nature Commun 3:934

20. Anh ND, Min OH, Thanh DN, Seungho B, Jun YS, Seok JM (2018) Highly enhanced photoresponsivity of a monolayer WSe_2 photodetector with nitrogen-doped graphene quantum dots. ACS Appl Mater Interf 10:10322–10329
21. Gerasimos K, Michela B, Louis G, Johann O, Maria B, Garcia de Arquer FP, Fabio G, Koppens FHL (2012) Hybrid graphene–quantum dot phototransistors with ultrahigh gain. Nat Nanotechnol 7:363–368
22. Son DI, Kwon BW, Park DH, Seo WS, Yi Y, Angadi B, Lee CL, Choi WK (2012) Emissive ZnO–graphene quantum dots for white-light-emitting diodes. Nat Nanotechnol 7:465–471
23. El Fatimy A, Myers-Ward RL, Boyd AK, Daniels KM, Gaskill DK, Barbara P (2016) Epitaxial graphene quantum dots for high-performance terahertz bolometers. Nat Nanotech 11:335–338
24. Abu Bakar E, Mohamed MA, Majlis BY, Hassan OH, Omar FC, Kudin TIT (2017) All-solution process flexible nanocomposite generator made of BaTiO3 nanoparticles and graphene quantum dots. In: IEEE regional symposium on micro and nanoelectronics (RSM). pp 54–57
25. Kim CO, Hwang SW, Kim S, Shin DH, Kang SS, Kim JM, Jang CW, Kim JH, Lee KW, Choi SH, Hwang E (2014) High-performance graphene-quantum-dot photodetectors. Scientific Rep 4:5603
26. Chen CC, Hong Q, Shenghuang L, Man LC, Yan L, Zaiquan X, Jingchao S, Yunzhou X, Delong L, Jian Y, Wenzhi Y, Chunxu P, Ping LS, Qiaoliang B (2015) Highly responsive MoS_2 photodetectors enhanced by graphene quantum dots. Sci Rep 5:11830

Chapter 11
Influence of Graphene Quantum Dots as Nano- and Bionano-Sensors

11.1 Introduction

Graphene quantum dots, or GQDs, are unique materials with good electrical, optical, and mechanical properties as well as chemical stability and biocompatibility. These qualities have made them promising for use in a variety of applications, including nanosensors [1]. GQDs are beneficial for optical sensing, and biomedical and catalytic applications due to their luminescent qualities combined with chemical and photostability, large surface area, biocompatibility, and low toxicity [2]. According to Mansuriya and Altintas [3], nanosensors are intriguing nanoscale instruments that measure physical quantities and translate them into perceptible signals. These small sensors are used in many industries, including health care, environmental monitoring, and defense [4]. To function as efficient nanosensors and bio-nanosensors, GQD has good photoluminescence, biocompatibility, high water solubility, quick electron mobility, and low cytotoxicity [5, 6]. As GQDs have numerous applications in nano- and nano-biosensors, we have consulted the articles in the tabulation form for nanosensors and nano-biosensors. Further, a few significant works have been discussed in detail.

11.2 Role of GQD in Nanosensors

A six-membered cyclized product and a nonradiative excited-state relaxation in DAN-GQDs proposed for phosgene sensing. When excited at 328 nm, the aqueous solution of DAN-GQDs (4.89×10^{-9} M) shows a significant emission peak at 423 nm. The initial fluorescence intensity of DAN-GQDs (ΦF 53.6 $\rightarrow$ 34.6%) is quenched by the addition of the phosgene molecule (0 $\rightarrow$ 88 μL) by the production of a stable six-membered cyclized product (Fig. 11.1a, b). When compared to other competing harmful contaminants in water, the DAN-GQDs demonstrated superior

N. Manjubaashini et al., *Graphene Quantum Dots*, Materials Horizons: From Nature to Nanomaterials, https://doi.org/10.1007/978-981-97-5722-0_11

selectivity and sensitivity for phosgene as $K_a = 3.84 \times 10^2\ M^{-1}$, and the limit of detection (LoD) is 2.26 ppb. Time-resolved fluorescence investigation verifies that nonradiative relaxing of excited electrons occurs before DAN-GQD quenching [7]. Utilizing highly stable nitrogen-doped GQD (N-GQD) functionalized with Pamoic Acid (PA@N-GQD), radioactive elements, uranium (VI) and thorium (IV), may be detected at the nanomolar level in pH $\pm$ 5.0. The PA@N-GQD aqueous solution has a spherical morphology and a size average of 6.5 nm. When uranium (VI) and thorium (IV) ions are added, PA@N-GQD shows a progressive drop in fluorescence intensity at 438 nm (λ_{ex} 344 nm). Using the photoluminescence approach, the selectivity, sensitivity, competitivity, pH, time effect, and reversibility of PA@N-GQDs have been studied (Fig. 11.1c). For uranium (VI) and thorium (IV) ions, the obtained fluorescence LoD of PA@N-GQD is 2.009×10^{-9} and 1.351×10^{-9} M, respectively [8].

To detect hazardous ions Ag^+, Ce^{2+}, Cd^{2+}, Cu^{2+}, Cr^{2+}, Pb^{2+}, Sn^{2+}, and F^+, as well as p-nitrophenol and thioacetamide in water, a photoluminescence (PL)-based nanosensor based on GQDs is constructed. The quenching of PL spectra suggests that the GQD/Fe (III) combination exhibits a high fluorescence quenching efficiency (~75%) in the direction of the hydrophobic GQD surface. With a high degree of linearity and sensitivity, the LoD for Fe (III) concentration was 40 $\pm$ 2 nM. The GQDs/Fe (III) complexes can be used as a turn-ON/turn-OFF type fluorescence probe for the detection of sulfur (turn-ON) at 7.6 $\pm$ 1 pM and p-nitrophenol (turn-OFF) at 5 $\pm$ 1 pM concentration (Fig. 11.2a). The suggested technique is simple, quick, and sensitive, making it perfect for using nanosensors to measure the water quality of ponds, rivers, or industrial wastes [9, 10].

Using sulfadimidine as a template molecule and a one-step electro-copolymerization of 3, 4-ethylenedioxythiophene (EDOT) and methacrylic acid (MAA) on a glassy carbon electrode (GCE) modified with GQDs-Pt nanoparticles (PtNPs), a novel molecularly imprinted polymer (MIP) electrochemical sensor was created (Fig. 11.2b). The electro-copolymerization of EDOT and MAA, according to the results, provides a significant quantity of functional groups for the binding of the template molecules, hence increasing the sensitivity of the MIP sensor. With a detection limit of 0.023 nM, the calibration curve showed linearity throughout a concentration range of 0.1 nM–0.1 mM [12]. α-MnO_2/GQD nanocomposites as a sensing material and microfluidic well as a portable resistive device with tiny electronics are created for the sensitive detection of Pb (II). The designed portable sensor was evaluated within the Pb (II) concentration range of 0.001 nM–1 μM, requiring a minimal sample volume of 200 μL. LoD and sensitivity were determined to be 1.05 $k\Omega/nM/mm^2$ and 0.81 nM, respectively, and were confirmed using a commercial impedance analyzer (Fig. 11.2c). It was discovered that the portable sensor had a roughly 45-day shelf life [11]. The GQD-based nanosensors used for various analyte sensing by fluorescence and electrochemical techniques are comprised in Table 11.1.

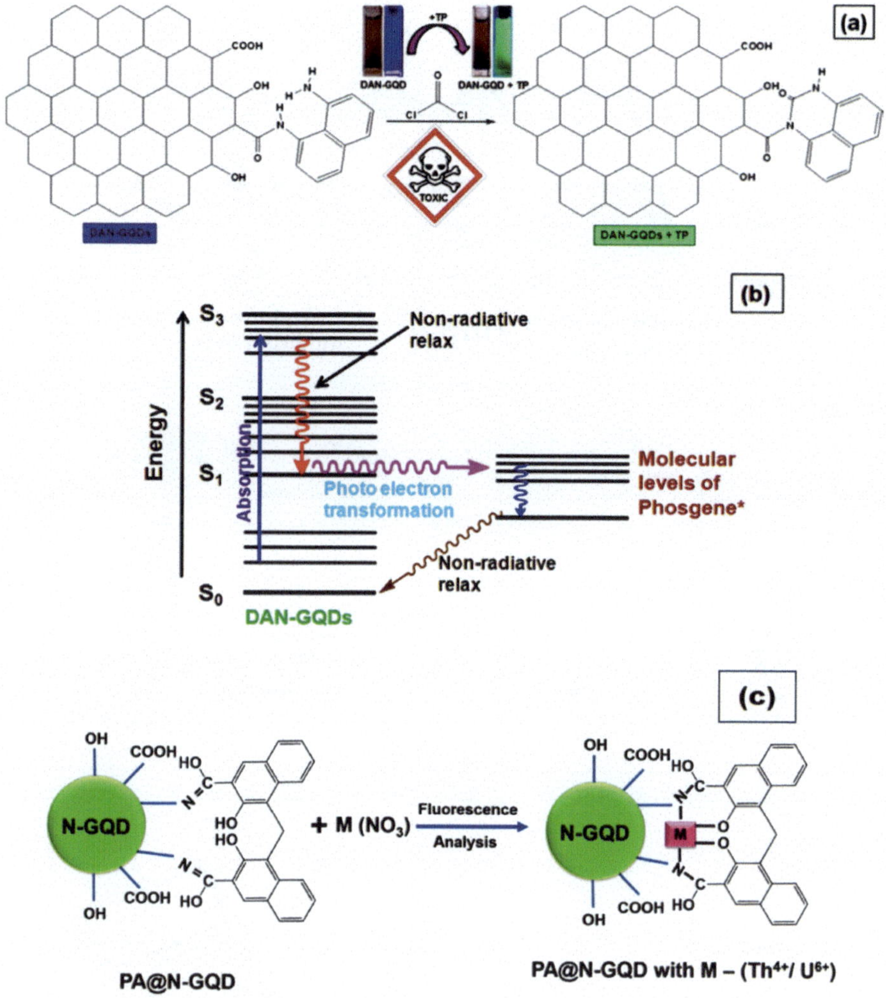

Fig. 11.1 **a** Phosgene sensing of DAN-GQDs by six-membered cyclized product. **b** Nonradiative excited-state relaxation in DAN-GQDs for phosgene sensing [7], and **c** PA@N-GQD for radioactive metal ion sensing [8]

11.3 Contribution of GQD in Bionanosensors

Using GQDs as a reductant and stabilizer of the AuNPs, a one-pot synthesis was used to produce the nanosensor of GQDs/AuNPs. The AuNP/GQD solution quickly turned blue due to the presence of cysteine (Cys), which also caused the absorption band's red shift from 525 to 645 nm during the aggregation phase. Conversely, in a PBS solution with a pH of 7.4, the AuNP/GQD nanosensor exhibits off–on fluorescence at 450 nm without causing a color change when lysine (Lys) is present. According

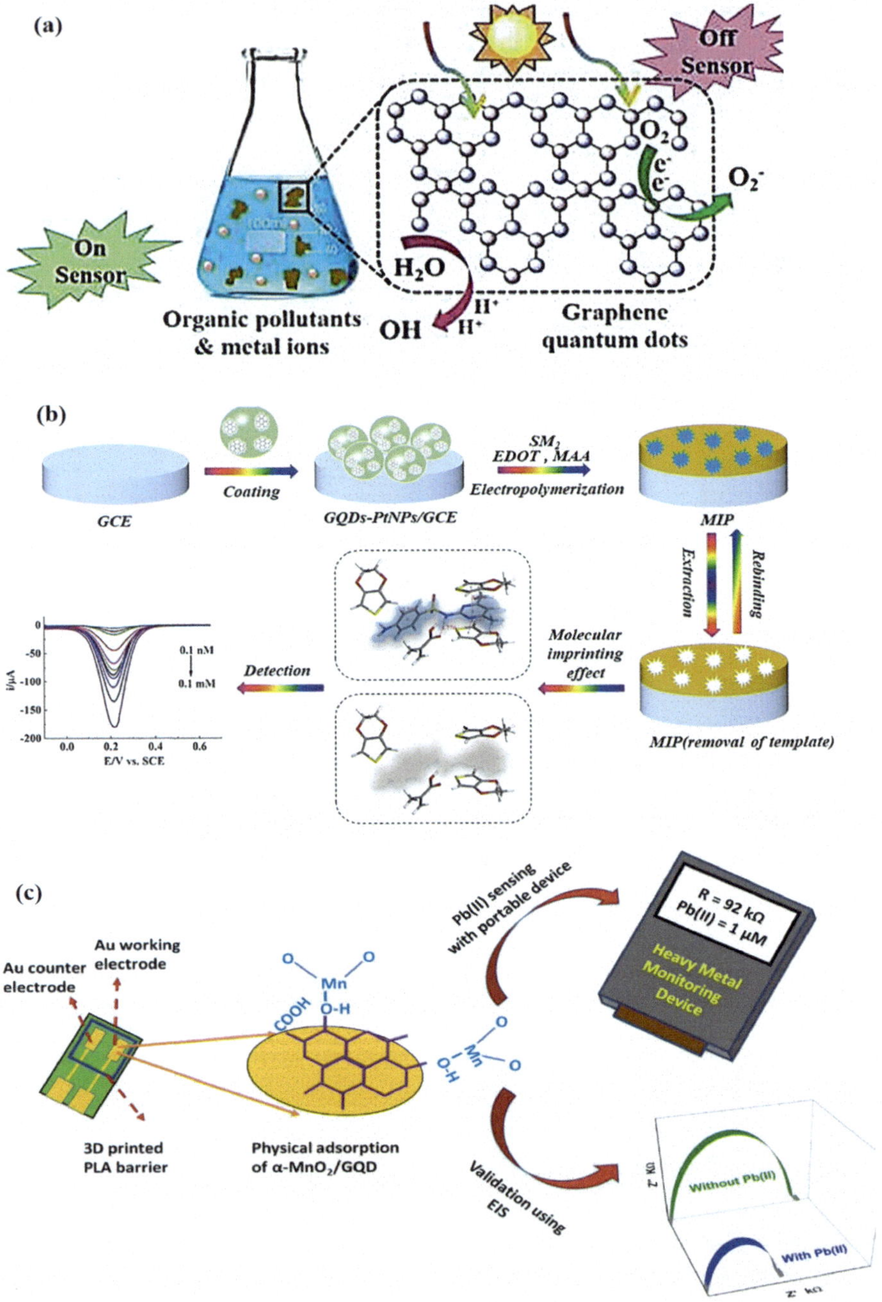

Fig. 11.2 **a** GQD as PL nanosensor. Reprinted with permission from @ Int. J. Environ. Sci. Technol. 21, 1637–1648 (2024). **b** MIP sensor to detect SM_2. Reprinted with permission from @ Food Chemistry 221, 1128–1134 (2017). **c** Schematic of nanosensor fabrication for Pb (II) detection [11]

Table 11.1 GQD-based nanosensors for various analyte sensing by fluorescence and electrochemical techniques

Material	Analyte	Methodology	LoD	References
Ionophore-based GQD sensor	Na^+ and K^+	Fluorescence technique	Na^+ 0.1 mM^{-1} M; K^{+3} μM^{-1} mM	Renjie et al. [13]
Graphene quantum dots–aptamer probe	Lead (II) ion	Fluorescence technique	0.6 nM	Sheng et al. [14]
Nitrogen-doped GQD	H_2O_2, ascorbic acid, acid phosphatase	Colorimetric, fluorescence technique	60 nM, 4.6 μU/ml, 3.6 μM	Yejian et al. [15]
GQD on glass carbon electrodes	Malathion	Electrochemical technique	0.62 nM	Sanju et al. [16]
GQD with Eriochrome Black T	Cobalt (II), copper (II)	Colorimetric and absorption techniques	0.014, 0.018 M	Vahab and Keshipour [17]
Mono-ethanolamine functionalized GQDs	Mercury, ethyl Xanthate	Fluorescence technique	10, 30 nM	Amini et al. [18]
GQD	Ag^+ ions	Fluorescence technique	250 nM	Xian-En [19]
GQD	Uranium	Fluorescence technique	1.5 ppb	Dewangan et al. [20]
Au–Nitrogen-doped graphene quantum dot	Caffeic acid	Electrochemical technique	0.03 μM	Zhao et al. [21]
Nitrogen-doped GQD	Fe^{3+}, Cu^{2+}, Hg^{2+}	Fluorescence	47.9 nM	Darwin et al. [22]
Palladium-functionalized GQD	Citrinin	Electrochemical technique	2.0×10^{-10} M	Onur et al. [23]
Sulfur and nitrogen-doped GQD	Heavy metal Ions Cd (II), Pb (II), Hg (II)	Electrochemical technique	1, 10, 1 pM	Saisree et al. [24]
Polyacrylonitrile/nitrogen-doped GQD	Methylene blue	Fluorescence technique	–	Saloni et al. [25]
PbS@CdS@CdS giant GQD	Temperature sensor, 281–313 K	Fluorescence technique	0.01 K^{-1}	Fanizza et al. [26]
L-Cysteine functionalized GQD	Mercury	Fluorescence technique	0.64 ppb	Pradeep Kumar [27]
GQDs@UiO-66-NH_2	Quercetin	Fluorescence technique	2.82 ng/mL	Sopan et al. [28]
Single-layered DQG	Fe^{3+} ion	Fluorescence technique	0.26 μM	Haoqiang et al. [29]

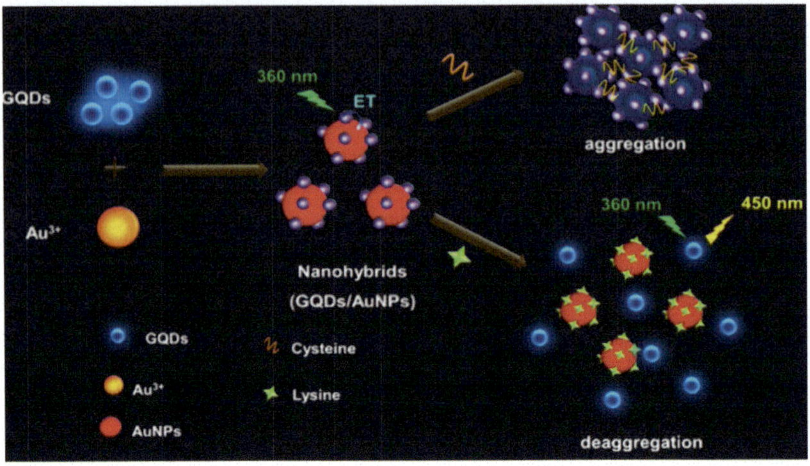

Fig. 11.3 Mechanism of AuNPs/GQD hybrid material with Cys and Lys. Reprinted with permission from @ Sensors and Actuators B 282, 936–944 (2019)

to Chiraporn et al. [30], the detection limits of Cys and Lys for sensing applications were 5.88 and 16.14 μM, respectively, with an acceptable percentage recovery of 101–106 in real samples (Fig. 11.3).

Nitrogen-doped GQD (N-GQD)-based switchable fluorescence probe to find tetracycline and L-Cysteine. Using urea and benzophenone hydrothermally, strong blue emissive N-GQDs were produced with a quantum yield of 34.9%. Because of their favorable biological compatibility and high optical stability, N-GQDs can be used for fluorescent labeling (Fig. 11.4a). The addition of tetracycline was shown to cause a significant dampening of fluorescence. The subsequent supplementation of L-cysteine has the potential to stimulate fluorescence regeneration. Because of the N-GQD nanosensor's pertinent sensitivity, a straightforward method for the detection of tetracycline and L-cysteine in natural water was created [31, 32]. Nitrogen-doped GQDs (N-GQDs) are made utilizing a bottom-up method with tris (hydroxymethyl) aminomethane (Tris) serving as the nitrogen source and citric acid as the carbon source. When the pH is raised from 3.0 to 12.0, the produced N-GQDs clearly show fluorescence quenching with a high yield of 61.50% and a quantum yield of 14.42%. Additionally, a strong linear association between the addition of TC in the range of 1–50 μM and the fluorescence intensity of N-GQDs is shown. It has been determined that 94 nM is the detection limit (Fig. 11.4b). Good linearity between RGB values and pH values is found by analyzing the RGB values from the fluorescence images ($R_2 = 0.996$), and the detection limit for TC detection is 97 nM [33].

To detect morphine, a new turn-ON fluorescence nanosensor was created by surface functionalizing GQDs with antimorphine. Following the addition of morphine, the antimorphine-GQDs' fluorescence intensity increased, enabling a sensitive detection with a detection limit of 0.06 μM (Fig. 11.5a). Additional research showed that this nanosensor does not interact with codeine or methamphetamine,

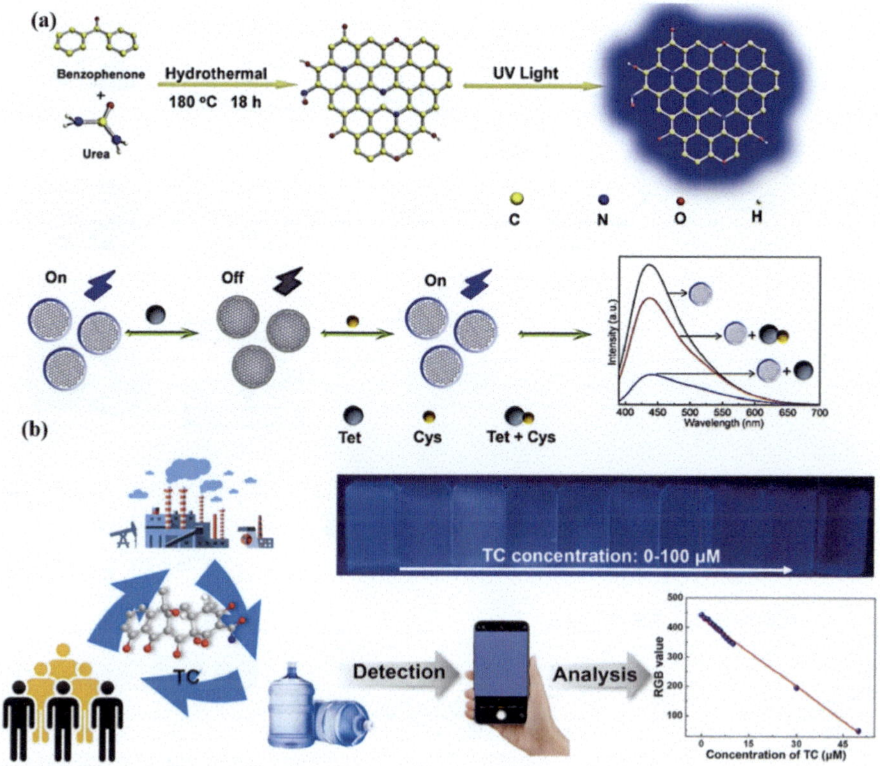

Fig. 11.4 **a** N-GQDs probe for Tet and L-Cys sensing. Reprinted with permission from @ Bulletin of Materials Science 45, 26 (2022). **b** Schematics of TC residue detection in real water samples by smartphone sensing procedure [34]

among other narcotics. As a result, it can function as an affordable and accurate nanosensor for determining low morphine concentrations [34]. Dopamine (DA) detection using an ultrasensitive electrochemical sensor based on GQDs and multi-walled carbon nanotubes (GQDs-MWCNTs) was created. The sensor exhibits high dopamine selectivity among other interfering bioanalytes, as predicted. With a detection limit of 0.87 nM (3S/N), this electrochemical sensor demonstrated optimal performance for DA measurement under ideal circumstances, displaying strong linearity across a wide linear range of 0.005–100.0 μM (Fig. 11.5b). Additionally, this electrochemical sensor was the first to measure DA produced from live PC12 cells with great performance, and it was effectively used to detect DA in human serum [35].

To detect gene mutation in pancreatic cancer (PCs), Preeti NJ et al. investigated the benefits of electrochemical biosensing and created a straightforward screen-printed DNA sensor. A tumor-suppressor gene called p16 has sequence changes that are linked to the start of PC. Developed an electrochemical based device for the

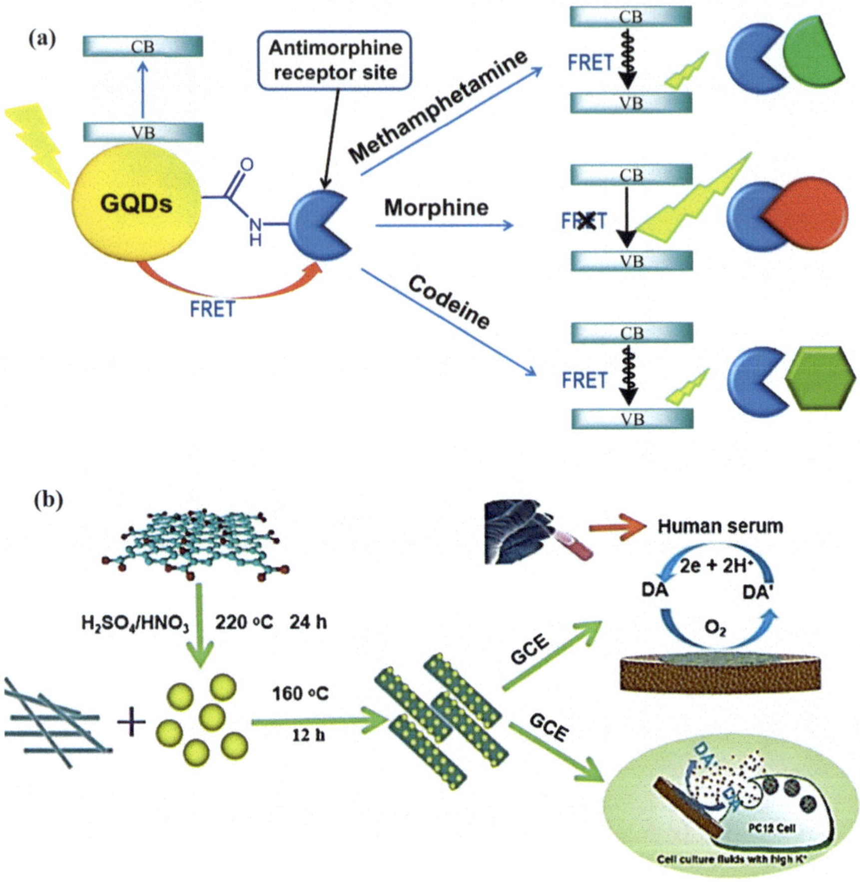

Fig. 11.5 **a** Nanosensor for the detection of morphine. Reprinted with permission from @ Materials Letter 241, 206–209 (2019). **b** Electrochemical Sensor for detecting dopamine [35]

early detection of PC and identified the p16 gene as a cancer marker. According to Preeti and Shobha [36], the biosensor demonstrated a 0.10 pM detection capacity (Fig. 11.6a). A single-digit GQD was reported by Sanju et al. utilizing hydrothermal and solvothermal processes. Three methods are used to study glucose oxidase (GOx) immobilized on glassy carbon (GC) electrodes modified with GQDs: differential pulse voltammetry, amperometry, and cyclic voltammetry (Fig. 11.6b). In comparison to rGO (0.025 $\mu A\ \mu M^{-1}\ cm^{-2}$, 4.16 μM) and GO (0.064 $\mu A\ \mu M^{-1}\ cm^{-2}$, 4.82 μM) nanosheets, the developed GOx-GQDs biosensor responds efficiently and linearly to the presence of glucose over concentrations ranging from 10 μM to 3 mM and LoD as 1.35 μM with the 0.00769 $\mu A\ \mu M^{-1}\ cm^{-2}$ sensitivity [37].

A new immunosensor technique that uses GQDs to simulate an enzyme in an electrochemical sensor offers a reliable way to identify *Y. enterocolitica*. The GQD-immunosensor allowed for the high sensitivity and specificity LoD of milk

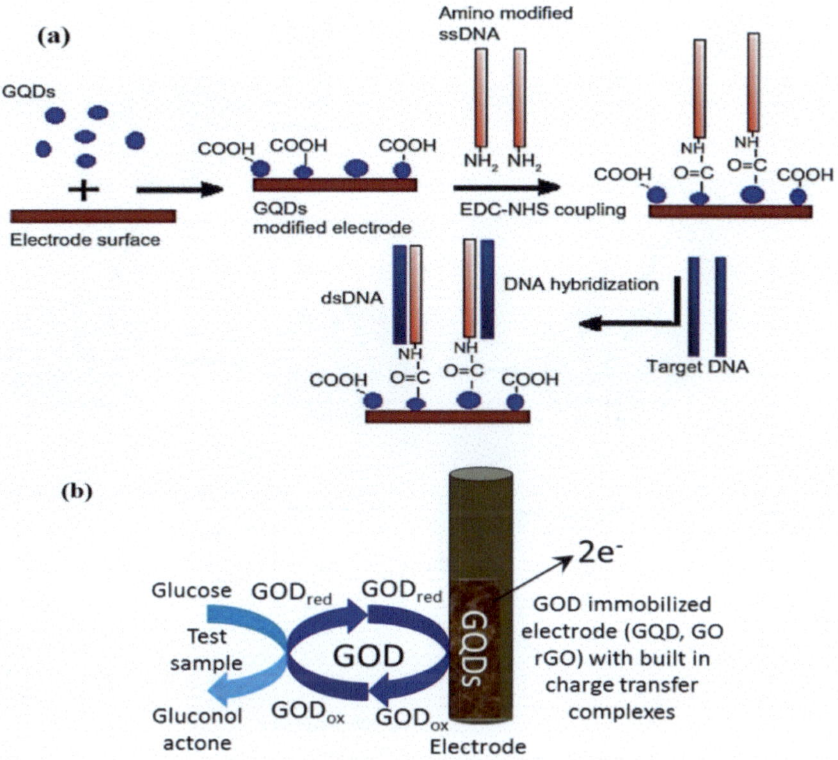

Fig. 11.6 **a** GQD as an electrochemical DNA hybridization sensor [36], and **b** Enzymatic glucose sensing on immobilized GQDs [37]

is 5 cfu mL^{-1}, and LoD of serum is 30 cfu mL^{-1} in the quantification of *Y. enterocolitica* over a broad concentration range (Fig. 11.7a). Without pre-sample treatment, the established approach can be used to detect any harmful bacteria in clinical and food samples [38]. Rather than using iron oxide nanoparticles as magnetic probes, magnetic GQDs were used and demonstrated excellent sensitivity. Using magnetic GQD as the magnetic probe and iron oxide as the magnetic carrier, magnetic separation-assisted MRS was created to detect endoglin. The test has a sensitivity LoD of 1.3 ng/mL, which is two orders of magnitude lower than that of MRS without magnetic separation. It also has a large linear region spanning from 5 ng/mL to 50 μg/mL. Endoglin (CD105) detection in real samples has demonstrated a high degree of accuracy and consistency [39]. After synthesizing and fabricating GQDs as a basic monodisperse platform, the biosensing performance was enhanced by heterogeneous electron transport (97.63×10^{-5} cm s^{-1}) and a time constant of 0.005 s. 213.88 Ω (ng mL^{-1}) $^{-1}$ cm^{-2} is the high sensitivity of the electrochemical immunosensor (Fig. 11.7b).

According to Bhardwaj et al. [40], the contaminated maize samples had a limit of detection of 0.05 ng g^{-1} and the reference samples had a LoD of 0.03 ng mL^{-1}

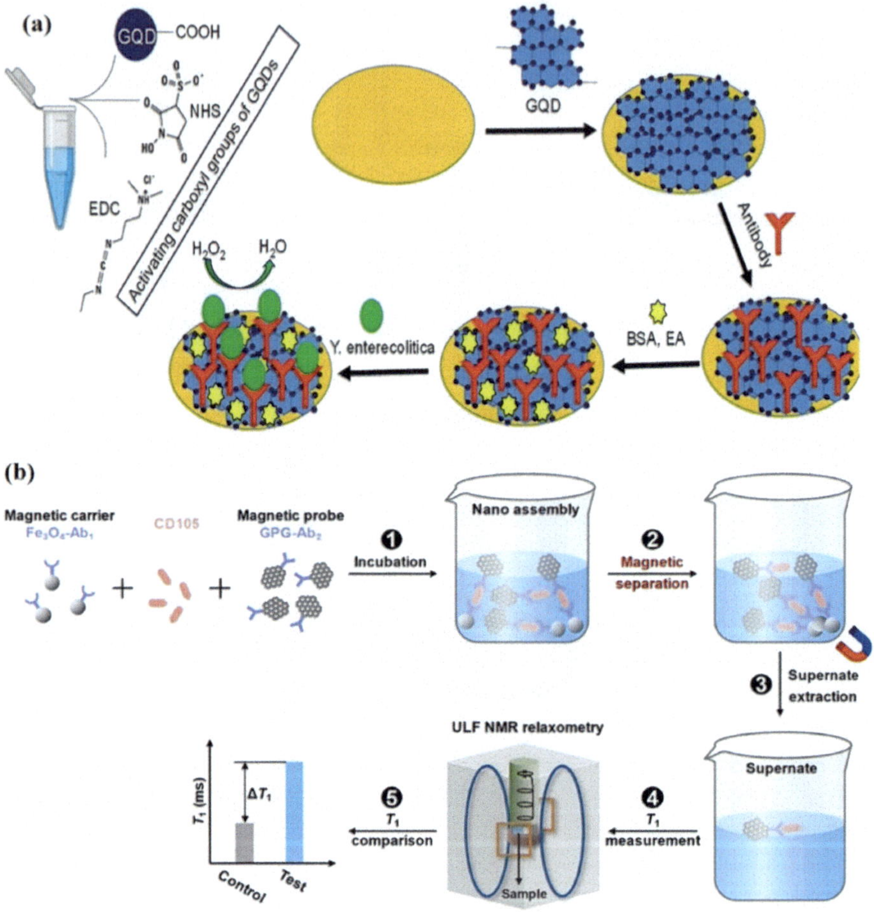

Fig. 11.7 **a** GQD-based immunosensor [38], and **b** View of MS-MRS assay. Reprinted with permission from @ Sensors and actuators B 380, 133389 (2023)

(Fig. 11.8). A GQD-based aptasensor can precisely identify minuscule intracellular concentrations of cytokine molecules. Both epitope-modified GQDs (Ep-GQDs) and GQDs modified with cytokine aptamers (Ap-GQDs) were produced. Aggregation between Ap-GQDs and Ep-GQDs causes the conjugates of Ap-GQDs to lose their fluorescence. Ap-GQD and Ep-GQD conjugates were employed as nanosensors to measure the very high sensitivity (2 pg mL^{-1}) of intercellular cytokine IFN-γ release. Similar nanosensors can be used for the detection of a wide range of cell-secreted compounds due to their straightforward and universal disaggregation-based sensing technique [41].

A ratiometric fluorescent nanosensor was disclosed that uses 2,3-diaminophenazine (DAP) as the specific response signal and amino-functionalized GQDs (af-GQDs) as the reference fluorophore for ultrasensitive and highly selective

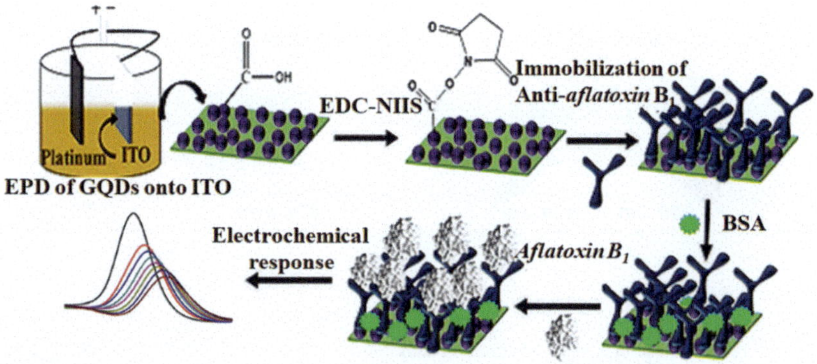

Fig. 11.8 Schematic representation of EPD process for coating GQD on ITO substrate and immobilization of antibody on the GQD-ITO substrate for AFB_1 detection. Reprinted with permission from @ Anal Bioanal Chem 410, 7313–7323 (2018)

identification of horseradish peroxidase (HRP) (Fig. 11.9a). With a detection limit of 0.21 fM (2.1 μU mL^{-1}), the fluorescence intensity ratio of DAP to af-GQDs rose linearly with an increase in HRP concentration throughout the range of 2–800 fM (0.02–8 μU mL^{-1}) [42]. The creation of an enzymatic impedimetric sensor may be used with an electrochemical paper-based analytical device (μ-EPADs) to detect these biomarkers several times. Using a sample volume of about 200 μL, electrochemical impedance spectroscopy has been used to quantify glucose and L-tyrosine within concentration ranges of 50–800 mg/dL and 1–500 μmol/L, respectively. With a shelf life of approximately one month, the impedance response showed a linear relationship over the spectrum of analyte concentrations, with detection limits of approximately 58 mg/dL and 0.3 μmol/L for glucose and tyrosine, respectively (Fig. 11.9b). By connecting the μ-EPADs to miniature electronics, the sensing technique was also applied to Arduino-based device applications [43].

In Nitrogen-doped graphene quantum dots (NGQDs), the nitrogen source were extracted from chitosan. Chitosan is a natural bioresource. Further, the atmospheric pressure microplasma system used for neurotransmitter biomarker detection. Multiple biomolecules, such as uric acid (UA), folic acid (FA), epinephrine (EP), and DA, can be detected simultaneously with excellent selectivity and sensitivity utilizing a single material by varying the pH conditions during the detection process. For UA, EP, FA, and DA, respectively, low LoDs of 123.1, 157.9, 80.5, and 91.3 nM are found, along with linear relationships between the biomarker concentration and the PL intensity ratio beginning at 0.8 and going up to 100 μM [44]. The purpose of the N-GQDs/PANI nanocomposite is to serve as a triboelectric layer for non-invasive glucose sensing. By adding electron-rich functional groups to the NGQDs, the surface electronegativity of PANI was altered. This increased surface charge aided in the charge transfer between PANI and intermedia, hence improving triboelectric outputs. In comparison to pristine PANI/GOx one (16.44 mM^{-1}), NGQDs/PANI/GOx-based TES delivers better sensitivity (23.52 mM^{-1}) in glucose detection

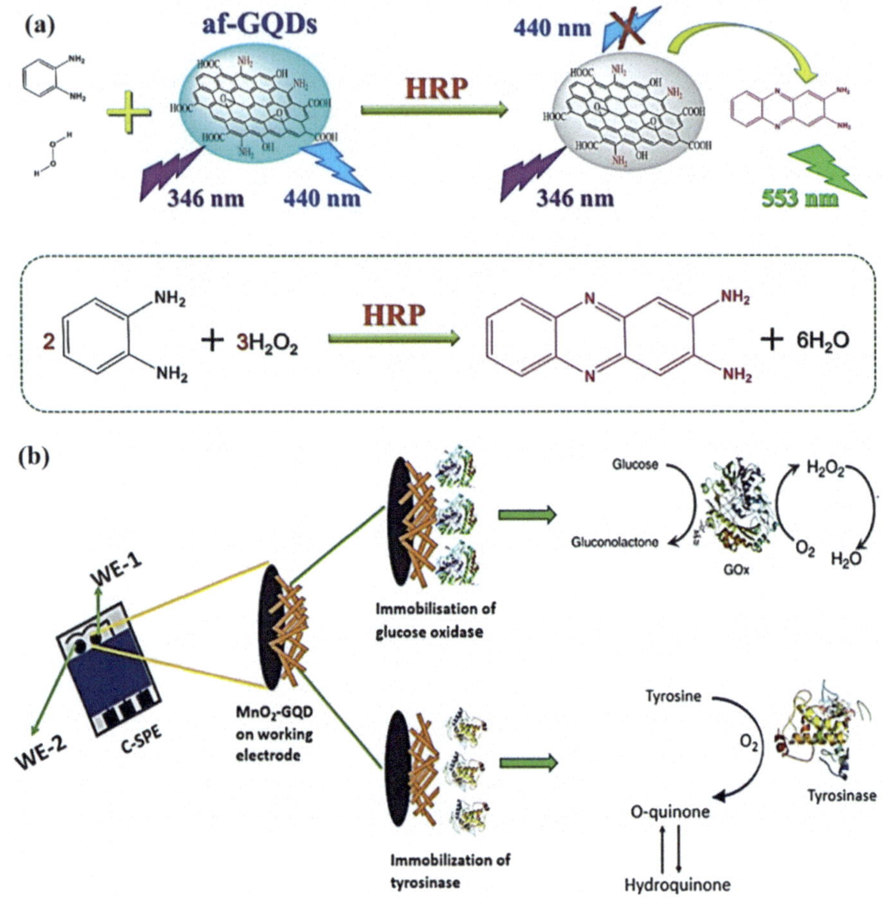

Fig. 11.9 **a** GQD as ratiometric fluorescence nanosensor. Reprinted with permission from @ Sensors and actuators B: Chemical 234, 255–263 (2016). **b** Fabrication of microfluidic multiplexed paper-based sensing strip [43]

due to the improved electron transport and surface charge by NGQDs (Fig. 11.10). The study shows that wearable TES are feasible to use on human skin and that they operate with a self-driving system that requires no external power source dependably and sensitively [45]. Table 11.2 lists the GQD-based bio-nanosensors that are utilized for various biomolecule sensing by fluorescence and electrochemical techniques.

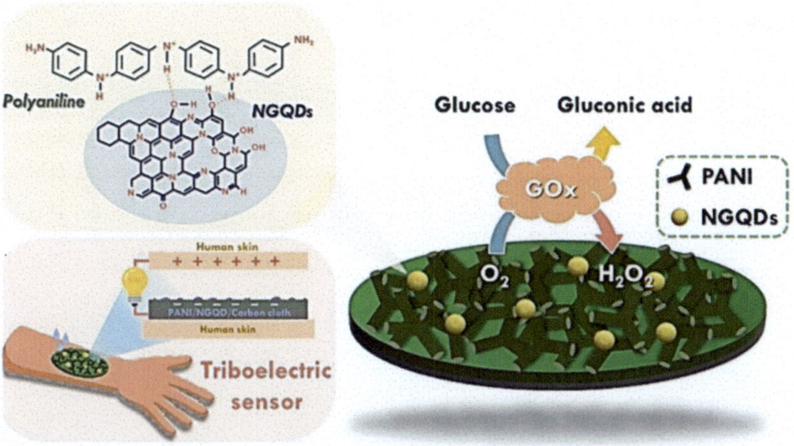

Fig. 11.10 Schematic for detection of human sweat TES using NGQDs/PANI [45]

Table 11.2 GQD-based bio-nanosensors using fluorescence and electrochemical techniques

Material	Analyte	Methodology	LoD	References
GQD with Phthalocyanines	ds-DNA	Fluorescence technique	30 pM	Ana et al. [46]
COOH- and NH_2-functionalized GQD	Antibiotics	Fluorescence technique	1 $\mu g\ L^{-1}$	Tongchang et al. [47]
Enzyme-generated GQD	Fenoxycarb	Enzymatic inhibition mechanisms	3.15 μM	Caballero-Díaz et al. [48]
Boron and nitrogen co-doped GQD	Tetracycline	Fluorescence technique	1 nM in PBS, 1.9 nM in urine, and 2.2 nM in human serum	Hai et al. [49]
AuNPs-GQD	Quercetin	Electrochemical technique	3.3×10^{-11} mol L^{-1}	Cristina et al. [50]
GQD	Acute myocardial infarction	Electrochemical technique	0.1–100 ng mL^{-1}	Tabish et al. [51]
GQD functionalized MWCNT	Dopamine	Electrochemical technique	95 nM	Shiva Kumar et al. [52]
GQD and 1-butyl pyridinium bromide (BPBr) as carbon paste electrode	5-Fluorouracil	Electrochemical technique	0.5 nM	Reyhane et al. [53]
Polyindole-derived nitrogen-doped GQDs	Dopamine	Electrochemical technique	0.15 nM	Anjitha et al. [54]
GQDs modified electrodes	Cortisol	Electrochemical technique	0.1 pg/ml	Sharma et al. [55]
GQD	C-reactive Protein	Electrochemical technique	176 pM	Xuehua and Guorong [56]
N, S-GQDs@ZIF-8@MIP	Octopamine	Fluorescence	0.062 mg L^{-1}	Ying et al. [57]
GQD	Metronidazole	Electrochemical technique	0.52 nmol L^{-1}	Ali et al. [58]
GQD-MnO_2 nanosheets	Glutathione detection	Fluorescence	150 nmol L^{-1}	Xu et al. [59]
GQD	Cell surface protein CD123	Electrochemical technique	1 cell/mL	Alireza et al. [60]
GQD-embedded silica molecular imprinted polymer	Ornidazole	Fluorescence technique	0.24 μM	Mina et al. [61]

(continued)

Table 11.2 (continued)

Material	Analyte	Methodology	LoD	References
GQD	Cephalexin	Electrochemical technique	0.53 fM	Pratik et al. [62]
Amino-functionalized GQD	Human telomere DNA	Fluorescence technique	25 fM	Shan et al. [63]
GQD	Adrenaline	Fluorescence technique	0.5 μM	Fatemeh et al. [64]
Sulfur-doped GQD @gold Nanostar probe	Prostate-specific antigen	Electrochemical technique	9.7 aM in PBS, 26 fg mL^{-1} in human serum	Mengke et al. [65]
N, Zn-GQDs	Cardiac troponin I (cTnI) in human serum	Electrochemical technique	4.59 pg L^{-1}	Hai et al. [66]
Covalent conjugation of GQDs	L-glutamine	Electrochemical technique Fluorescence technique	360 μM 9.42 μM	Mingxing et al. [67]
GQD by 3D nanomesh graphene frameworks	Cu^{2+} in rat brain	Fluorescence technique	0.067 μM	Pooja et al. [68]
Crown ether-functionalized GQD	Potassium and sodium ions	Electrochemical, Fluorescence technique	–	Chunxia et al. [69]
GQD	Aflatoxin M1 in milk samples	Electrochemical technique	2 μM	Iannazzo et al. [70]
GQDs@ZrO_2	Ochratoxin-A from the spiked coffee sample	Electrochemical technique	0.38 ng/mL	Shadjou et al. [71]
GQD	Hepatitis B virus DNA	Electrochemical technique	1 nM	Pramod et al. [72]
Gold-loaded boron-doped GQD	Guanine (GUA), and adenine (ADE) biomolecules	Electrochemical technique	GUA - 1.71 μM, ADE - 1.84 μM	Qian et al. [73]
Coal-derived GQDs	Glutathione	Fluorescence technique	27 μM	Reshma et al. [74]
GQDs encased in molecularly imprinted polymer	Biotin	Fluorescence technique	315 nmol L^{-1}	Aayushi et al. [75]
Nitrogen-doped GQD	Breast cancer cells	Electrochemical technique	1, 2 cells mL^{-1}	Parizad et al. [76]

(continued)

Table 11.2 (continued)

Material	Analyte	Methodology	LoD	References
GQD-polyfluorene (PF) nanocomposite	Mitomycin C-DNA interaction sensing	Electrochemical technique	0.039 ng	Hai et al. [77]
GQDs@La^{3+}@ZrO_2	Flutamide in urine samples	Electrochemical technique	0.00082 μM	Deniz et al. [78]
N-GQDs-H-ZIF-8 colloidosomes	Glucose	Fluorescence technique	5.4 μM	Thota et al. [79]
N-GQDs-CoOOH nanoflakes	Tetracycline	Fluorescence technique	0.95 ng mL^{-1}	Meng et al. [80]
Nitrogen-doped GQDs/ Tb–Phen NPs	Metformin	Fluorescence technique	0.76 nM	Zhang et al. [81]
Arginine-functionalized GQD	Thiamine	Fluorescence technique	53 nM	Masoud et al. [82]
GQD nanoprobe	Propofol	Fluorescence technique	5.34 mg l^{-1}	Fatemeh et al. [83]
1,8-diaminonaphthalene functionalized GQD	Vitamin B9 and B12	Fluorescence technique	1.73×10^{-15} M	Pavithra et al. [84]
Nitrogen-doped GQDs	Glycine	Fluorescence technique	21.04 pM	Sravya et al. [85]

References

1. Leyla K, Goksu SV, Sibel AO (2021) An overview on quantum dot-based nanocomposites for electrochemical sensing on pharmaceutical assay. Iranian J Pharma Res 20:187–203
2. Murilo HMF, Rodrigo S, Jessica BSL, Luiza AM, Daniel SC (2021) Graphene quantum dots-based nanocomposites applied in electrochemical sensors: a recent survey. Electrochem 2:490–519
3. Mansuriya BD, Altintas Z (2020) Applications of graphene quantum dots in biomedical sensors. Sensors 20:1072
4. Darwin K, Ren-Jie W, Yan-Yi C, Michael Ryan R, Zetitty Cesario N, Wei-Hung C (2022) Recent advances in the graphene quantum dot-based biological and environmental sensors. Sens Actua Report 4:100130
5. Anas NAA, Fen YW, Omar NAS, Daniyal WMEMM, Ramdzan NSM, Saleviter S (2019) Development of graphene quantum dots-based optical sensor for toxic metal ion detection. Sensors 19:3850
6. Stephen RB, Eli José MRJ (2022) Graphene-Based electrochemical sensors for detection of environmental pollutants. Curr Opinion Environ Sci Health 29:100381
7. Pavithra VR, Daniel Thangadurai T, Nataraj D, Senthilkumar K, Manonmani G, Nandakumar K, Sabu T, Praveen G (2019) Graphene nanobuds: a new second-generation phosgene sensor with ultralow detection limit in aqueous solution. ACS Appl Mater Interf 11:19339–19349
8. Shilpa AS, Daniel Thangadurai T, Bhalerao GM, Siuli M (2024) Tailor-designed carbon-based novel fluorescent architecture for nanomolar detection of radioactive elements U(VI) and Th(IV) in pH $\pm$ 5.0. Talanta 272:125783
9. Mandal D, De P, Khatun S et al (2024) Highly fluorescent graphene quantum dots as "turn off–on" nanosensor for detecting toxic metal ions to organic pollutant. Int J Environ Sci Technol 21:1637–1648
10. Ashish K, Sachin K, Rangadhar P, Gaurav M, Gopinath P (2021) Recent advances in graphene quantum dot-based optical and electrochemical (bio)analytical sensors. Mater Adv 2:5513–5541
11. Amit KG, Mansi K, Souradeep R, Pankaj SN, Ranjit K, Shikha W, Ashish M (2021) Design and development of a portable resistive sensor based on α-MnO_2/GQD nanocomposites for trace quantification of Pb(II) in water. IET Nanobiotechnol 15:505–511
12. Lihua Z, Lihua H, Qi W, Qianlin T, Fei L (2020) Theoretical and experimental studies of a novel electrochemical sensor based on molecularly imprinted polymer and GQDs-PtNPs nanocomposite. Microchemical J 158:105196
13. Renjie W, Xinfeng D, Yaotian W, Jingying Z, Xiaojiang X (2018) Graphene quantum dots integrated in ionophore based fluorescent nanosensors for Na and K. ACS Sens 3:2408–2414
14. Sheng QZ, Yue SX, Jing CL, Rong CJ, Feng H (2015) A fluorescent nanosensor based on graphene quantum dots–aptamer probe and graphene oxide platform for detection of lead (II) ion. Biosens Bioelectro 68:225–231
15. Yejian W, Jingwen Z, Xiaochun D, Jie C, Xi F, Xiaobo W (2021) Colorimetric and fluorescent dual-modality sensing platform based on fluorescent nanozyme. Front Chem 9:774486
16. Sanju T, Aditi S, Dhirendra M (2023) A graphene quantum dots–glassy carbon electrode-based electrochemical sensor for monitoring malathion. Beilstein J Nanotechnol 14:701–710
17. Vahab L, Keshipour S (2022) Novel nanosensor of cobalt (II) and copper (II) constructed from graphene quantum dots modified with Eriochrome Black T. Sci Rep 12:13179
18. Amini MH, Faridbod F, Ganjali MR et al (2017) Functionalized graphene quantum dots as a fluorescent "off–on" nanosensor for detection of mercury and ethyl xanthate. Res Chem Intermed 43:7457–7470
19. Xian-En Z, Cuihua L, Yue G, Han G, Shuyun Z, Xue Y, Jinmao Y, Hua W (2017) A ratiometric fluorescent nanosensor for the detection of silver ions using graphene quantum dots. Sens Actua B 253:239–246
20. Dewangan PK, Khan F, Shrivas K et al (2019) Determination of uranium in environmental sample by nanosensor graphene quantum dots. J Radioanal Nucl Chem 320:757–763

21. Zhao Q, Zhou L, Li X, He J, Huang W, Cai Y, Wang J, Chen T, Du Y, Yao Y (2020) Au–Nitrogen-doped graphene quantum dot composites as “On–Off” nanosensors for sensitive photo-electrochemical detection of caffeic acid. Nanomaterials 10:1972
22. Darwin K, Ren-Jie W, Owen S, Kostya (Ken) O, Wei-Hung C (2021) Microplasma nano-engineering of emission-tuneable colloidal nitrogen-doped graphene quantum dots as smart environmental-responsive nanosensors and nanothermometers. Carbon 185:501–513
23. Onur A, Faruk K, Murat B, Haydar Y, Necip A, Mehmet Lütfi Y (2017) Palladium nanoparticles functionalized graphene quantum dots with molecularly imprinted polymer for electrochemical analysis of citrinin. J Mol Liq 243:677–681
24. Saisree S, Arya Nair JS, Sandhya KY (2023) Graphene quantum dots doped with sulfur and nitrogen as versatile electrochemical sensors for heavy metal ions Cd (II), Pb (II), and Hg (II). ACS Appl Nano Mater 6:1224–1234
25. Saloni S, Rajesh K, Ram Manohar Y (2023) Polyacrylonitrile/N-doped graphene quantum dots nanocomposite activity as SERS nanosensors for detection of methylene blue. Mater Today Commun 36:106860
26. Fanizza E, Zhao H, De Zio S, Depalo N, Rosei F, Vomiero A, Curri ML, Striccoli M (2020) Encapsulation of dual emitting giant quantum dots in silica nanoparticles for optical ratiometric temperature nanosensors. Appl Sci 10:2767
27. Pradeep Kumar D, Fahmida K, Vinayak S, Komal K, Chandra C, Khilawan P (2020) Fabrication of L-Cysteine functionalized Graphene quantum dots as a fluorescent nanosensing probe for determination of mercury. IOP Conf Ser: Mater Sci Eng 798:012030
28. Sopan N, Sayali P, Kalyani C, Zamir K, Ashwini P, Pravin P (2023) Graphene quantum dots incorporated UiO-66-NH_2 based fluorescent nanocomposite for highly sensitive detection of quercetin. Nano Biomed Eng 15:1–13
29. Haoqiang W, Xingxing W, Weilong D, Shern-Long L, Qunhui Y, Wei G (2020) One-step preparation of single-layered graphene quantum dots for the detection of Fe^{3+}. Spectrochimica Acta Part A 226:117626
30. Chiraporn C, Thawatchai T, Vinich P, Boosayarat T (2019) Effective GQD/AuNPs nanosensors for selectively bifunctional detection of lysine and cysteine under different photophysical properties. Sens Actua B 282:936–944
31. Feng X, Ting W, Xiameng L, Da C, Changxing W, Gang W, Xiaohu Z, Qinglei G (2022) A fluorescence labelling and switchable nanosensor based on nitrogen-doped graphene quantum dots. Bull Mater Sci 45:26
32. Seyyed MM, Seyyed AH, Ahmad G, Sargol M, Wei-Hung C, Omid A, Navid O, Aziz B (2021) Precise blood glucose sensing by nitrogen-doped graphene quantum dots for tight control of diabetes. J Sens, Article ID 5580203: 14 p
33. Xiameng L, Yun Z, Tingting L, Guoqiang G, Yiru W, Xu W, Jingzheng Y, Da C (2023) Large-scale synthesis of dual sensing nitrogen-doped graphene quantum dots for real-time detection of pH and antibiotic. Bull Mater Sci 46:77
34. Majid MF, Nazanin M (2019) Functionalization of graphene quantum dots with antimorphine: design of selective nanosensor for detection of morphine. Mater Lett 241:206–209
35. Qitong H, Xiaofeng L, Lili T, Qing-Xiao T (2020) Graphene quantum dots/multiwalled carbon nanotubes composite-based electrochemical sensor for detecting dopamine release from living cells. ACS Sust Chem Eng 8:1644–1650
36. Preeti NJ, Shobha W (2016) Graphene quantum dot-based on-chip electrochemical DNA hybridization sensor for pancreatic cancer. Rep Electrochem 6:31–40
37. Sanju G, Tyler S, Alexander B, John B (2018) Graphene quantum dots electrochemistry and development of ultrasensitive enzymatic glucose sensor. MRS Adv 3:831–847
38. Sumeyra S, Zeynep A (2019) Graphene quantum dots as nanozymes for electrochemical sensing of yersinia enterocolitica in milk and human serum. Materials 12:2189
39. Yongqiang L, Zhifeng S, Liuyang S, Quan T, Qisheng T, Hans-Joachim K, Siwei Y, Guqiao D, Hui D (2023) Graphene quantum dots-based magnetic relaxation switch involving magnetic separation for enhanced performances of endoglin detection using ultra-low-field nuclear magnetic resonance relaxometry. Sens Actua B 380:133389

40. Bhardwaj H, Singh C, Kotnala RK et al (2018) Graphene quantum dots-based nano-biointerface platform for food toxin detection. Anal Bioanal Chem 410:7313–7323
41. Guozhen L, Kai Z, Ke M, Andrew C, Mark RH, Ewa MG (2017) Graphene quantum dot based "switch-on" nanosensors for intracellular cytokine monitoring. Nanoscale 9:4934–4943
42. Shan H, Lumin W, Chusheng H, Wei S, Qi X (2016) Amino-functionalized graphene quantum dots based ratiometric fluorescent nanosensor for ultrasensitive and highly selective recognition of horseradish peroxidase. Sens Actua B 234:255–263
43. Mathur A, Nayak HC, Rajput S, Roy S, Nagabooshanam S, Wadhwa S, Kumar R (2021) An enzymatic multiplexed impedimetric sensor based on α-MnO_2/GQD nano-composite for the detection of diabetes and diabetic foot ulcer using micro-fluidic platform. Chemosensors 9:339
44. Yan-Yi C, Darwin K, Seyyed MM, Pavel VF, Elena DO, Wei-Hung C (2022) Bioresource-derived colloidal nitrogen-doped graphene quantum dots as ultrasensitive and stable nanosensors for detection of cancer and neurotransmitter biomarkers. J Mater Chem B 10:9654–9661
45. Yu-Hsin C, Ching-Cheng C, Ling-Yu C, Pang-Chen W, Pawisa K, Min-Hsin Y (2023) Self-powered triboelectric sensor with N-doped graphene quantum dots decorated polyaniline layer for non-invasive glucose monitoring in human sweat. Nano Energy 112:108505.
46. Ana MS, Carla IMS, Leandro MOL, Ines FAM, Joao PCT, Ermelinda M (2022) Graphene quantum dots and phthalocyanines turn-OFF-ON photoluminescence nanosensor for ds-DNA. Nanomaterials 12:1892
47. Tongchang Z, Arnab Halder ID, Yi S (2018) Fluorescent nanosensor based on molecularly imprinted polymers coated on graphene quantum dots for fast detection of antibiotics. Biosensors 8:82
48. Caballero-Díaz E, Benítez-Martínez S, Valcárcel M (2017) Rapid and simple nanosensor by combination of graphene quantum dots and enzymatic inhibition mechanisms. Sens Actua B 240:90–99
49. Hai LT, Win D, Ruey-An D (2020) Ultrasensitive detection of tetracycline using boron and nitrogen co-doped graphene quantum dots from natural carbon source as the paper-based nanosensing probe in difference matrices. Nanomaterials 10:1883
50. Cristina S, Catalina CN, Livia ADG, van Staden JF (2020) Gold nanoparticle-graphene quantum dots nanozyme for the wide range and sensitive electrochemical determination of quercetin in plasma droplets. Microchim Acta 187:611
51. Tabish TA, Hayat H, Abbas A, Narayan RJ (2022) Graphene quantum dots-based electrochemical biosensing platform for early detection of acute myocardial infarction. Biosensors 12:77
52. Shiva Kumar A, Saravanan G, Kyusik Y (2020) Electrochemical sensor for detecting dopamine using graphene quantum dots incorporated with multiwall carbon nanotubes. Appl Sur Sci 508:145294
53. Reyhane E, Mahmoud E, Hassan KM (2020) A sensitive sensor for nano-molar detection of 5-fluorouracil by modifying a paste sensor with graphene quantum dots and an ionic liquid. J Nanostruct 10:230–238
54. Anjitha T, Dipin T, Yahya AI, Pradeepan P (2022) Polyindole-derived nitrogen-doped graphene quantum dots-based electrochemical sensor for dopamine detection. Biosensors 12:1063
55. Sharma V, Tarun Kumar S, Inderpeet K (2021) Electrochemical detection of cortisol on graphene quantum dots modified electrodes using a rationally truncated high affinity aptamer. Appl Nanosci 11:2577–2588
56. Xuehua B, Guorong W (2017) Label free c-reactive protein detection based on an electrochemical sensor for clinical application. Int J Electrochem Sci 12:6304–6314
57. Ying G, Guanqing Y, Xuelian H, Jinni Z, Guozhen F (2022) A high-luminescence biomimetic nanosensor based on N, S-GQDs-embedded Zinc-based metal-organic framework@molecularly imprinted polymer for sensitive detection of octopamine in fermented foods. Foods 11:1348
58. Ali AE, Parisa NE, Rezaei B (2018) Metronidazole determination with an extremely sensitive and selective electrochemical sensor based on graphene nanoplatelets and molecularly imprinted polymers on graphene quantum dots. Sens Actua B 270:192–199

59. Xu Y, Yang S, Chengzhou Z, Junhua S, Dan D, Xingguang S, Yuehe L (2016) Graphene quantum Dot–MnO_2 nanosheet based optical sensing platform: a sensitive fluorescence "Turn Off–On" nanosensor for glutathione detection and intracellular imaging. ACS Appl Mater Interf 8:21990–21996
60. Alireza S, Balal K, Mahdi M, Amir Reza A, Alireza K, Sajad Razavi B, Majid Ebrahimi W, Mohammad Reza R, Majid M (2021) An efficient graphene quantum dots-based electrochemical cytosensor for the sensitive recognition of CD123 in acute myeloid leukemia cells. IEEE Sens J 21:16451–16463
61. Mina MS, Farnoush F, Mohammad Reza G (2019) A luminescence nanosensor for Ornidazole detection using graphene quantum dots entrapped in silica molecular imprinted polymer. Spectrochimica Acta Part A 206:430–436
62. Pratik K, Akanksha R, Sonu (2023) Fabrication of an ultrasensitive electrochemical immunosensor coupled with biofunctionalized zero-dimensional graphene quantum dots for rapid detection of cephalexin. Food Chem 398:133846
63. Shan H, Lumin W, Chusheng H, Wei S, Qi X (2017) Label-free and ratiometric fluorescent nanosensor based on amino-functionalized graphene quantum dots coupling catalytic G-quadruplex/hemin DNAzyme for ultrasensitive recognition of human telomere DNA. Sens Actua B 245:648–655
64. Fatemeh N, Hadi H, Mojtaba R, Amir Mohammad S, Kourosh A, Mehdi RN, Hermann E (2022) A glassy carbon electrode modified with N-TiO_2@AgNPs@GQDs for electrochemical determination of dopamine. Diam Related Mater 127:109120
65. Mengke M, Yan L, Lei W, Xing GS (2017) A label-free fluorescence nanosensor for the determination of adrenaline based on graphene quantum dots. Anal Methods 9:4434–4438
66. Hai LT, Win D, Ruey-An D (2021) Electrochemical immunosensor for ultra-sensitive detection of attomolar prostate specific antigen with sulfur-doped graphene quantum dot@gold nanostar as the probe. Electrochim Acta 389:138700
67. Mingxing L, Rongrong J, Meie Z, Mengjiao L, Qingjie Y, Hongda Z, Huiling G, Hongmei S (2022) A sensitive ratiometric biosensor for determination cardiac troponin I of myocardial infarction markers based on N, Zn-GQDs. Talanta 249:123577
68. Pooja D, Deepak K, Manpreet K, Anupma T, Ki-Hyun K, Preeti K, Kamalpreet K, Harsimran K (2021) Conjugate of graphene quantum dots and glutaminase for the sensing of L-glutamine: electrochemical vs. fluorescent sensing approaches. Inorg Chem Commun 130:108745
69. Chunxia W, Fan Y, Yushu T, Wang Y, Huangliang Z, Changchun Y, Ranjia L, Hongjun Z, Yongfeng L, Lanqun M (2018) Graphene quantum dots nanosensor derived from 3D nanomesh graphene frameworks and its application for fluorescent sensing of Cu^{2+} in rat brain. Sens Actua B 258:672–681
70. Iannazzo D, Espro C, Ferlazzo A, Celesti C, Branca C, Neri G (2021) Electrochemical and fluorescent properties of crown ether functionalized graphene quantum dots for potassium and sodium ions detection. Nanomaterials 11:2897
71. Shadjou R, Hasanzadeh M, Heidar-poor M, Shadjou N (2018) Electrochemical monitoring of aflatoxin M1 in milk samples using silver nanoparticles dispersed on α-cyclodextrin-GQDs nanocomposite. J Mol Recognit 31:e2699
72. Pramod KG, Deepika C, Zishan HK, Pratima RS (2020) ZrO_2 nanoflowers decorated with graphene quantum dots for electrochemical immunosensing. ACS Appl Nano Mater 3:2506–2516
73. Qian X, Jingyun H, Huiyao H, Weiwei M, Zhizhen Y (2018) A label-free electrochemical platform for the highly sensitive detection of hepatitis B virus DNA using graphene quantum dots. RSC Adv 8:1820–1825
74. Reshma K, Patel NM, Belqasem A, Sambandam A, Muthupandian AK (2022) Ultrasound-aided synthesis of gold-loaded boron-doped graphene quantum dots interface towards simultaneous electrochemical determination of guanine and adenine biomolecules. Ultrason Sonochem 83:105921
75. Aayushi K, Banibrata M, Soumen B (2022) Coal-derived graphene quantum dots with a Mn^{2+}/Mn^{7+} nanosensor for selective detection of glutathione by a fluorescence switch-off–on assay. New J Chem 46:7545–7556

76. Parizad M, Seyed Mohamadreza MH, Beheshteh S (2024) The graphene quantum dots encased in the molecularly imprinted polymer as a new fluorescent nanosensor for the detection of biotin. Sens Actua Report 7:100187
77. Hai LT, Van DD, Naresh Kumar D, Sih-Min L, Yu-Fen H, Rueyan D (2022) Ultrasensitive detection of breast cancer cells with a lectin-based electrochemical sensor using N-doped graphene quantum dots as the sensing probe. Sens Actua B 368:132233
78. Deniz E, Nuray D, Dilsat OA, Ali B, Omer S, Fatih A, Selehattin Y (2024) Graphene quantum dots-polyfluorene hybrid nanobiosensor for mitomycin C-DNA interaction sensing. Mater Sci Eng B 299:116944
79. Thota T, Harisankar K, Tummala A, Kalli SB, Shanmukha Kumar JV, Pradeep Kumar B (2020) Synthesis and characterization of nanocomposite material based on graphene quantum dots and lanthanum doped zirconia nanoparticles: an electrochemical sensing application towards flutamide in urine samples. Diam Related Mater 110:108143
80. Meng L, Chenxi L, Liang Q, Hongsheng L (2022) Nitrogen-doped graphene quantum dots anchored on hollow zeolitic imidazolate framework-8 colloidosomes for fluorescence detection of glucose. ACS Appl Nano Mater 5:5425–5438
81. Zhang L, Wang J, Deng J et al (2020) A novel fluorescent "turn-on" aptasensor based on nitrogen-doped graphene quantum dots and hexagonal cobalt oxyhydroxide nanoflakes to detect tetracycline. Anal Bioanal Chem 412:1343–1351
82. Masoud G, Gholamreza D, Jafar S (2022) A ratiometric fluorescent sensor for detection of metformin based on terbium–1,10-phenanthroline–nitrogen-doped-graphene quantum dots. RSC Adv 12:22255–22265
83. Fatemeh N, Rouholah ZD, Morteza H, Mohammad RG (2018) Fluorescence turn-on sensing of thiamine based on Arginine–functionalized graphene quantum dots (Arg-GQDs): central composite design for process optimization. Sens Actua B 255:2078–2085
84. Pavithra VR, Daniel Thangaduraia T, Nataraj D (2020) Ultra-sensitive detection of commercial vitamin B9 and B12 by graphene nanobuds through inner filter effect. J Photochem Photobio A 400:112691
85. Sravya V, Pavithra VR, Daniel Thangadurai T, Nataraj D, Sathish Kumar N (2022) Excitation-independent and fluorescence-reversible N-GQD for picomolar detection of inhibitory neurotransmitter in milk samples—an alleyway for possible neuromorphic computing application. Talanta 239:123132
86. Hakimi M, Salehi A, Boroumand FA, Mosleh N (2018) Fabrication of a room temperature ammonia gas sensor based on polyaniline with N-doped graphene quantum dots. IEEE Sens J 18:2245–2252
87. Pavithra VR, Daniel Thangadurai T, Nehru K, Lee YI, Nataraj D, Sabu T, Nandakumar K, Jiya J (2020) Surface and morphology analyses, and voltammetry studies for electrochemical determination of cerium (III) using a graphene nanobud-modified-carbon felt electrode in acidic buffer solution (pH 4.0 $\pm$ 0.05). RSC Adv 10:37409

Chapter 12
Graphene Quantum Dots in Energy Sectors

12.1 Introduction

Graphene, a two-dimensional (2D) single-atom carbon sheet, has garnered significant attention because of its remarkable characteristics, including its semi-metallic electrical structure, high carrier mobility, and exceptional flexibility [1]. Its strong conductivity and eco-friendliness, in particular, fulfill crucial needs for the creation of optoelectronic devices. The intriguing nanomaterial known as graphene quantum dots, or GQDs, has a lot of potential uses in a lot of different industries, including energy [2]. A few glimpses of GQD energy applications are given below.

Solar Cells: GQDs have demonstrated potential for raising solar cell efficiency. They have good light absorption qualities and can efficiently create and transfer charge carriers, making them useful as sensitizers or electron acceptors in solar cells. Moreover, GQDs can be included in solar cell electrodes to increase electron transfer kinetics and boost power conversion efficiency [3].

Energy Storage Devices: The application of GQDs in batteries and supercapacitors has been investigated. They are appropriate materials for enhancing the functionality of energy storage devices due to their large surface area, superior electrical conductivity, and adjustable bandgap [4, 5]. In batteries and supercapacitors, GQDs can be added to electrodes to improve charge storage capacity, cycling stability, and rate capability [6, 7].

Fuel Cells: Due to their great stability, strong catalytic activity, and huge surface area, GQDs are being researched for application as fuel cell catalysts. They can function as effective electrocatalysts for a range of electrochemical reactions, including the vital fuel cell processes of hydrogen evolution reaction (HER) and oxygen reduction reaction (ORR). Fuel cell electrode performance and durability can be increased by adding GQDs [8].

N. Manjubaashini et al., *Graphene Quantum Dots*, Materials Horizons: From Nature to Nanomaterials, https://doi.org/10.1007/978-981-97-5722-0_12

12.2 Supercapacitors Based on GQDs

For supercapacitor applications, N-doped GQD (N-GQDs) coated on Fe_3O_4-halloysite nanotubes (Fe_3O_4-HNTs) have been developed as active anode materials. In neutral electrolyte solutions, N-GQD@Fe_3O_4-HNTs maintain strong rate capability and show a high specific capacitance of 418 F g^{-1} (Fig. 12.1a). Furthermore, according to Akhilesh Babu et al. [9], the anode materials exhibit outstanding electrochemical performance, with energy and power densities of 10.4^{-29} W h kg^{-1} and 0.25–5.2 kW kg^{-1}, respectively. In symmetrical supercapacitors, the capacitive characteristics of the GQD–3DG composites are assessed [10]. With a high specific capacitance of 268 F g^{-1}, the supercapacitor constructed from the GQD–3DG composite is very stable and performs more than 90% better than the supercapacitor made from pure 3DG electrodes (136 F g^{-1}) (Fig. 12.1b). Graphite oxide was chemically oxidized to produce GQDs. Using GNSs as representative electrode materials, KOH neutralized the GQDs, significantly improving their conductivity and ion-donating capacity (Fig. 12.1c–d). When compared to supercapacitors using GQD electrolytes, those utilizing the solution and solid-type NGQD electrolytes exhibited noticeably better capacitance and rate performance [11].

When measured at 0.5 A g^{-1}, the electrodes with negative (AC) and positive (AC@GQDs) polarity show a significant specific capacitance of 220 and 265 F

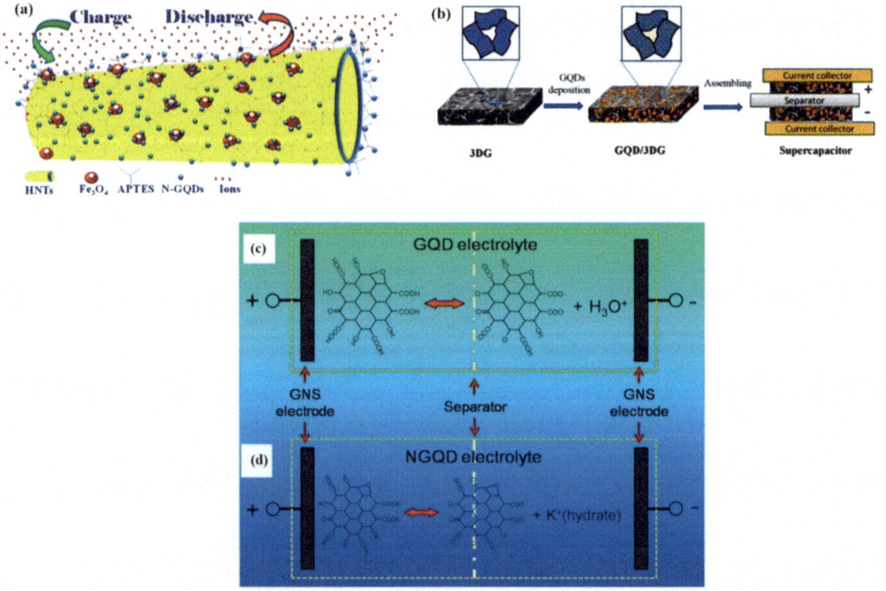

Fig. 12.1 **a** Schematic of N-GQD@Fe_3O_4-HNTs supercapacitor applications [9]; **b** Fabricating a symmetrical supercapacitor based on the GQD–3DG composite material [10]; **c** GQD electrolyte supercapacitor mechanism; (d) NGQD electrolyte supercapacitor mechanism [11]

g^{-1}, respectively (Fig. 12.2a–c). Furthermore, galvanostatic charge–discharge experiments demonstrate that these electrodes possess high-rate capabilities of 165 and 230 F g^{-1} when measured at 5 A g^{-1}. The asymmetric AC/AC@GQD supercapacitor device has a specific capacitance of 118 F g^{-1}. In addition, the asymmetric device has outstanding cycling behavior; even after 10,000 cycles, and its capacitance retention is an amazing 92% [12]. The GQD/Co $(OH)_2$ composite was used as the positive electrode and active carbon as the negative electrode to create a wire-shaped, quasi-solid-state hybrid supercapacitor device. After 10,000 cycles at a current density of 2 A cm^3, the device exhibits an excellent capacitance retention rate of 99.5% and can withstand bending tests almost without degrading the electrochemical performance. We further investigated the atomic electronic characteristics of GQD/Co $(OH)_2$ related to the interface by doing systematic first-principles calculations on it. We observed enhanced electronic states that extend throughout the whole Co $(OH)_2$ energy gap as a result of the inclusion of GQDs, and these states are directly associated with the electrical conductivity and device performance [13].

The SnO_2/N-GQD NTs composite's supercapacitor performance was examined using electrochemical investigations, which included galvanostatic charge–discharge (GCD) measurement in a 2 M KOH solution used as the electrolyte. SnO_2/N-GQD NTs were subjected to electrochemical studies at various scan rates and current

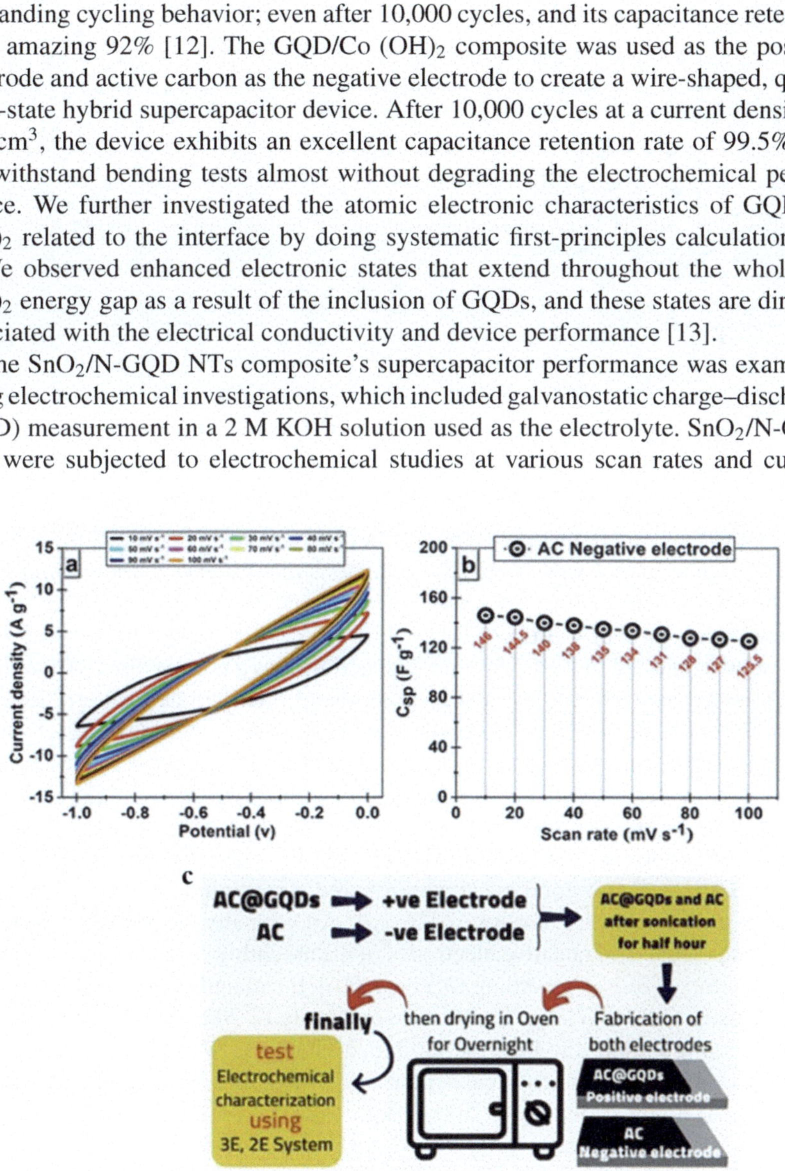

Fig. 12.2 **a** AC@GQDs Electrodes CV curve. **b** AC@GQDs Electrodes calculated C_{sp}, and **c** Preparation of AC and AC@GQDs Electrodes [12]

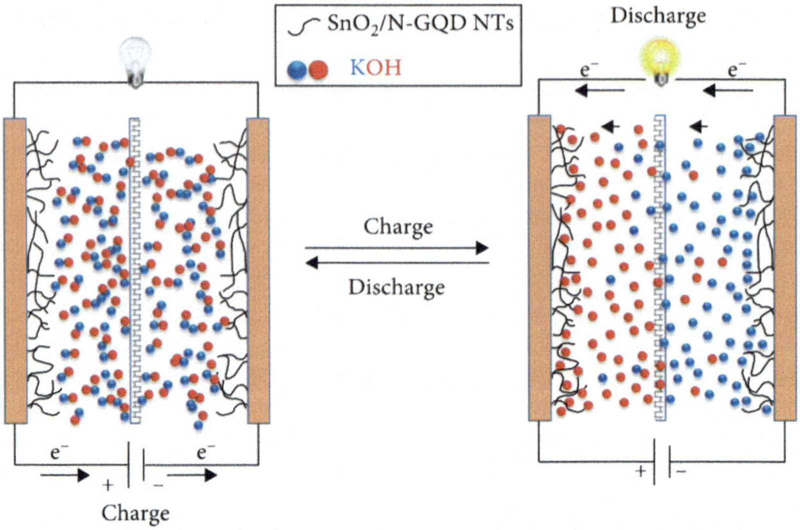

Fig. 12.3 Mechanism of SnO_2/N-GQD NT composite electrodes during the charge–discharge processes [14]

densities [14]. Compared to pure SnO_2 NTs (230 mF cm^{-2}) (Fig. 12.3), SnO_2/N-GQD NTs made with 3 wt% of N-GQDs demonstrated superior capacity retention even after 5000 GCD cycles and a maximum specific capacitance of 420 mF g^{-1} at a current density of 8 mA cm^{-2}.

The active material in the micro-supercapacitors was a porous film made of interdigitated graphene (ipG GQDs) and its consequent interdigitated pattern. Remarkably, these supercapacitor devices demonstrated high energy storage (9.09 μF cm^{-2}), quick relaxation time (8.55 ms) (Fig. 12.4A(a–d)), steady cycle retention, great stability even under extreme bending angle 45° with 10,000 cycles, and high transparency (92.97% at 550 nm) [15]. The electrochemical performances of GQDs/$Ni(OH)_2$ composites on carbon cloth (G-NH/CC) electrodes are improved. In particular, the G-NH//CC has good cycle stability and a maximum specific capacitance of 1825 F g^{-1} at a current density of 1 A g^{-1}. Additionally, G-NH/CC composites are used as positive and negative electrodes in the assembly of solid-state symmetric supercapacitors (SSCs). According to Yiwen [16], the manufactured SSC has a high energy density of 80.8 Wh kg^{-1} and a power density of 2021 W kg^{-1}. In the three-electrode configuration, the heteroatom-doped GQD (S, N-GQDs) nanocomposite exhibits the maximum specific capacitance of 645 F g^{-1} at a current density of 0.5 A g^{-1}. At a power density of 500 W kg^{-1}, the S, N-GQDP2 composite-based symmetric cell energy density is 17.25 Wh kg^{-1} (Fig. 12.4B(a–f)). At a volumetric power density of 525 W L^{-1}, a high volumetric energy density of 18.11 Wh L^{-1} is achieved. Furthermore, with a capacitance retention of 90%, the S, N-GQDP2 nanocomposite-based symmetric device has good cycling stability for 1000 cycles [17].

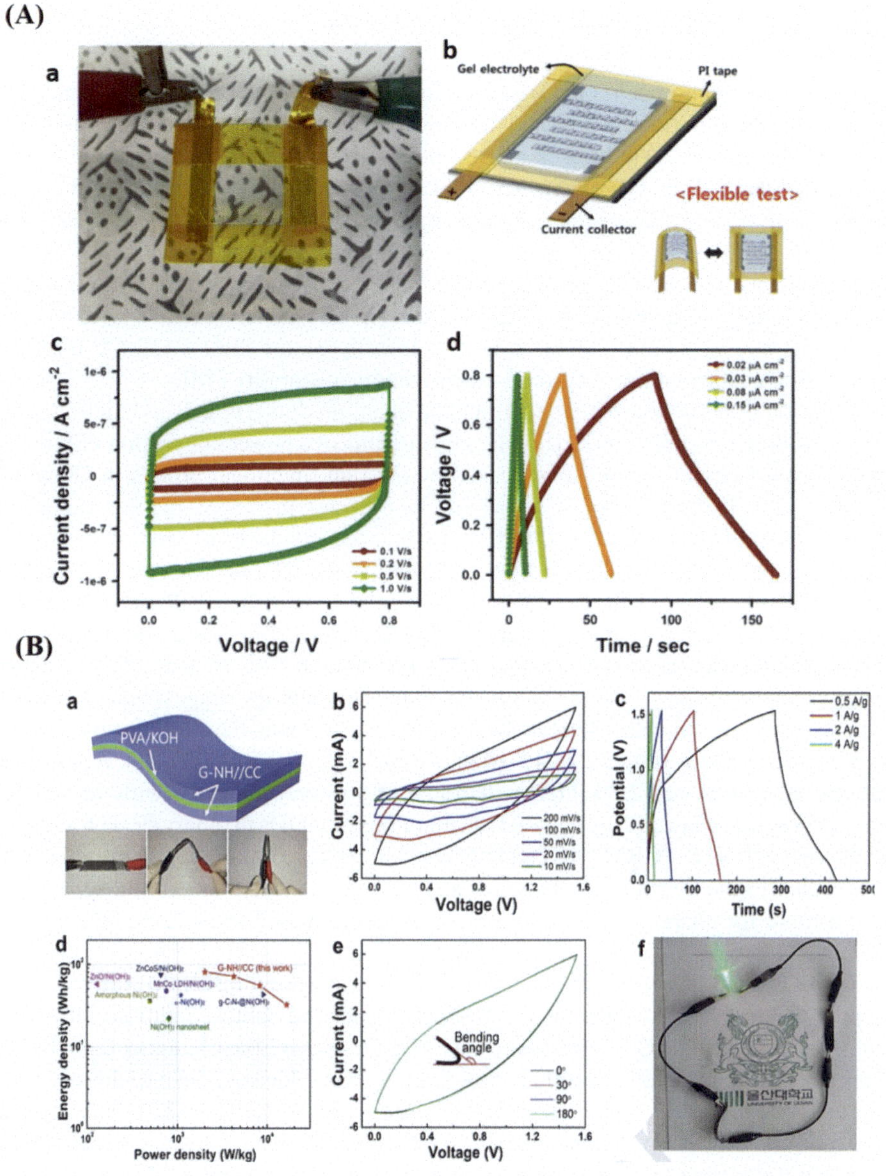

Fig. 12.4 **A** (**a**) Optical image of ipG-GQDs-MSC. (**b**) Illustration for the fabricated structure of ipG-GQDs-MSC during bending test. (**c**) Cyclic voltammetry (CV) curves, and (**d**) Galvanostatic charge–discharge curves. Reprinted with permission from @ Nano energy 26, 746–754 (2016). **B** (**a**) SSS supercapacitor using G-NH//CC electrodes and bending state image. (**b**) CV curves at different scan rates. (**c**) Galvanostatic charge–discharge curves at different current densities. (**d**) Ragone plots of G-NH//CC SSC and reported $Ni(OH)_2$-based supercapacitors. **e** CV curves of the G-NH//CC SSC at different bending states, and **f** Images of lighting LED bulb powered by two G-NH//CC SSCs in series. Reprinted with permission from @ Journal of Materials Science and Technology 58, 73–79 (2020)

12.3 Photovoltaic Cells Based on GQDs

Using a dipping approach, GQDs possessing exceptional inherent catalytic activity and high conductivity were adorned onto the $SrRuO_3$ (SRO) counter electrode (CE), creating an SRO–GQD hybrid. Increased electrocatalytic activity toward I3—reduction is the result of the synergistic action of $SrRuO_3$ and GQDs, which leads to more active catalytic sites, faster ion diffusion, and electron transfer than a pure $SrRuO_3$ CE. The result is an outstanding power conversion efficiency (PCE) of 8.05% for the produced dye-sensitized solar cells (DSSCs) based on the optimized SRO–GQD CE, which is significantly higher than the reference device constructed with a traditional platinum (Pt) CE (7.44%). In $I^{3-}/^{I-}$ electrolytes, the SRO–GQD CE likewise demonstrates outstanding long-term electrochemical stability [18].

Creation of reduction-regulated graphene quantum dots (GQDs) and their utilization in bulk heterojunction (BHJ) solar cells to improve power conversion efficiency (PCE). Short-circuit current is increased by the rich functional groups in GOQDs' greater optical absorptivity, whereas fill factors are increased by the enhanced conductivity of decreased GQDs [19]. To balance conductivity and absorptivity, the reduction level of GQDs must be moderate. When BHJ devices were compared to a reference device without GQDs (6.70%), the BHJ devices' PCE was remarkably enhanced by 7.60% due to the partially reduced GQDs. Power conversion efficiency (PCE) was increased in crystalline silicon solar cells by using GQDs as an energy-downshift layer on the silicon substrate. The findings suggest that GQDs are crucial in raising the cells' optical absorptivity. At 30 mm/s, the short-circuit current density increased by roughly 2.94% (0.9 mA/cm^2) (Fig. 12.5a–c). The PCE of p-type silicon solar cells was 2.7% higher than that of a reference device without a GQD energy downshift layer [20]. The various functions of GQDs in binary (P3HT: GQDs) and ternary (P3HT: PCBM: GQDs) hybrid solar cells are methodically investigated. According to the findings, GQDs offer an effective interface for exciton separation and a route for charge transfer, which can enhance hybrid solar cells (Wentian W 2019).

GQD is used as a counter electrode (CE) in a tungsten disulfide–molybdenum copper oxide (WM) composite for DSSCs. Under the same conditions, the efficiency of the WM@GQDs composite was 10.38% (Fig. 12.6a–c), higher than that of the costly platinum CE [22]. When compared to other devices, the maximum open-circuit voltage (VOC = 360 mV), short-circuit current (JSC = 1.84 mA cm^{-2}), and power conversion efficiency (0.293%) of sulfur- and nitrogen-doped GQD (SNGQD)-decorated nanorod-based solar cells were demonstrated [23].

Binder-free TiO_2 paste containing DQDs has been investigated as a potential photoelectrode material for flexible DSSCs built on ITO/PEN substrates. When compared to those with DSSC built with just a TiO_2 photoanode (open-circuit output potential difference (Voc) of 0.68 V and short-circuit density (Jsc) of 10.67 mA cm^{-2}), DSSC based on GQDs-TiO_2 exhibited an increment in power conversion efficiency of 5.48%. Additionally, the DSSC exhibited an open-circuit output potential difference

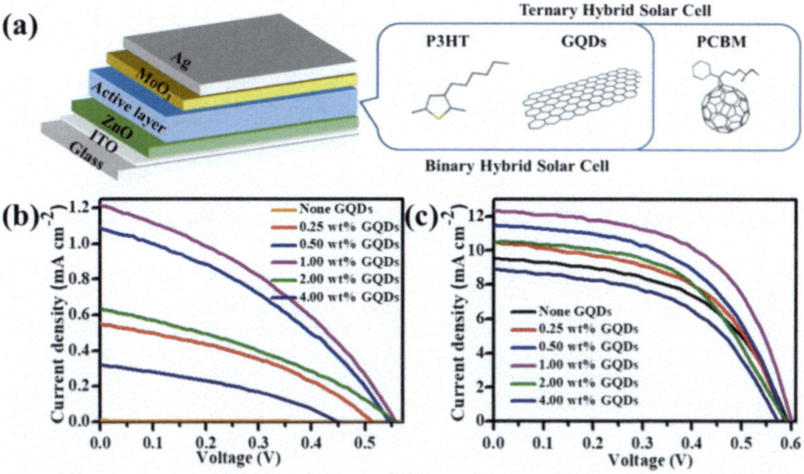

Fig. 12.5 **a** Formation of binary and ternary hybrid solar cells with P3HT: GQDs and P3HT:PCBM: GQDs as active layers; chemical structure of P3HT, GQDs, and PCBM, *J-V* characteristics of **b** binary and **c** ternary hybrid solar cells with different contents of GQDs under AM 1.5G 100 mW cm^{-2} illumination [21]

(Voc) of 0.73 V and short-circuit current density (Jsc) of 11.54 mA cm^{-2} (Fig. 12.7a–b). The integration of GQDs has been understood to result in decreased recombination losses and increased charge extraction [24]. Nitrogen and fluorine co-doped GQDs (N, F-GQDs) are used for solar cell applications. Using additive-free electrophoretic deposition (EPD), a homogeneous and stable layer of N, F-GQDs was formed on the TiO_2/FTO electrode, which served as the photoanode. The manufactured solar cells' photovoltaic characteristics were investigated in an AM 1.5 G synthetic sun irradiation. It was discovered that the development of the EPD layer along with a suitable ratio of N and F doping in GQDs significantly increased the efficiency of the solar cell (Fig. 12.7c–d). The efficiency, open-circuit voltage, and short-circuit current density of the best cell with N, F-GQD0.07/TiO_2/FTO photoanode ($1 \times 1 \times 0.2$ cm^3) increased from 0.07 to 0.51, 370 to 699 mV, and 0.3 to 1.2 mA/cm^2, respectively [25]. The incident photon-to-current conversion efficiency (IPCE) of DSSCs with various sensitization affects their photoresponse properties (Fig. 12.7e). The results showed a 20% improvement in the photoelectric conversion efficiency of DSSCs utilizing the GQD + N719 co-sensitized photoanode, to 5.14% [26].

12.4 Hydrogen Production and Storage Based on GQDs

Highly effective H_2 formation from water splitting was achieved by using N-doped titanate nanotubes (N-TNTs) coated with graphene quantum dots (GQDs). N-doping in N-TNT/GQD composites can extend light responsiveness to visible light range,

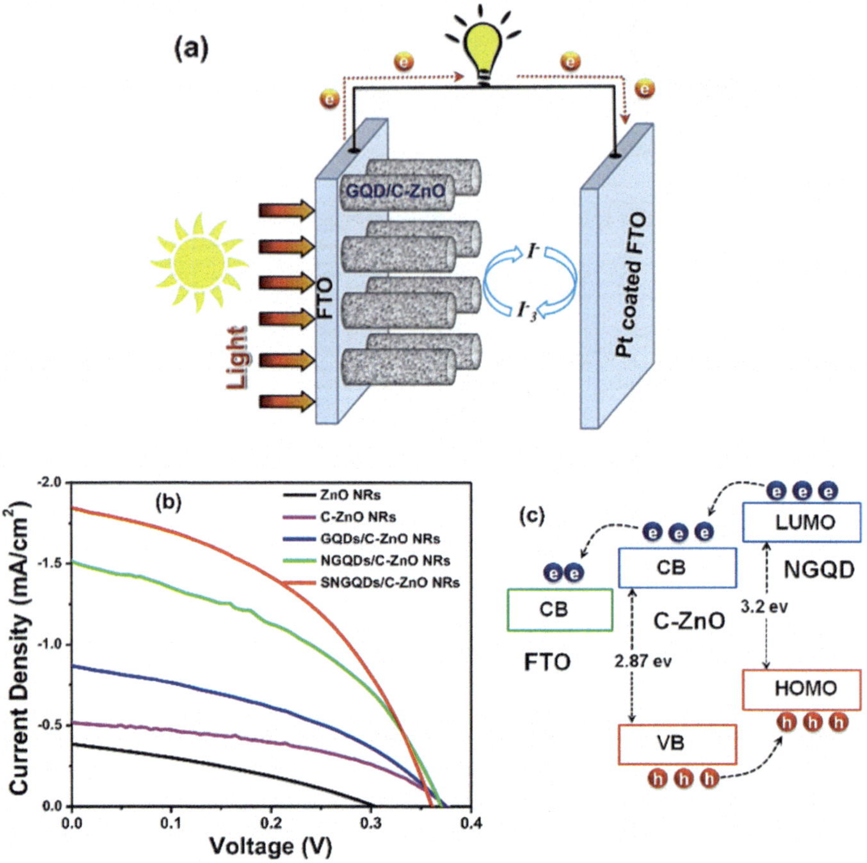

Fig. 12.6 **a** A typical GQD-sensitized C-ZnO NR-based QDSSC. **b** Current density *versus* voltage (*J-V*) characteristics of pristine ZnO NR, C-ZnO NR, GQD/C-ZnO NR, NGQD/C-ZnO NR, and SNGQD/C-ZnO NR-based QDSSCs under 100 mW cm^{-2} white light illumination (AM 15), and **c** Band diagram of charge transportation and separation mechanism in a typical NGQD-sensitized solar cell. Reprinted with permission from @ Bulletin of Materials Science 42, 65 (2019)

as demonstrated by photoelectrochemical studies, and coupling with GQDs not only enhances visible light absorption but also promotes interfacial charge carrier transfer. The closely integrated N-TNT/GQD composite shows a substantially greater photocatalytic H_2 production performance under UV–vis irradiation, being 2.1 times higher than that of pure TNTs. This is due to the synergistic impact between N-doping and GQD coupling. When combined with P25 (TiO_2), S, N co-doped graphene quantum dots (S, N-GQDs) and S, N-GQD/P25 demonstrated exceptional photocatalytic hydrogen production capabilities [27]. When N-GQDs' photoelectrochemical characteristics were examined concerning water splitting (Fig. 12.8a–c), the findings revealed that 2N-GQDs made with 2 g of urea generated the most photocurrent. Examined under AM 1.5G illumination, the photocatalytic activity of 2N-GQDs

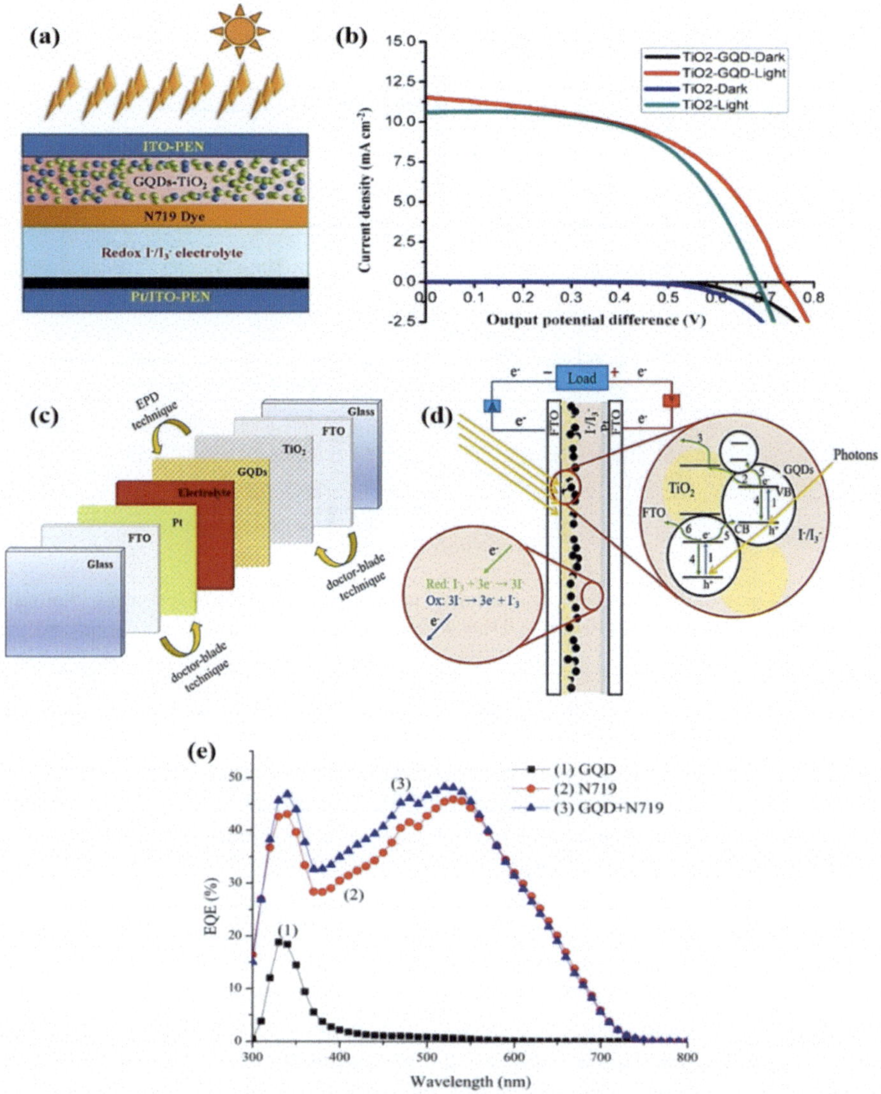

Fig. 12.7 **a** Formation of DSSC. **b** J-V characteristics of flexible solar cell made with binder-free GQD-TiO_2 paste. Reprinted with permission from @ Electrochimica Acta 305, 278–284 (2019). **c** Schematic illustration of photoanode and photocathode in QSSC assembly. **d** GQD explosion with sunlight irradiation and possible paths for photogenerated electrons. Reprinted with permission from @ Ind. Eng. Chem. Res. 62, 7481–7491 (2023). **e** Incident photon-to-current conversion efficiency (IPCE) curves with different photoanodes. Reprinted with permission from @ IEEE Transactions on Nanotechnology, 22, 20–27 (2015)

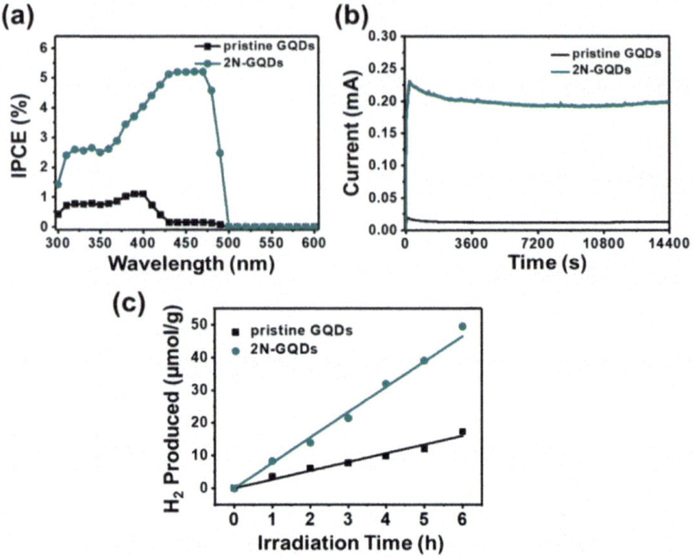

Fig. 12.8 **a** Curve of IPCE. **b** Photocurrent-time of pristine GQDs and 2N-GQDs, and **c** Photocatalytic hydrogen production of pristine GQDs and 2N-GQDs [28]

powder photocatalyst for hydrogen production demonstrated a significant increase over that of pristine GQDs [28].

When it comes to PEC hydrogen production, the ternary TiO_2/CdSe/GQDs photoanode exhibits superior photostability and more photocurrent. The improved surface reaction kinetics and improved charge carrier separation provided by the added GQDs also helped to explain TiO_2/CdSe/GQDs' higher PEC efficiency (Fig. 12.9a). For long-lasting solar-to-fuel conversion applications, the ground-breaking finding offers a potentially global approach for developing GQD-modified chalcogenide-sensitized semiconductor photoelectrodes [29]. When compared to pure TNTs and binary GQDs-TNTs photocatalysts, the CdS-GQD-titanate nanotube (CdS-GQDs-TNT) nanocomposite shows much better photocatalytic performance toward H_2 production. This is likely due to the combination of factors such as faster interfacial charge transfer rate, faster and long-distance electron transport pass, and stronger visible light harvesting (Fig. 12.9b). Based on photoelectrochemical experiments, the interfacial charge transfer mechanism of the CdS-GQDs-TNTs ternary composite is postulated [30].

In photoelectrochemical (PEC) hydrogen fuel generation powered by solar light, GQD onto metal oxide electron transport layer is receiving a lot of attention. By using electrostatic attraction, surface-modified GQD (2 nm) are directly assembled onto a TiO_2 hollow nanowire, and the photocarrier accumulation and recombination mechanisms that lead to device operation are examined. GQDs absorbed light photons at visible light wavelengths up to 600 nm, according to optical characterization. Using sacrificial-free electrolytes, hybrid TiO_2-GQD heterostructures demonstrate

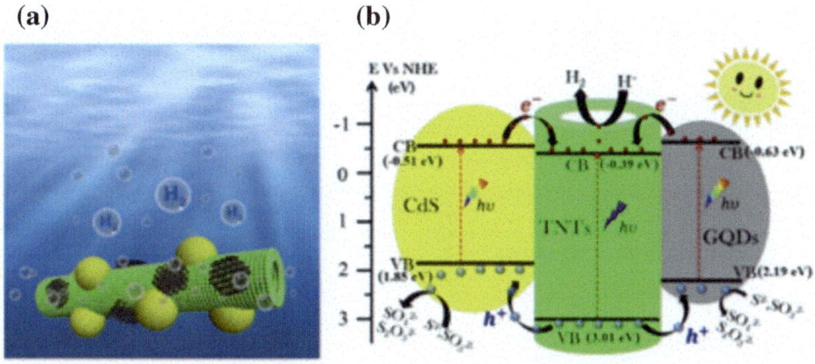

Fig. 12.9 **a** Structure of CdS-GQDs-TNTs. **b** Charges transfer mechanism of the CdS-GQDs-TNTs ternary photocatalyst. Reprinted with permission from @ International Journal of Hydrogen energy 45, 6476–6486 (2020)

a 70% photocurrent improvement for water oxidation compared to pristine TiO_2 (Fig. 12.10a–b). This is further supported by incident photon to current efficiency [31]. Magnesium hydride (MgH_2) improved hydrogen storage behavior, thanks to the catalytic action of GQDs. Nearly 5 weight percent of hydrogen is reabsorbed by the 7% GQDs-MgH_2 sample in 2.5 min at 300 °C and 15 atm hydrogen pressure. The MgH_2 sample that was catalyzed by GQDs shows good cyclic stability during de/re-hydrogenation (Fig. 12.10c(i–ii)). Even after 25 rounds of de- and re-hydrogenation, there was no discernible decrease in hydrogen capacity. The production enthalpy of 7% GQDs-MgH_2 was calculated from Van't Hoff plot to be 58 $kJmol^{-1}$ [32].

12.5 Batteries and Fuel Cells Based on GQDs

The idea is to create a 3D nanoflower-like GQDs@ZnxMnO$_2$ composite cathode for aqueous zinc-ion batteries (ZIBs) by integrating GQDs with Zn-intercalated MnO_2 nanosheets (Fig. 12.11a–b). The GQDs@ZnxMnO$_2$ composite cathode has good structural stability, outstanding electrochemical kinetics, and an exceptional aqueous Zn^{2+} storage capacity (~403.6 mAh · g^{-1}). The produced aqueous ZIBs have the potential to be effective electrochemical storage devices for a variety of energy-related fields because they can provide significant energy density (226.8 W · h · kg^{-1}), impressive power density (650 W · kg^{-1}), and long-term cycle performance in real applications [33]. Hydrophilic GQDs are introduced to manipulate the nucleation process and successfully build an electrochemically stable zinc anode. The GQDs with oxygen-containing groups stimulate interfacial hydrogen bonding, which is advantageous for enhancing the kinetics of the Zn ion reaction and reducing the impact of water-induced side reactions (Fig. 12.11c–e). Significantly, symmetric cell polarization voltage drops from 80 to 50 mV at 0.8 mA cm^{-2}, resulting in a longer lifespan of 2200 h [34]. Consequently, after 600 cycles at 1 A g^{-1}, the constructed

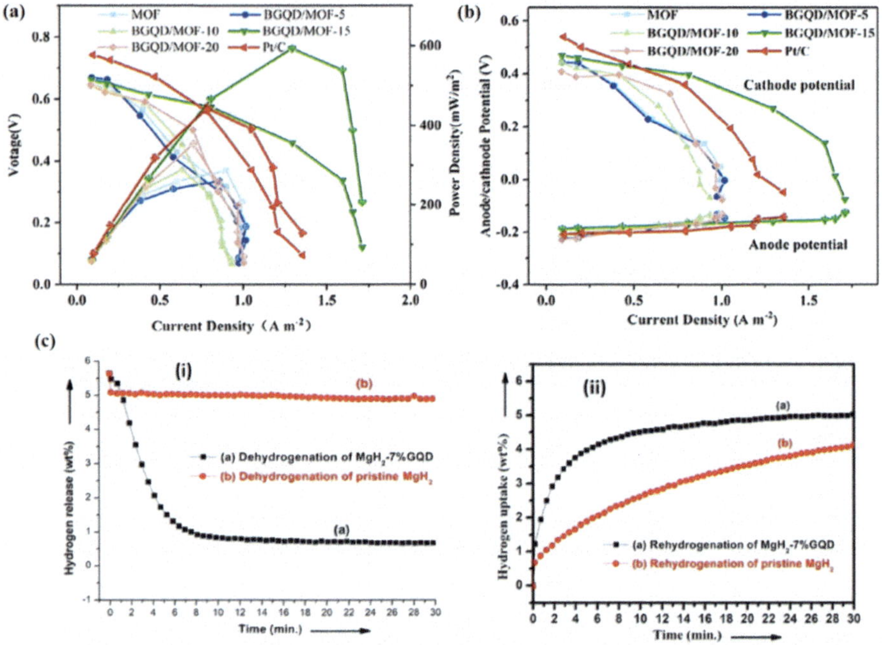

Fig. 12.10 **a** Polarization curves and power generations obtained of MFC saturated O_2. **b** electrode potentials (vs. SCE) [31], **c-i** Dehydrogenation kinetics MgH_2@GQD, and **c-ii** Re-hydrogenation kinetics of dehydrogenated sample. Reprinted with permission from @ International Journal of Hydrogen Energy (2024)

ZIBs with vanadium cathode perform better, displaying a capacity of 164.3 mAh g^{-1}.

New carbonaceous LIB anodes are based on porous carbon nanotube (CNT) sponge that incorporate GQD as a freestanding all-carbon electrode. A high specific capacity (700 mAh $\cdot$ g^{-1} at 100 mA $\cdot$ g^{-1} after 100 cycles) and rate performance were achieved by this hybrid structure (Fig. 12.12A(a–f)), which combined conductive CNTs with highly active GQDs. GQDs with predesigned functional groups provided massive storage sites for Li-ions, and the 3D CNT frameworks prevented the agglomeration of GQDs. The reversible specific capacity remains at 483 mAh $\cdot$ g^{-1} with high current density of 1000 mA $\cdot$ g^{-1} after 350 cycles [35]. To further improve the electrochemical qualities, GQDs are applied to the VO_2 surfaces as a highly effective surface sensitizer and protective layer. At 100 mA/g, the integrated electrodes provide a Na storage capacity of 306 mAh/g, while at 18 A/g after 1500 cycles, the capacity increases to over 110 mAh/g (Fig. 12.12B(a–c)). The Na-ion battery research could open the door for the development of post-lithium batteries of the future [36].

A deeper comprehension of the critical influence of regulated redox site distributions on battery performance is made possible by the subdomain GQD-based

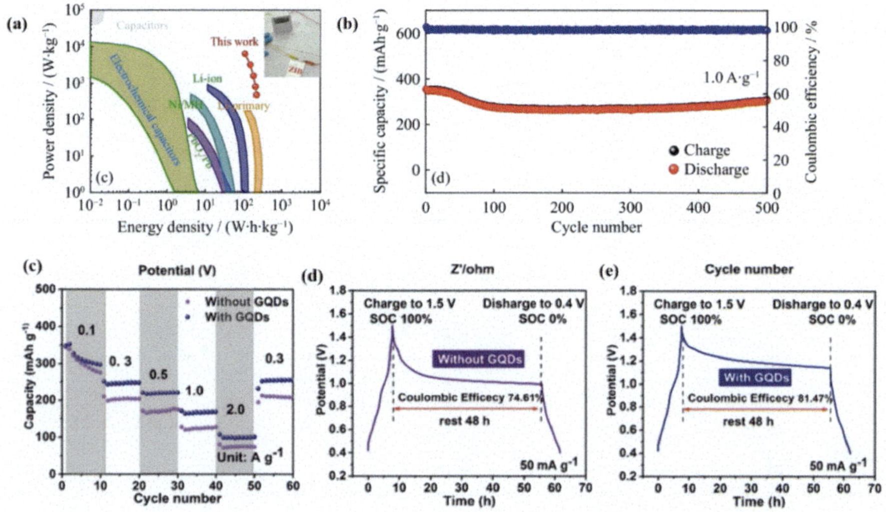

Fig. 12.11 **a** Ragone plot and **b** long-term cycle performance of the aqueous ZIBs based on GQDs@Zn_xMnO_2cathode. Reprinted with permission from @ International Journal of Minerals, Metallurgy and Materials 30, 25 (2023). **c** Rate capabilities of the Zn||NVO at different current densities from 0.1 to 2 A g^{-1}; self-discharge of Zn||NVO: **e** without and **f** with GQDs. Reprinted with permission from @ Nano Energy 92, 106752 (2022)

anode, which exhibits stable cyclability over 1000 cycles. The GQDs are also used in the cathode as a platform to fully use the phenoxazine bioinspired redox-active organic motif's inherent electrochemical activity (Fig. 12.13a–c). An all-GQD battery employs the GQD-derived cathode and anode to reach a high energy density of 290 Wh kg/cathode, showing how to enhance the energy density and reaction reversibility of metal-free, sustainable batteries in a practical manner [37]. Phosphating nano-sulfur (P-NS) particles disperse and anchor on monolayer graphene oxide (MGO) to form a hierarchical sandwich super-structure (P-NS/GQDs/MGO), with GQDs serving as the intermediate phase (Fig. 12.14a–d). To speed heterogeneous nucleation and efficiently avoid P-NS agglomeration and volume change, GQDs can help generate particular bonding conditions and interfacial interactions. This enhances sulfur consumption. After 1000 charge–discharge cycles at 1.68A g^{-1}, the capacity decay rate per cycle is 0.041% [38].

Novel ZnO/NiO anode materials for lithium and sodium-ion batteries have been N-doped, GQD-modified, and have a clearly defined yolk-shell structure. With an average delithiation capacity of 525 mAh/g at 372 mA g^{-1} over 400 cycles, the NG-ZnO/NiO_s anode produced by the solvothermal method demonstrated outstanding cycling performance and a high reversible delithiation capacity of 912 mAh/g at 18.6 mA g^{-1}. In addition, the NG-ZnO/NiO_s electrode demonstrated respectable electrochemical performance in Na-ion batteries, with capacity values up to 235 mAh/g at 18.6 mA g^{-1} (Fig. 12.15a–d). Crucially, the combination route of NG QDs with ZnO/NiO microspheres determined the surface characteristics, morphology, and

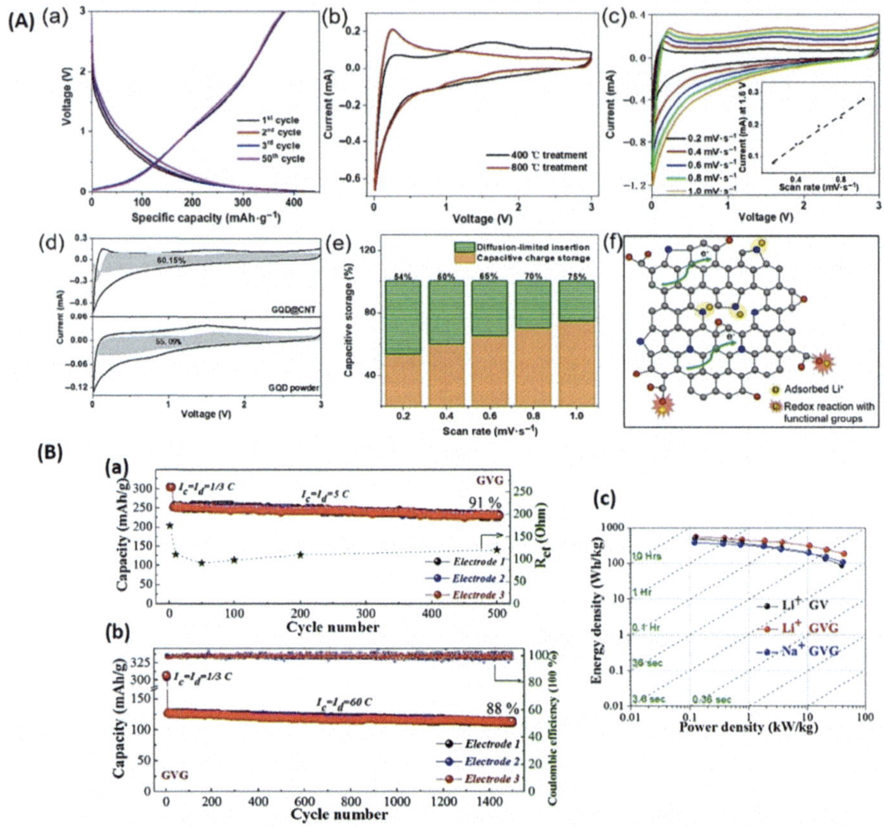

Fig. 12.12 **A** (**a**) Discharge–charge profiles. **b** CV measurements. **c** CV measurements under numerous scan rates. **d** CV profiles for 400 °C annealed GQD@CNT and GQD powder electrodes. **e** Capacity contribution composition of 400 °C annealed GQD@CNTs under different CV scan rates, and **f** Schematic illustration of the interaction between the Li ions and the functionalized GQD. Reprinted with permission from @ Nanoresearch 13: 1044–1052 (2020). **B** (**a**) Cycling performance of three different GVG electrodes at 5 °C for 500 cycles, (**b**) Cycling performance of three different electrodes at 60 C for 1500 cycles, and (**c**) Ragone plot based on the total mass of the whole electrode [36]

electrochemical behavior of the resulting NG-ZnO/NiO hybrids. This suggests that the quality of the heterojunction between composite components has a significant impact on the electrode performance [39].

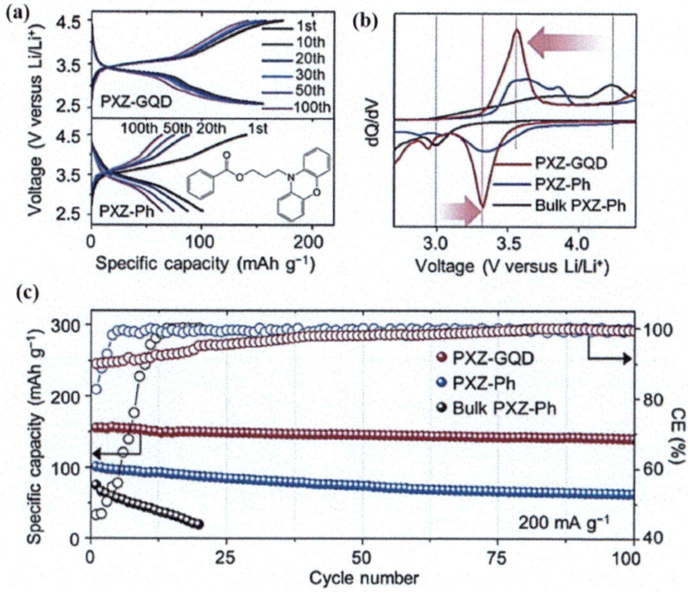

Fig. 12.13 **a** Charge–discharge profiles of PXZ-GQD and PXZ-Ph cathodes. **b** dQ/dV curves for PXZ-GQD, PXZ-Ph, and bulk PXZ-Ph cathodes, and **c** Cycling performance and CE over 100 cycles at a current density of 200 mA g^{-1}. Reprinted with permission from @ Small 19, 2,303,432 (2023)

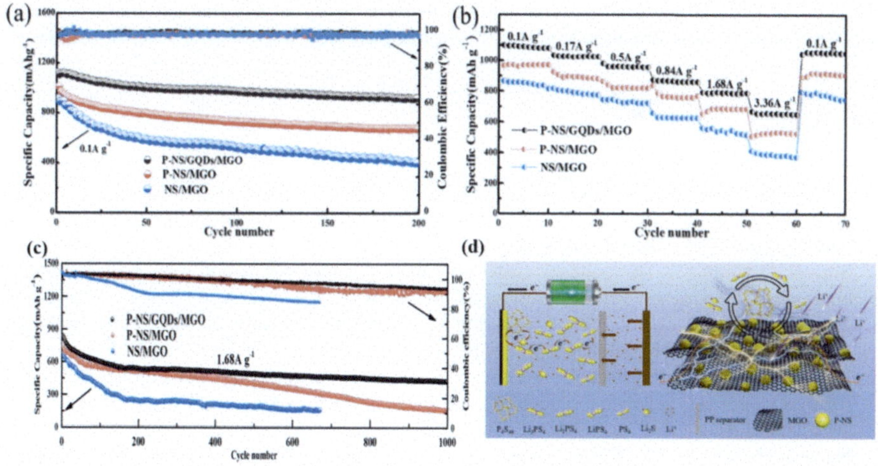

Fig. 12.14 **a** Cycling performance of each electrode under constant current charge and discharge at a low current density. **b** Rate performance. **c** Electrode under constant current charge–discharge long cyclic curve, and **d** Mechanism of P-NS during battery charging and discharging. Reprinted with permission from @ Journal of Power Sources 553, 232,269 (2023)

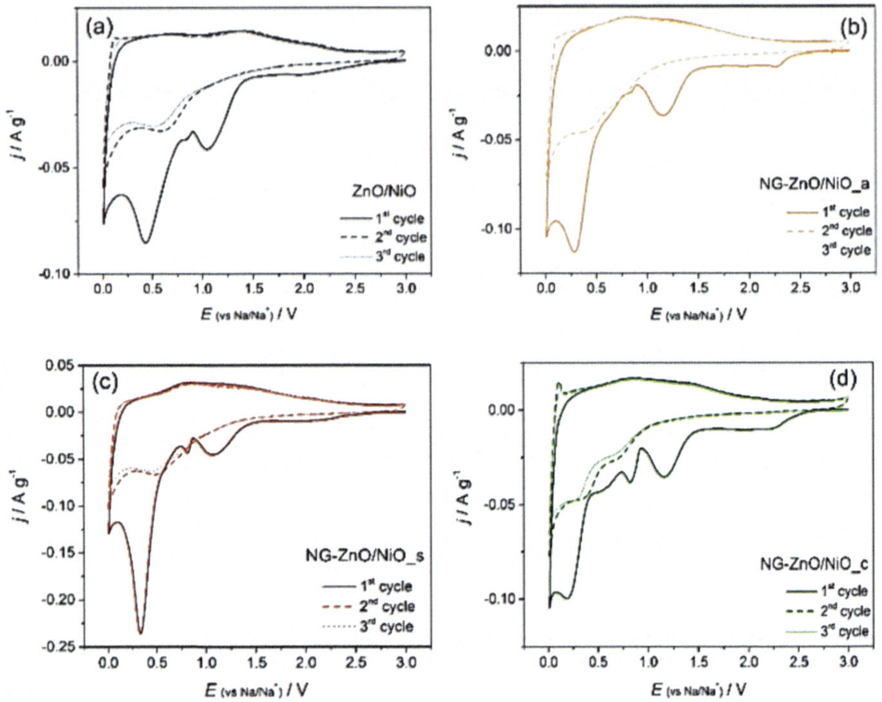

Fig. 12.15 Electrochemical activity of pristine ZnO/NiO material and NG-ZnO/NiO_a, NG-ZnO/NiO_c, NG-ZnO/NiO_s hybrids toward sodium-ions; and (**a–d**) cyclic voltammograms, scan rate 0.1 mV s^{-1}. Reprinted with permission from @ Applied Surface Science 655, 159,702 (2024)

References

1. Qianwen L, Jianhan S, Kun G, Nan C, Xiaotong S, Dan T, Congcong B, Ranran C, Liangti Q (2020) Graphene quantum dots for energy storage and conversion: from fabrication to applications. Mater Chem Front 4:421
2. Sora B, Doyoung K, Hyoyoung L (2016) Graphene quantum dots and their possible energy applications: a review. Curr Appl Phys 16:1192–1201
3. Mohammad D, Mohamad MM, Meysam G (2016) Dye-sensitized solar cells based on nanocomposite of polyaniline/graphene quantum dots. J Mater Sci 51:2964–2971
4. Shuo Z, Lina S, Hongzhou D, Wenbo H, Lifeng D, Liyan Y (2018) High-performance supercapacitor of graphene quantum dots with uniform sizes. ACS Appl Mater Interf 10:12983–12991
5. Nanasaheb MS, Qi XX, Pritamkumar VS, Je MY, Rajaram SM, Kwang HK (2019) Sulphur source-inspired self-grown 3D NixSy nanostructures and their electrochemical supercapacitors. ACS Appl Mater Interfaces 11:4551–4559
6. Xu Q, Niu Y, Li J et al (2022) Recent progress of quantum dots for energy storage applications. Carbon Neutrality 1:13
7. Ghiyasiyan-Arani M, Salavati-Niasari M (2022) Decoration of green synthesized S, N-GQDs and $CoFe_2O_4$ on halloysite nanoclay as natural substrate for electrochemical hydrogen storage application. Sci Rep 12:8103

8. Norazuwana S, Siti KK, Raihana B (2020) Carbon and graphene quantum dots in fuel cell application: an overview. Int J Energy Res 45:1–29
9. Akhilesh Babu G, Ankan Dutta C, Rueyan D (2017) Nano assembly of N-doped graphene quantum dots anchored Fe_3O_4/halloysite nanotubes for high performance supercapacitor. Electrochim Acta 245:912–923
10. Qing C, Yue H, Chuangang H, Huhu C, Zhipan Z, Huibo S, Liangti Q (2014) Graphene quantum dots–three-dimensional graphene composites for high-performance supercapacitors. Phys Chem Chem Phys 16:19307
11. Zhang S, Li Y, Song H et al (2016) Graphene quantum dots as the electrolyte for solid-state supercapacitors. Sci Rep 6:19292
12. Huda SA, Khadijah MSK, Mona SB, Soha Talal AG, Ahmed MW (2023) Fabrication of asymmetric supercapacitors (AC@GQDs//AC) with high electrochemical performance utilizing activated carbon and graphene quantum dots. ACS Omega 43:40808–40816
13. Qiuyu X, Wen Z, Fengzhen J, Xinman C, Yong Z, Faling L, Wei H, Liang F, Said Nasir K, Miao Z (2019) Graphene quantum dot/$Co(OH)_2$ electrode on nanoporous Au–Ag alloy for superior hybrid micro-supercapacitors. J Mater Chem C 7:11441
14. Seshendra RC, Ravindra NB, Jeevan Kumar RM, Murali G, Reddy SA, Jiyeong K, Eunji P, Byungin K, Insik I, Arup Kumer R, Kiruthiga R, Udayabhaskar R (2022) Tin oxide/nitrogen-doped graphene quantum dots composite nanotubes: an efficient electrode for supercapacitors. J Nanomater 2022, Article ID 3167809
15. Keunsik L, Hanleem L, Yonghun S, Yeoheung Y, Doyoung K, Hyoyoung L (2016) Highly transparent and flexible supercapacitors using graphene-graphene quantum dots chelate. Nano Energy 26:746–754
16. Yiwen H, Jingli X, Jin Suk C, Won MC (2020) Graphene quantum dots/$Ni(OH)_2$ nanocomposites on carbon cloth as a binder-free electrode for supercapacitors. J Mater Sci Tech 58:73–79
17. Hemalatha K, Sornalatha M, Suresh Kannan B, Manik Clinton F, Hee-Je K, Karthick SN (2021) Sulfur and nitrogen-doped graphene quantum dots/PANI nanocomposites for supercapacitors. New J Chem 45:4101
18. Tao L, Kun Y, Lina G, Hui C, Ning W, Luhan H, Tingxi L, Hongcai H, Zhanhu G (2017) Graphene quantum dots decorated $SrRuO_3$ mesoporous film as an efficient counter electrode for high-performance dye-sensitized solar cells. J Mater Chem A 5:17848–17855
19. Jung KK, Myung JP, Sang JK, Dong HW, Sung PC, Sukang B, Jong HP, Byung HH (2013) Balancing light absorptivity and carrier conductivity of graphene quantum dots for high-efficiency bulk heterojunction solar cells. ACS Nano 7:7207–7212
20. Kyung DL, Myung JP, Do-Yeon K, Soo MK, Byungjun K, Seongtak K, Hyunho K, Hae-Seok L, Yoonmook K, Sam SY, Byung HH, Donghwan K (2015) Graphene quantum dot layers with energy-down-shift effect on crystalline-silicon solar cells. ACS Appl Mater Interf 7:19043–19049
21. Wentian W, Haixia W, Min Z, Shouwu G (2019) Dual role of graphene quantum dots in active layer of inverted bulk heterojunction organic photovoltaic devices. ACS Omega 4:16159–16165
22. Areerob Y, WC O, Hamontree C et al (2023) A novel of WS2–$MoCuO_3$ supported with graphene quantum dot as counter electrode for dye-sensitized solar cells application. Sci Rep 13:7762
23. Majumder T, Mondal SP (2019) Graphene quantum dots as a green photosensitizer with carbon-doped ZnO nanorods for quantum-dot-sensitized solar cell applications. Bull Mater Sci 42:65
24. Kishore KD, Suazo-Davila D, García-Torres D, Nathan PC, Aruna I, Min-Hung H, Angel AM, Carlos RC, Baixin C, Nick B, Hari MU (2019) Low-temperature titania-graphene quantum dots paste for flexible dye-sensitized solar cell applications. Electrochim Acta 305:278–284
25. Mehraneh F, Shohreh F (2023) Electrophoretic deposition of fluorine-nitrogen graphene quantum dots with highly luminescent upconversion properties for solar cell applications. Ind Eng Chem Res 62:7481–7491
26. Jung-Chuan C, Ruei-Hong S, Po-Hui Y, Po-Yu K, Yu-Hsun N, Chih-Hsien L, Po-Feng C, Yi-Ting W, Shang-Wen Z (2015) Graphene quantum dots as a co-sensitizer with improving light absorption for dye-sensitized solar cells. IEEE Trans Nanotechnol 22:20–27

27. He X, Chengyi H, Hongzhi W, Qinghong Z, Yaogang L (2017) S, N Co-doped graphene quantum dot/TiO_2 composites for efficient photocatalytic hydrogen generation. Nanoscale Res Lett 12:400
28. Kai-An T, Ping-Yen H, Ting-Hsuan L, Chun-Wen T, Hui P, Yan-Gu L, Yung-Jung H (2020) Nitrogen-doped graphene quantum dots for remarkable solar hydrogen production. ACS Appl Energy Mater 3:5322–5332
29. Yung-Shan C, Ping-Yen H, Tso-Fu Mark C, Chun-Yi C, Masato S, Yung-Jung H (2020) Incorporating graphene quantum dots to enhance the photoactivity of CdSe-sensitized TiO_2 nanorods for solar hydrogen production. J Mater Chem A 8:13971–13979
30. Jiayan X, Lizhen L, Lin Z, Hongtian L, Li Y, Fuchi L, Fengzhen L, Wenjie K, Jun L (2020) Enhanced H_2 evolution and the interfacial electron transfer mechanism of titanate nanotube sensitized with CdS quantum dots and graphene quantum dots. Int J Hydrogen Energy 45:6476–6486
31. Sudhagar P, Herraiz-Cardona I, Park H, Song T, Hyun NS, Gimenez S, Mora SI, Fabregat-Santiago F, Bisquert J, Terashima C, Paik U, Soo KY, Fujishima A, Hee HT (2016) Exploring graphene quantum dots/TiO_2 interface in photoelectrochemical reactions: solar to fuel conversion. Electrochim Acta 187:249–255
32. Rashmi K, Ashish B, Satish KV, Sterlin Leo, Hudson M, Shaz MA (2024) Enhancement in hydrogen sorption behavior of MgH_2 catalyzed by graphene quantum dots. Inter J Hydrogen Energy (in Press)
33. Minjie S, Hangtian Z, Cong C, Jintian J, Liping Z, Chao Y (2023) Synergistically coupling of graphene quantum dots with Zn-intercalated MnO_2 cathode for high-performance aqueous Zn-ion batteries. Int J Miner Metall Mater 30:25–32
34. Hao Z, Ruiting G, Shuo L, Cheng L, Haoyang L, Guoqiang Z, Jiugang H, Hongshuai H, Xiaobo J (2022) Graphene quantum dots enable dendrite-free zinc ion battery. Nano Energy 92:106752
35. Xuewei Z, Yizeng W, Yunsong W, Huaisheng W, Yawei Y, Zhipeng W, Linxiu D, Yuanyuan S, Anyuan C (2020) High-performance Li-ion batteries based on graphene quantum dot wrapped carbon nanotube hybrid anodes. Nano Res 2020:1044–1052
36. Chao D, Zhu C, Xia X, Liu J, Zhang X, Wang J et al (2015) Graphene quantum dots coated VO_2 arrays for highly durable electrodes for Li and Na Ion batteries. Nano Lett 15:565–573
37. Youngjin H, Chungryeol K, Donghan S, Il-Doo K, Kisuk K, Youn JJ, Lee D, Seokwoo J (2023) All-graphene quantum dot-derived battery: regulating redox activity through localized subdomains. Small 19:2303432
38. Bin W, Xinyu L, Yong C, Yaping Z, Ling L, Yong C, Ming L, Jianrong X (2023) Boosting lithium–sulfur battery performance using a phosphating nano-sulfur anchored on graphene framework anode. J Power Sourc 553:232269
39. Beata B, Monika WZ, Wojciech L, Andrzej Z, Tomasz K (2024) N-doped graphene quantum dot-decorated MOF-derived yolk-shell ZnO/NiO hybrids to boost lithium and sodium ion battery performance. Appl Surf Sci 655:159702

Chapter 13
Impact of Graphene Quantum Dots as Catalyst

13.1 Introduction

Graphene quantum dots (GQD) are prospective catalysts due to their special qualities like a wide range of uses that could increase chemical process sustainability, selectivity, and catalytic efficiency [1]. Because of these unique qualities, GQDs have many benefits in catalysis. (i) Large Surface Area: The high surface area-to-volume ratio of GQDs makes them an excellent source of active sites for catalytic processes. When compared to bulk graphene or other carbon-based compounds, this greater surface area enables increased catalytic activity [2]. (ii) Tunable Properties: By adjusting the size, shape, and surface chemistry of GQDs, it is simple to modify their electrical structure. By modifying the surface functional groups or doping elements, this tunability allows the catalytic performance to be fine-tuned for particular processes [3]. (iii) Electron Transfer Ability: Because of their distinct quantum confinement effects and edge states, GQDs have exceptional electron transfer properties. By facilitating effective charge transfer during catalytic processes, these characteristics improve selectivity and catalytic activity. (iv) Chemical Stability: GQDs can catalyze a variety of chemical transformations, including those involving harsh environments or corrosive reactants because they are chemically stable under a range of reaction conditions. (v) Versatility: GQDs are adaptable catalysts that can be used in a variety of reactions, such as heterogeneous catalysis, photocatalysis, and electrocatalysis. They have been investigated in several catalytic applications, including chemical synthesis, CO_2 reduction, hydrogen evolution processes, and oxygen reduction reactions [4]. (vi) Environmental friendliness: GQDs are more economical and environmentally benign than conventional catalysts made of noble metals or other non-renewable resources. Sustainable and scalable techniques are frequently used in their synthesis, such as top-down tactics that make use of inexpensive carbon sources or bottom-up procedures that use precursors obtained from biomass. (vii) Effects of Nanoconfinement: Reactant molecules can be contained within the porous structure of GQDs, acting as nanoscale reaction chambers. By aiding molecular adsorption,

N. Manjubaashini et al., *Graphene Quantum Dots*, Materials Horizons: From Nature to Nanomaterials, https://doi.org/10.1007/978-981-97-5722-0_13

encouraging particular reaction routes, and reducing side reactions, this nanoconfinement effect can improve the efficiency and selectivity of catalytic reactions [5].

13.2 Electrocatalysis Applications of GQDs

As a highly active electrocatalyst for OER, a novel oxygen plasma-treated graphene quantum dots embedded Ni–Fe Prussian blue analog (O-GQD-NiFe PBA) is proposed (Fig. 13.1a–d). As-assembled O-GQD-NiFe PBA demonstrates remarkable long-term stability for 100 h in an alkaline solution and outstanding electrocatalytic efficacy toward OER with a low overpotential of 259 mV for reaching a current density of 10 mA cm^{-2} [6].

Potentially functionalized GQDs that can increase bulk MoS_2's hydrogen evolution reactions (HER) activity. The resulting ALQDs are thinner and more active

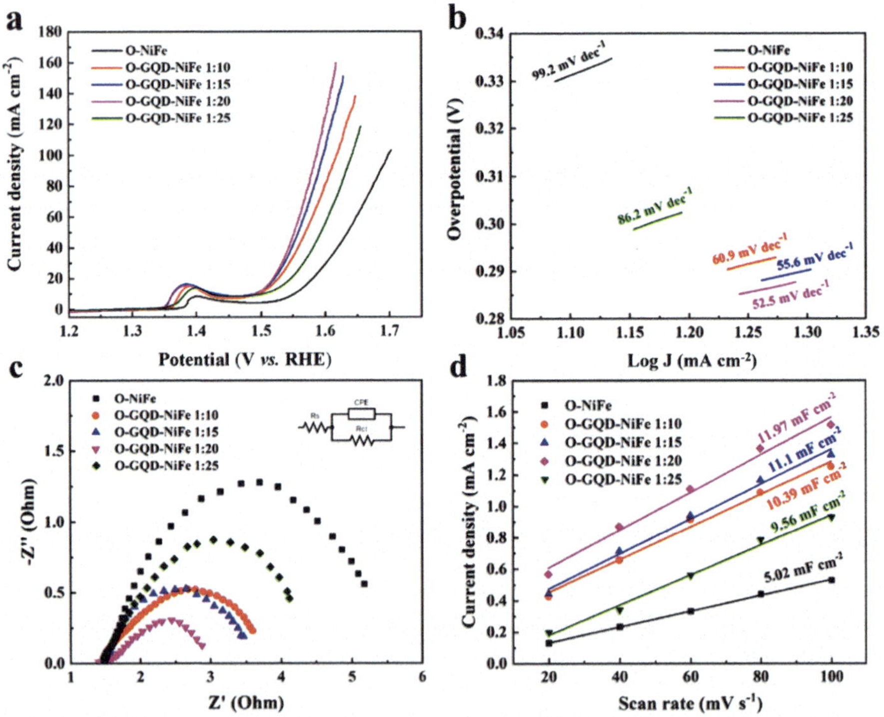

Fig. 13.1 (**a**) LSV curves, (**b**) Tafel slope plots, (**c**) EIS Nyquist plots, and (**d**) current density of O-NiFe PBA, O-GQD-NiFe PBA 1:10, O-GQD-NiFe PBA 1:15, O-GQD-NiFe PBA 1:20, O-GQD-NiFe PBA 1:25. Reprinted with permission from @ Journal of Colloid and Interface Science 648, 193–202 (2023)

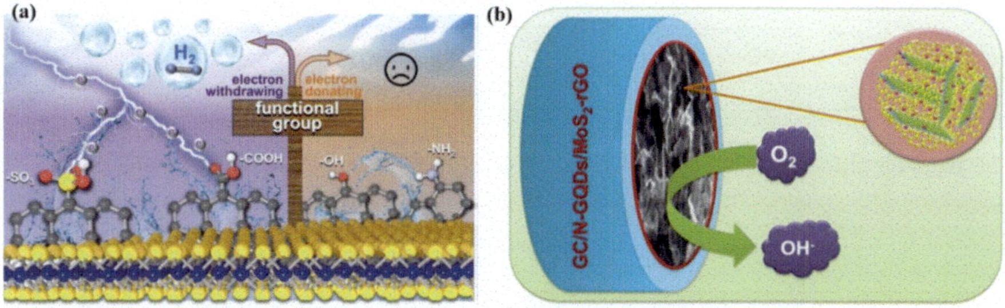

Fig. 13.2 (**a**) Functionalized GQDs facilitate the formation of nanosheet architectures of MoS_2 compared to an electron-donating group (Bingjie H 2023), and (**b**) Representation of ORR on N-GQDs/MoS_2 – rGO nanohybrid-modified electrode [8]

when the concentration and intensity of electron-withdrawing functional groups on GQDs are higher (Fig. 13.2a). Surprisingly, the artificially created near atom-layer ALQD-SO_3 exhibits noticeably better HER performance [7]. Better ORR stability was demonstrated by the N-GQDs/MoS_2 – rGO nanohybrid for up to 3000 cycles, with hardly any variation in the half-wave potential ($E_{1/2}$) (Fig. 13.2b). Most notably, [8] reported that the N-GQDs/MoS_2 – rGO nanohybrid showed an improved capacity to tolerate methanol, even at high methanol concentrations of 3.0 M in an alkaline medium. The creation of a low-cost, highly methanol-tolerant graphene quantum dot-based 3D nanohybrid with no metal may pave the way for the development of innovative cathode electrocatalyst designs for direct methanol fuel cell applications.

Using various approaches, the newly synthesized ZIF-67/GQD@Au was characterized and used as a capable heterogeneous catalyst for the reduction of aromatic nitro compounds. This catalyst facilitates the efficient reactions of structurally distinct alkynes, amines, and aldehydes at 60 °C, resulting in good to very good yields (76–88%) of the required propargylic amine in water in just one day [9]. In both reactions, the catalyst demonstrated outstanding stability and recyclability. It was recycled for six runs with only a little drop in activity. The Heck and decarboxylative cross-coupling processes were then catalyzed by the as-prepared Pd-GQD, which demonstrated more catalytic activity than other reference systems (Fig. 13.3). Further evidence that the GQD is a useful support for creating new heterogeneous catalysts with enhanced catalytic performances came from the extended substrate scope of several substituted aryl iodides [10].

A novel composite catalyst system for effective HER that is doped with highly defective graphene quantum dots (HDGQDs) in 1 T/2H-MoS_2. The increased conductivity, more active sites in 1 T-MoS_2 than in 2H-MoS_2, and extra defects brought about by HDGQDs are the causes of the high electrocatalytic activity, which is indicated by an overpotential of 136.9 mV and a Tafel slope of 57.1 mV/decade. According to all experimental data, the composite catalyst system's greatly increased electrocatalytic activity for HER is the result of each component's distinct advantages working together synergistically [11]. A Composite of NiCo-MOF and N-doped

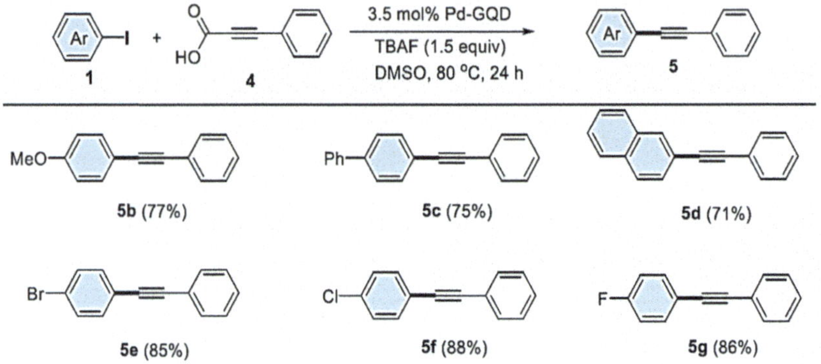

Fig. 13.3 Cross-coupling reactions of substituted aryl iodides with phenyl propionic acid [10]

GQD (N-GQDs) with a hierarchical structure was synthesized via a controlled electrodeposition strategy as an exemplary non-noble metal catalyst for alkaline water splitting. The increased active sites by synergy effect between Ni and Co ions and the enhanced local electrical conductivity through uniform incorporation of N-GQDs into porous NiCo-MOF matrix offer a catalyst that only demanded overpotentials of 218 and 359 mV to deliver a current density of 100 mA·cm^{-2} toward the HER and OER, respectively, and also exhibited drastically enhanced electrochemical stability for over 150 h in alkaline media (Fig. 13.4). In particular, the assembled two-electrode setup for overall water electrolysis yields a current density of 10 mA·cm^{-2} at a low cell voltage of 1.62 V [12].

Here, a prepared vanadium-based electrolyte is mixed with a given volume of GQD, approximately 300 μL, to act as a molecular catalyst. The electrochemical activity of the VO^{2+}/VO^{2+} redox reaction was shown to vary significantly on the GQD integrated electrolyte's CV curve. Comparing the resistive behavior of GQD with the bare electrolyte, the results of the electrochemical impedance spectroscopy

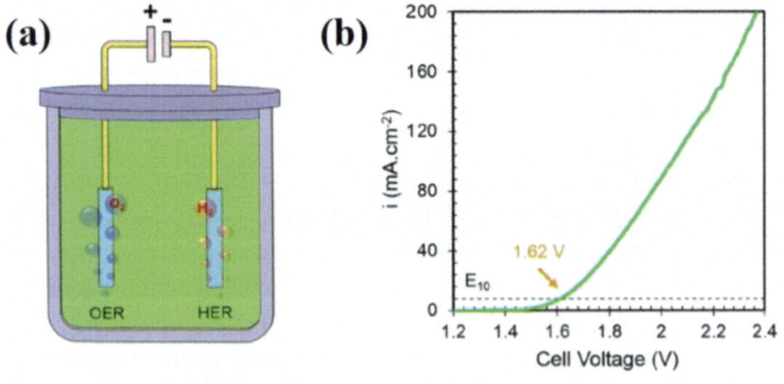

Fig. 13.4 Representation of the two-electrode system, and (**b**) LSV curve at 5 mV·s^{-1}. Reprinted with permission from @ ACS Appl Nano Mater 7:2086 – 2099 (2024)

(EIS) investigation revealed much higher values. The observed kinetics metrics, such as exchange current density, rate constant, and others, demonstrate a significant improvement in the VO^{2+} to VO^{2+} redox reaction kinetics following the addition of GQD to the active electrolyte. The exchange current density value of $i_0 = 6.175 \times 10^{-4}$ A cm^{-2} for the GQD added electrolyte exhibited good improvisation, and this value is two times greater than that of the bare electrolyte ($i_0 = 2.365 \times 10^{-4}$ A cm^{-2}) [13].

13.3 Photocatalysis Applications of GQDs

When TiO_2 was co-doped with GQD-0.1Fe, the photocatalytic decolorization efficacy of RB5 dye was significantly increased. With a progressive drop in Kapp values with an increase in RB5 concentration, the photocatalytic reaction proceeded as the pseudo-first-order reaction did. The results showed that the heterogeneous reaction system followed the Langmuir–Hinshelwood isotherm and concurrently occurred adsorption and photocatalytic processes. The KC value was determined to be 2.45 mg L^{-1} min^{-1}, while the KLH value was 0.45 L mg^{-1}. Studies on the photocatalytic reaction mechanism revealed that the primary active species in the GQD-0.1Fe-TiO_2-300 that caused the decolorization of RB5 were holes and OH radicals (Fig. 13.5 a). The suggested reaction route demonstrated the significant roles that Fe-TiO_2 and GQDs play in the production of electrons and holes. GQD-0.1Fe-TiO_2-300 was also robust for up to four cycles [14]. The results of the phytotoxicity assay showed that the germination of Lycopersicon esculentum seeds was unaffected by treated water or the best photocatalysts. When converting N_2 to NH_3, the X% (Co_3O_4)/GQDs composites are used as reliable photocatalysts (Fig. 13.5 b). These composites are used to study the kinetics and the concentration of the ammonia generated [15].

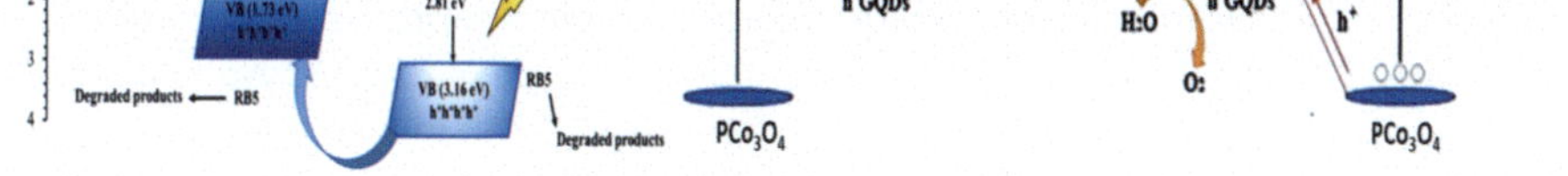

Fig. 13.5 (a) Charge transfer mechanism for GQD-Fe-TiO_2 photocatalysts for the degradation of RB5 dye. Reprinted with permission from @ Ecotoxicology and environmental safety @ 226, 112,855 (2021), and (b) Mechanism of NH_3 in the presence of X% (Co_3O_4)/GQDs. Reprinted with permission from @ Journal of Molecular Liquids 391, 123223 (2023)

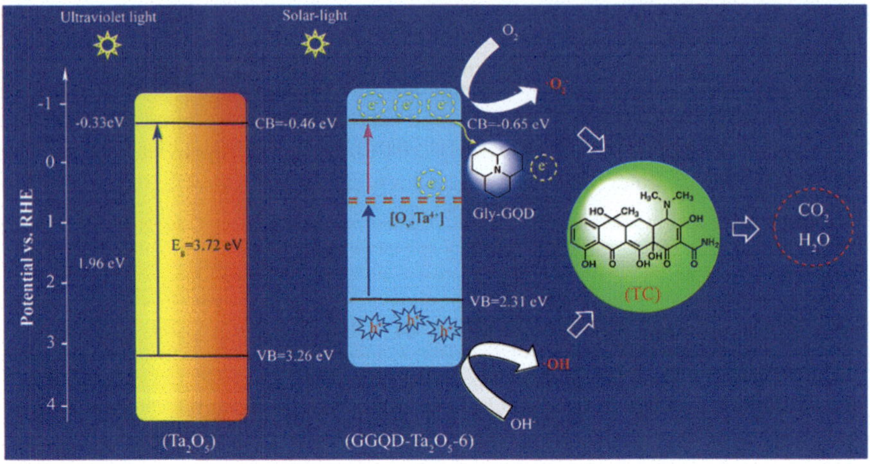

Fig. 13.6 Energy band structures of GGQD-$Ta_2O_5{}^{-6}$ and Ta_2O_5 and transfer mechanism of photogenerated carriers in photocatalytic process with GGQD-$Ta_2O_5{}^{-6}$ under solar light irradiation. Reprinted with permission from @ Journal of Alloys and Compounds 968, 171757 (2023)

Tetracycline degradation is facilitated by the introduction of glycine-functionalized GWD (GGQD), which exhibits exceptional photocatalytic performance as a Ta_2O_5 photocatalyst. Ta^{4+} self-doped oxygen vacancies optimize the band position, produce new defect energy levels, and reduce the band gap of Ta_2O_5. These enhance the ability to absorb visible light and encourage the transfer of charge created by photosynthesis. The recombination of photogenerated charge carriers is inhibited by graphene acting as an electron trap. By acting as a light sensitizer, graphene increases absorption of visible light (Fig. 13.6). Ta_2O_5 exhibits high photocatalytic activity for solar light-driven tetracycline photodegradation due to the synergy of these various structure engineering [16]. Compared to Ta_2O_5, the photodegradation efficiency is 6.6 times greater, the apparent kinetic constant is 19.3 times higher, and the photocurrent density is 13.4 times higher after 60 min of solar light irradiation.

To act as catalysts in the photodegradation of methylene blue (MB) under visible light illumination, plasmonic Au nanoparticles (NPs) are placed on top of the hybrid nanosystem consisting of graphitic carbon nitride (GCN) and GQD [17]. Plasmonic Au serves as antennas at the interface with GQD-GCN, accumulating and enhancing the localized electromagnetic field. This encourages the photogeneration of many carriers on GQD-GCN. Charge transfer migration from hybrid to Au NPs separates the carriers. Ultimately, in visible light, the carriers on the plasmonic Au nanostructures start the destruction of MB. Methyl Orange and Methylene Blue dyes were degraded by visible light irradiation using the photocatalytic activity of manufactured GQDs at pH 10 [18]. It was shown how some parameters, including the kinetic model, contact time, GQD dosage, and pH of the GQD, have an impact. The tenable process by which the harmful dyes under visible light illumination degrade is based on GQDs (Fig. 13.7).

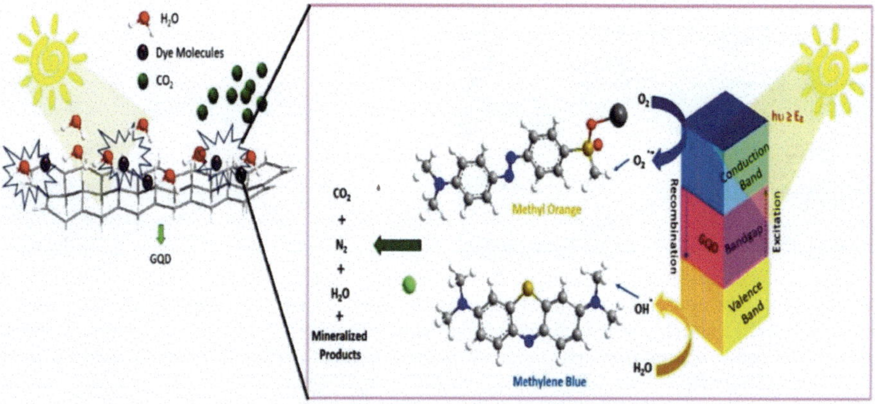

Fig. 13.7 Mechanism of photocatalysis on the surface of the GQDs [18]

First-ever fluorescent ink and novel graphene quantum dot (S, N-GQD) based photocatalysts that are co-doped with sulfur and nitrogen are reported (Fig. 13.8a). As green nanomaterials, the generated S and N-GQDs' catalytic behavior were examined, and organic dyes were employed as standard dyes. The photocatalytic mechanism was investigated using a variety of radical scavenging agents [19]. Consequently, the new photocatalyst presents a viable substitute for the degradation of organic contaminants. Under visible light irradiation, the degradation of MB was used to assess the photocatalytic activity of GQDs/ZnPor. It was possible to increase the photocatalytic activity of GQDs/ZnPor in comparison to pure zinc porphyrin (Fig. 13.8b). Additionally, photoluminescence measurements and tests involving free radicals and hole scavenging demonstrated the photocatalytic degradation mechanism of GQDs/ZnPor [20].

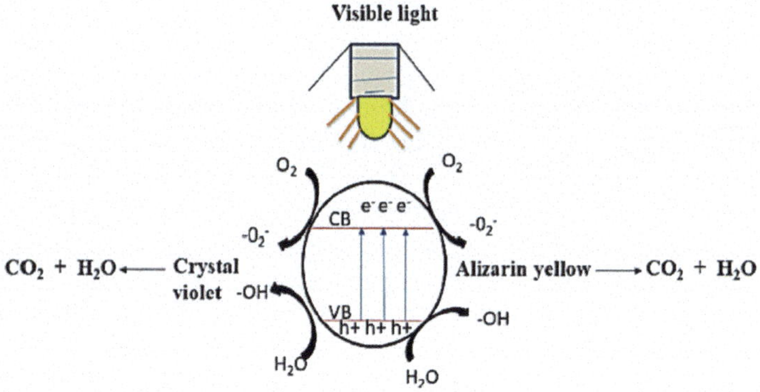

Fig. 13.8 Photocatalytic pathway of S, N-GQDs under visible light irradiation [19]

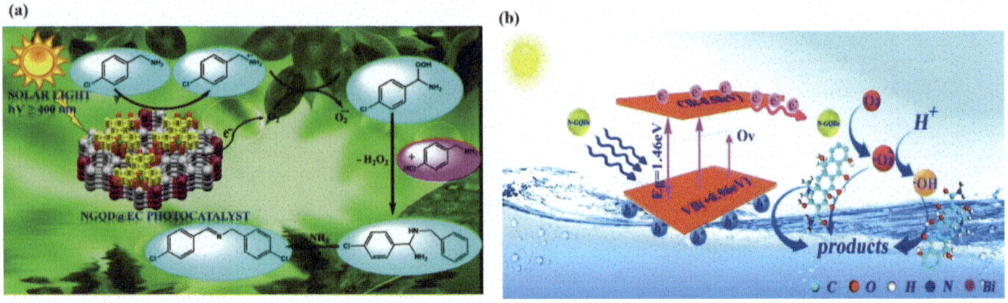

Fig. 13.9 (**a**) Represents the coupling of benzylamine in the presence of oxygen and N-GQDS@EC photocatalyst [21], and (**b**) photocatalytic mechanism over the GBO-0.4 composite toward TC degradation under full-spectrum light irradiation. Reprinted with permission from @ Applied Catalysis B 277, 119218 (2020)

Utilizing graphene-based materials that respond to sunlight to activate C–N bonds and promote coenzyme renewal. Utilizing a highly efficient and stereospecific response to solar light, the N-GQDs@EC composite photocatalyst system converts imine 98.5 percent and regenerates NADH 55 percent more than N-GQDs alone (Fig. 13.9a). Under solar illumination, the current study work demonstrates the advancements in the application of N-GQDs@EC composite photocatalyst for stereospecific NADH regeneration and imine conversion [21]. To create 0D/2D nanoscale heterostructure photocatalysts with enhanced full-spectrum photoactivity for environmental applications, nitrogen-doped GQD (N-GQDs) modified BiO_{2-x} nanosheets were successfully fabricated and demonstrated superior performance in light-harvesting, electron–hole pair separation, and full-spectrum driven molecular oxygen activation (Fig. 13.9b). The increased photocatalytic performance was attributed to the combined impacts of the produced nanojunction and the electron collection effect and upconversion photoluminescence capabilities of the N-GQDs. By creating a bulk-to-surface conduit between BiO_{2-x} and N-GQDs, effective molecular oxygen activation was accomplished [22].

The hybrid catalysts comprise hollow TiO_2/poly (diallyl-dimethylammonium chloride)/GQD. For directed photo-charge migration, the composite catalysts' multilayered nanostructure guarantees a synchronous charge transfer chain [23]. In addition, the hybrids were coated with GQDs varying in hydroxyl content to adjust the catalyst-reactant interfacial interaction (Fig. 13.10a-c). Group-specific absorption and activation dominated the interfacial photo-charge transport dynamics between the catalysts and reactants, according to transient-state surface photovoltage studies. The utilization of Ni2P nanoparticles in conjunction with OH-functionalized GQDs (OH-GQDs) photosensitizer for photocatalytic H_2 generation at wavelengths greater than 420 nm. Compared to naked OH-GQDs, the rate of H_2 production is approximately 94 times higher, and it is even on par with OH-GQDs with 1.0 wt% Pt cocatalyst (Fig. 13.10d). This improvement is most likely the result of the contact between Ni_2P and OH-GQDs at the semiconductor-cocatalyst interface, which facilitates an effective charge transfer mechanism [24].

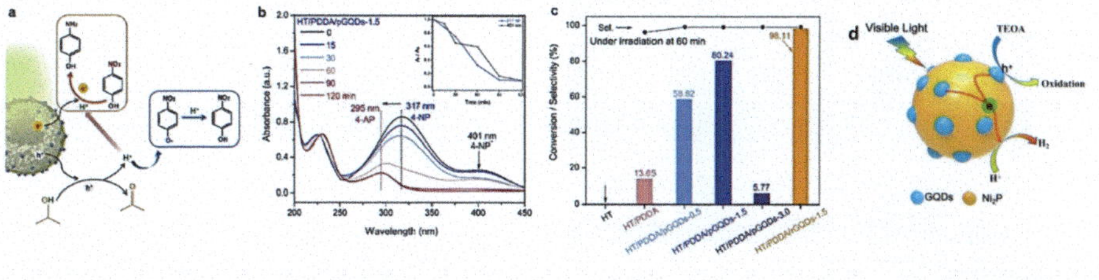

Fig. 13.10 (**a**) Formation of hydrogenation, (**b**) Absorbance of photocatalysis of 4-NP by HT/PDDA/pGQD-1.5, (**c**) Photocatalytic hydrogenation performance of HT, HT/PDDA, HT/PDDA/pGQD-x ($x = 0.5, 1.5,$ and 3.0), and HT/PDDA/rGQD-1.5 NPs under irradiation at 60 min. Reprinted with permission from @ ACS Catalysis 13, 10824–10834 (2023), and (**d**) Mechanisms for the photocatalytic H_2 evolution over Ni_2P/OH-GQDs under visible light irradiation Reprinted with permission from @ Chinese Journal of Catalysis 39, 1753–1761 (2018)

The synthesized GQDs exhibit extensive absorption in the visible spectrum between 400 and 700 nm. They may also be used as smart photosensitizers under visible light irradiation (≥420 nm) with substantial singlet oxygen ($^1O_2$) production (Fig. 13.11a). The produced GQDs have the potential to be employed as a metal-free, visible light-active, sensitized photocatalyst through an energy transfer mechanism. This mechanism involves the GQDs converting light energy into $^1O_2$, which can degrade organic dyes and kill adjacent microbes [25]. For the first time, functionalized GQDs (fGQDs) are investigated as visible light-collecting photocatalysts for the manufacture of solar fuel from CO_2. After being integrated with the biocatalyst, the highly effective light-harvesting fGQDs photocatalyst converts CO2 to formic acid in a very selective manner (Fig. 13.11b). The findings provide a model for the use of GQD-based materials as photocatalysts in the selective synthesis of solar fuels from CO_2, and they also create opportunities for the more environmentally friendly synthesis of additional high-value compounds [26].

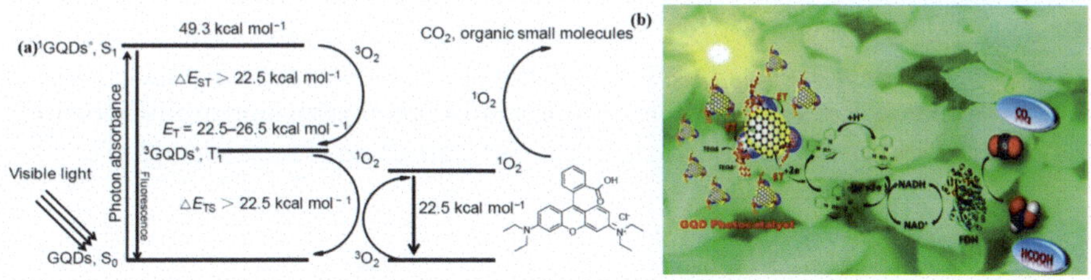

Fig. 13.11 (**a**) Photosensitized degradation of dyes over GQDs under visible light irradiation. Reprinted with permission from @ Sci China Mater 59, 12–19 (2016), and (**b**) fGQDs photocatalyst-biocatalyst integrated system for exclusive formic acid production from CO_2. Reprinted with permission from @ ChemCatChem 8, 3389 (2016)

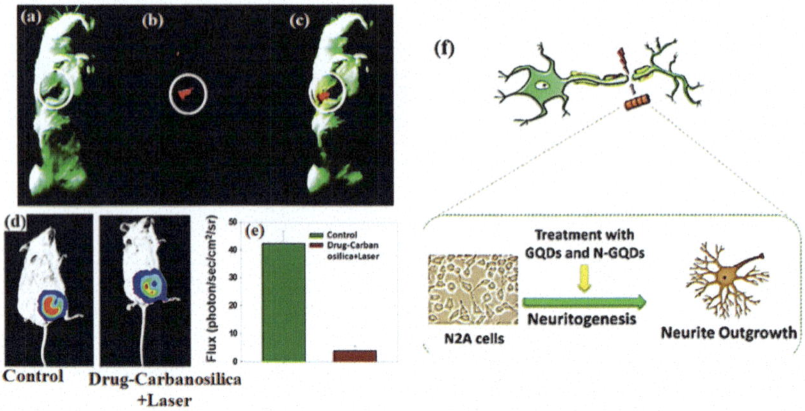

Fig. 13.12 Cu (II)/GQDs/$NiFe_2O_4$ nanocatalyst in the synthesis of propargyl amines. Reprinted with permission from @ Catalysis Lett 151, 1444–1455 (2021)

13.4 GQDs as a Catalyst for Various Synthesis

Using GQDs to modify $NiFe_2O_4$ NPs, Cu (II) nanoparticles are stabilized as part of a new magnetically retrievable catalytic system (Cu (II)/GQDs/NiFe2O4) for the solvent-free, green production of propargyl amines (Fig. 13.12). Propargyl amines were synthesized by reacting several aldehydes/ketones, amines, and phenylacetylene at 100 °C with the catalyst present [27]. With the aid of an external magnet, the produced catalyst can be separated and recovered for five courses without experiencing a discernible decrease in efficiency.

A generic procedure for the production of single-atom catalysts that can achieve high loadings of transition metal atoms up to 40 weight percent or 3.8 percent, which is many times better than standards found in the literature. The utilization of graphene quantum dots, which were subsequently woven into a carbon matrix, served as a support by offering a multitude of anchoring points. This enabled the production of high densities of transition metal atoms with enough distance between them to prevent aggregation. On a Ni single-atom catalyst with increased Ni loading, a notable rise in activity in electrochemical CO_2 reduction (Fig. 13.13) [28]

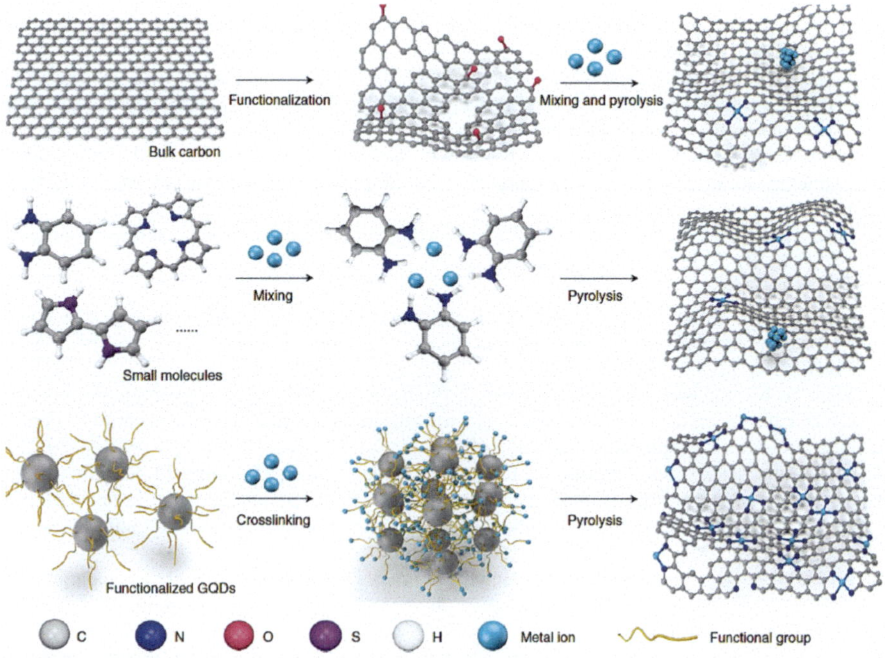

Fig. 13.13 Top-down and Bottom-up approach for the synthesis of single-atom catalysts GQDs. Reprinted with permission from @ Nat Chem 13, 887–894 (2021)

References

1. Zheng D, Shu-Ling S, Zhi-Hong T, Jun-He Y (2021) Graphene quantum dots-based heterogeneous catalysts. New Carbon Mater 36:449–467
2. Hari Prakash S, Mohana S (2023) A comprehensive review on recent developments in the graphene quantum dot framework for organic transformations. J Organometallic Chem 997:122790
3. Wenwen L, Matthew L, Gaopeng J, Gaoran L, Jianbing Z, Meiling X, Yanfei Z, Rui G, Aiping Y, Ming F, Zhongwei C (2020) Graphene quantum dots-based advanced electrode materials: design, synthesis and their applications in electrochemical energy storage and electrocatalysis. Adv Energy Mater 10:2001275
4. Qin X, Ola O, Zhao J, Yang Z, Tiwari SK, Wang N, Zhu Y (2022) Recent progress in graphene-based electrocatalysts for hydrogen evolution reaction. Nanomaterials 12:1806
5. Chunyu C, Qinghua L, Ming Y, Zhifeng L, Qingyun H, Ting W, Songhao L, Yuan P, Chenhui Z, Yang L (2022) Advances in preparation, mechanism and applications of graphene quantum dots/semiconductor composite photocatalysts: A review. J Hazardous Materials 424:127721
6. Yin-Chen L, Sofiannisa A, Min-Hsin Y, Li-Yin H, Angelina Melanita T, Kuo-Chuan H (2023) Graphene quantum dots induced defect-rich NiFe Prussian blue analogue as an efficient electrocatalyst for oxygen evolution reaction. J Colloid and Interface Sci 648:193–202
7. Bingjie H, Kai H, Bijun T, Zhendong L, Zeming W, Huazhang G, Cheng L, Zheng L, Liang W (2023) Graphene quantum Dot-Mediated Atom-Layer semiconductor electrocatalyst for hydrogen evolution. Nano-Micro Lett 15:217
8. Ramalingam V, Indrajit MP, Alagarsamy P, Bhalchandra AK, Nay MH, Dionysiou DD, Bernaurdshaw N (2016) Synergistically enhanced electrocatalytic performance of an N-Doped

graphene Quantum Dot-Decorated 3D MoS_2−Graphene nanohybrid for oxygen reduction reaction. ACS Omega 1:971–980
9. Zhwan N, Mohammad G, José MS, Mahboobe E (2024) Graphene quantum dots incorporated ZIF-67 for stabilization of Au nanoparticles: Efficient catalyst for A3- coupling and nitroarenes reduction reactions. Appl Organomet Chem 2024:e7400
10. Jisoo K, Jeongah L, Ji Dang K, Myong YC, Sunwoo L, Hyun CC (2022) Preparation, characterization, and catalytic properties of Pd-Graphene quantum dot catalysts. Catalysts 12:619
11. Sheng-Fu C, Tai-Sing W, Yun-Liang S (2023) Highly defective graphene quantum dots-doped 1T/2H-MoS_2 as an efficient composite catalyst for the hydrogen evolution reaction. Sci Rep 13:15184
12. Hamed SG, Milad R (2024) N-Doped graphene quantum dots incorporated into NiCo Bimetallic-Organic framework as an electrocatalyst for Alkaline water splitting. ACS Appl Nano Mater 7:2086–2099
13. Mani U, Karuppusamy M, Subramanian S, Pitchai R (2020) Graphene quantum dot beyond electrocatalyst: an in situ electrolyte catalyst towards improved reaction kinetics of VO^{2+}/VO^{2+} redox couples. J Electrochem Soc 167:140540
14. Muhammad SK, Nadia R, Ahson JS, Jehanzeb AS, Jamshaid H, Muhammad I, Saifullah Awan M, Asad S, Jean K, Muhammad A, Muhammad B (2021) Graphene quantum dot and iron co-doped TiO_2 photocatalysts: Synthesis, performance evaluation and phytotoxicity studies. Ecotoxicol Environ Saf 226:112855
15. Mokhtari A, Khatamian M (2023) Tuning the photocatalytic activity of graphene quantum dots via decorating the X% (Co_3O_4) as modern photocatalysts to produce ammonia. J Molecular Liquids 391:123223
16. Li R, Wen T, Sun X, Shen Y, Li Z, Yang Y (2023) Synthesis of Ta_2O_5 by introducing graphene quantum dot with excellent photocatalytic performance for degradation of tetracycline. J Alloys and Compounds 968:171757
17. Gone R, Biswajit C, Giri PK (2017) In situ decoration of plasmonic Au nanoparticles on graphene quantum dots-graphitic carbon nitride hybrid and evaluation of its visible light photocatalytic performance. Nanotechnology 28:395703
18. Peetam M, Kandarpa KN, Mitali S (2021) Efficient blue luminescent graphene quantum dots and their photocatalytic ability under visible light. Biointerface Research in Appl Chem 11:8171–8178
19. Tamizharasan S, Muralidharan R, Abirami N, Leelavathi H, Kumarasamy A, Arulmozhi R (2022) Highly efficient sulfur and nitrogen codoped graphene quantum dots as a Metal-Free green photocatalyst for photocatalysis and fluorescent ink applications. ACS Omega 2022:12825–12834
20. Qian L, Yuanjian Z, Songqin L (2015) Graphene quantum dots enhanced photocatalytic activity of zinc porphyrin toward the degradation of methylene blue under visible-light irradiation. J Mater Chem A 3:8552
21. Ruchi S, Rajesh KY, Ravindra KS, Satyam S, Atul PS, Dilip KD, Ahmad U, Navneet KG (2023) Highly selective Nitrogen-Doped Graphene Quantum Dots/Eriochrome Cyanine composite photocatalyst for NADH regeneration and coupling of benzylamine in aerobic condition under solar light. Catalysts 13:199
22. Fei C, Lian-Lian L, Ying-Jie Z, Jing-Hang W, Gui-Xiang H, Qi Y, Jie-Jie C, Han-Qing Y (2020) Enhanced full solar spectrum photocatalysis by nitrogen-doped graphene quantum dots decorated BiO_{2-x} nanosheets: Ultrafast charge transfer and molecular oxygen activation. Applied Catalysis B 277:119218
23. Linjia L, Rui Z, Pan H, Yanhong L, Dejun W, Xie T (2023) Boosting hydrogenation of graphene quantum Dot-Modified photocatalysts: specific functionalized modulation at active sites. ACS Catal 13:10824–10834
24. Liang Z, Qiudi Y, Daochuan J, Huanlin C, Rana MI, Pingwu D (2018) Metal-free graphene quantum dots photosensitizer coupled with nickel phosphide cocatalyst for enhanced photocatalytic hydrogen production in water under visible light. Chinese J Catalysis 39:1753–1761

25. Jiechao G, Minhuan L, Weimin L, Qingyan J, Liang G, Bingjiang Z, Xiangmin M, Guangle N, Pengfei W (2016) Graphene quantum dots as efficient, metal-free, visible-light-active photocatalysts. Sci China Mater 59:12–19
26. Dolly Y, Rajesh KY, Abhishek K, No-Joong P, Baeg J-O (2016) Functionalized graphene quantum dots as efficient visible light photocatalysts for selective solar fuel production from CO_2. ChemCatChem 8:3389
27. Soudabeh M, Farid M, Fatemeh SMS (2021) Investigation into the catalytic performance of Cu (II) supported graphene quantum dots modified $NiFe_2O_4$ as a proficient Nano-Catalyst in the synthesis of propargylasmines. Catalysis Lett 151:1444–1455
28. Chuan X, Yunrui Q, Yang X, Peng Z, Graham K, Xiao Z, Zhenyu W, Jung Yoon K, David AC, Dongxing Z, Peng L, Mohsen S, Emilio H, Peixin C, Husam NA, Yongfeng H, Haotian W (2021) General synthesis of single-atom catalysts with high metal loading using graphene quantum dots. Nat Chem 13:887–894

Chapter 14
Industrial Application of Graphene Quantum Dots

14.1 Introduction

Graphene quantum dots, or GQDs, are widely used in many different industries. GQDs' special optical qualities are employed to improve imaging methods in bioimaging. GQDs are used in photothermal therapy and targeted medication delivery in cancer therapeutics. Because of their size-dependent fluorescence, which varies with temperature, GQDs are also used in temperature sensing. Furthermore, GQDs are being investigated for light-converting LEDs and drug delivery systems, allowing for lighter and more energy-efficient lighting options.

Furthermore, GQDs may find use in other industries including environmental monitoring and energy storage. Quantum dots of graphene are showing great promise as a material with a broad range of uses across multiple industries. There are two primary kinds of graphene quantum dots on the market: liquid and powder. Since liquid graphene quantum dots are usually distributed in a liquid solution, solution-based technologies like optoelectronics and electronics can easily employ them. Conversely, since powdered graphene quantum dots are solid particles, they can be used in energy storage devices, solar cells, and composite materials. Graphene quantum dots are available in liquid and powder forms, each of which has special benefits that are propelling the market's expansion.

GQDs are graphene particles at the nanoscale that have special chemical, optical, and electrical characteristics. They are widely sought after for a variety of applications in industries like electronics, energy, medicine, and optoelectronics because of their great thermal stability, high electrical conductivity, and high chemical reactivity. Though it is still in its early stages, the GQD market has a lot of room to develop. To fully realize the promise of GQDs, the market is seeing significant investments made in R&D. The market is anticipated to develop because of the rising demand for sophisticated electronic and energy storage devices as well as the expanding use of renewable energy sources. The demand for GQDs is also being

N. Manjubaashini et al., *Graphene Quantum Dots*, Materials Horizons: From Nature to Nanomaterials, https://doi.org/10.1007/978-981-97-5722-0_14

driven by the increased emphasis on environmental sustainability and the requirement for energy-efficient solutions. Compared to traditional materials, these dots offer improved performance, durability, and efficiency and are utilized in solar cells, batteries, sensors, and displays.

Further driving the market expansion is expected to come from their use in drug delivery systems, illness detection, and biomedical imaging. Asia Pacific is anticipated to lead the global market for graphene quantum dots due to the existence of significant electronics and energy storage industries in nations like South Korea, Japan, and China. Because more emphasis is being placed on adopting sophisticated technologies, growth in North America and Europe is also anticipated to be substantial. In general, the market for graphene quantum dots appears to have a bright future. On the other hand, issues with large-scale manufacturing, high production costs, and strict restrictions can limit market expansion somewhat. Notwithstanding, the market is anticipated to expand at a compound annual growth rate (CAGR) of % throughout the projected period due to further research and technological improvements.

14.2 Market Strategy of GQDs (2023–2028)

Global Market Revenue for Graphene Quantum Dots, 2017–2022, 2023–2028, in millions of dollars. Worldwide Sales of Graphene Quantum Dots, 2017–2022, 2023–2028, (Ton). Top five firms in the world for graphene quantum dots in 2021 (%). With a compound annual growth rate (CAGR) of % between 2022 and 2028, the global market for graphene quantum dots, which was estimated at million in 2021, is expected to reach US$ million by that time. Merck KGaA, NANOCHEMAZONE, Sisco Research Laboratories Pvt. Ltd., American Elements, ACS Material, LLC, Graphene Square Inc., Strem Chemicals, Inc., Dotz Nano Limited, and Green Science Alliance Co., Ltd. are some of the major global manufacturers of GQD. In terms of revenue, the top five players worldwide own a share of roughly % in 2021.

To gather information on sales, revenue, demand, price changes, product types, current developments, and plans, industry trends, drivers, challenges, obstacles, and potential risks, Market Monitor Global, INC. (MMG) surveyed GQD manufacturers, suppliers, distributors, and industry experts. The global graphene quantum dot market by type, 2017–2022, 2023–2028 ($ Millions) & (Ton), as well as the segment percentages for the global graphene quantum dot market by type, 2021 (%), make up the total market.

Dotz Nano Ltd. is among the businesses producing GQDs for biomedical applications utilizing coal as a precursor. Global sales of semiconductors hit USD 595 billion in 2021, and it is anticipated that the semiconductor market will expand once more to USD 676 billion in 2022. In the next years, memory semiconductors and 5G chips will drive the semiconductor market's expansion. Some of the major players in the global market that are driving the growth of the graphene quantum dots market are Quantag Nanotechnologies, Dotz Nano Ltd., Green Science Alliance Co. Ltd., Zen Graphene Solutions Ltd., Sisco Research Laboratories Pvt. Ltd., Jiangsu Xfnano Raw

Materials Tech Co., Ltd., Graphene Square Inc., Nanochemazone, Merck KGaA, and Strem Chemicals, Inc.

GQDs are novel carbon-based fragments that are used in the biomedical industry to help create composites that are stronger and more biocompatible than polymers. The excellent mechanical and electrical properties of carbon in its various forms and configurations are advantageous to graphene quantum dots (GQDs), which are nanoscale, crystalline graphene fragments with a size of 2–20 nm. The primary component of GQDs is sp2 hybridized carbon, and its zigzagged edges, surface imperfections, and quantum confinement make them luminous. Because of their greater surface areas, non-toxicity, improved cross-linking with the polymer matrix, and fluorescent qualities, GQD-polymer-based composites have recently emerged as viable alternatives for biomaterial and biosensor applications.

GQDs are also used in bioimaging, orthopedic applications for replacing bone tissue, and optoelectronics. GQDs have been synthesized using both top-down and bottom-up techniques. Less complex top-down techniques entail employing various techniques to reduce bigger chunks of graphite, charcoal, coal, anthracite, carbon fibers, etc., into nano-sized particles. Small polycyclic hydrocarbons or fullerenes are typically used in more intricate bottom-up methods to create planar GQD structures through chemical synthesis.

14.3 Various Industrial Applications of GQDs

Optoelectronics: GQDs have remarkable optical characteristics, such as adjustable absorption spectra and potent photoluminescence. Because of these characteristics, they can be used in optoelectronic devices such as solar cells, photodetectors, displays, and light-emitting diodes (LEDs) [7].

Biomedical Imaging: Because of their outstanding biocompatibility, little cytotoxicity, and adjustable photoluminescence, graphene quantum dots (GQDs) have garnered a lot of attention for biomedical imaging applications. They can be employed as contrast agents in bioimaging methods such as drug delivery systems, fluorescence imaging, and bioimaging.

Sensors: Due to their large surface area, superior electrical conductivity, and sensitivity to a wide range of analytes, graphene quantum dots (GQDs) have demonstrated considerable promise in sensing applications. They are useful for environmental monitoring, medical diagnostics, food safety, and the detection of gases, heavy metals, biomolecules, and other chemical species [8].

Energy Storage: To enhance the performance of energy storage systems like lithium-ion batteries and supercapacitors, GQDs can be added to the electrodes and electrolytes. The development of next-generation energy storage systems is advanced by their high surface area and electrical conductivity, which improve charge storage capacity, cycling stability, and rate capability [9].

Catalysis: Because of its many edge sites, large surface area, and adjustable electronic structure, graphene quantum dots (GQDs) have shown great promise as catalysts for a range of chemical reactions. Their ability to catalyze significant reactions like the hydrogen evolution reaction (HER), oxygen reduction reaction (ORR), and photocatalytic degradation of pollutants makes them potentially useful for industrial processes, environmental remediation, and energy conversion.

Flexible Electronics: To create flexible electronics, such as bendable displays, wearable technology, and electronic skins, GQDs can be incorporated into transparent, flexible substrates. They are appropriate for applications requiring conformal and lightweight electronic components because of their remarkable mechanical flexibility, electrical conductivity, and optical transparency.

Water Purification: GQDs can be functionalized or altered using techniques like photocatalysis and membrane filtration to selectively adsorb and break down contaminants in water. By eliminating impurities like bacteria, heavy metals, and organic dyes from water sources, they provide viable answers to problems with water pollution [10].

GQDs are used as smart nano-adsorbents of uranium (^{238}U) from radioactive industry waste. The GQDs were developed in a size range of 160–220 nm using a green route. The results showed that the GQDs were capable of adsorbing almost 40% of the uranium (238U) in alamine 3366 solution. Also, the results demonstrated that using GQDs treatment-like smart nanomaterials for radioactive waste in a volume reduction of almost 90% is achieved, helping the storage process as the final disposal of this material. GQDs are smart devices for the treatment of radioactive waste as an alternative to absorbent in the radioactive industry [11].

References

1. Graphene Quantum Dots Market Research Report Provides thorough Industry Overview, which offers an In-Depth Analysis of Product Trends and New Market D (linkedin.com)
2. Graphene Quantum Dots Market Research Reports. Get Customized Market Research Report (reliableresearchreports.com)
3. The Global Market for Graphene Quantum Dots 2023–2033: Superior Stability Compared to Non-Carbon QDs Drives Adoption (yahoo.com)
4. Global Graphene Quantum Dots Market Report 2023: Increasing Semiconductor Memory Demand is Driving Growth (yahoo.com)
5. Graphene Quantum Dots Market–Global Industry Size, Share, (globenewswire.com)
6. Nanomechanical Mapping of Graphene Quantum Dot-Epoxy Composites Used in Biomedical Application
7. Solomon SD, Abolanle SA, Lukman OO, John AOO, Odunayo TO, Samuel OO (2022) A review on graphene quantum dots for electrochemical detection of emerging pollutants. J Fluoresc 32:2223–2236
8. Qianwen L, Jianhan S, Kun G, Nan C, Xiaotong S, Dan T, Congcong B, Ranran C, Liangti Q (2020) Graphene quantum dots for energy storage and conversion: from fabrication to applications. Mater Chem Front 4:421–436
9. Zahir N, Magri P, Luo W, Gaumet JJ, Pierrat P (2022) Recent advances on graphene quantum dots for electrochemical energy storage devices. Energy Environ Mater 5:201–214

10. Sandy Subala A, Anand KV, Antony Sibi S (2020) Quantum dots in waste water Treatment-Review article. Int J Innov Sci Res Technol 5:1449–1455
11. Frederico Duarte de Menezes, Luciana Magalhães Rebelo Alencar, Clenilton Costa dos Santos, Maria Isabel Barbosa da Silva & Ralph Santos-Oliveira (2020) Using graphene quantum dots for treating radioactive liquid waste. Environ Sci Pollut Res 27:3508–3512

Chapter 15
Tailor-Made Graphene Quantum Dots for Textile Applications

15.1 Introduction

Graphene quantum dots (GQDs) are nanoscale fragments of graphene that display special electrical, optical, and chemical capabilities. Their characteristics render them auspicious contenders for an array of uses, encompassing textiles [1]. There are a few issues taken into account while tailoring GQDs for textile applications:

Size and shape: GQD performance in textile applications is significantly influenced by their size and shape. To guarantee uniform dispersion and integration into the textile matrix, GQDs must be tailored to have sizes compatible with textile fibers. GQDs can be produced in a variety of sizes and forms [2].

Surface Chemistry: GQDs must be surface functionalized to increase their compatibility with textile fibers and their dispersibility in textile matrices. To improve GQDs' interaction with textile materials, functional groups like carboxyl, hydroxyl, or amine can be added to their surface [3].

Optical Properties: GQDs have special optical qualities, such as photoluminescence, which can be used in textile applications for luminous or smart textiles, sensing, and other purposes. Certain capabilities in textiles can be made possible by customizing the optical properties of GQDs through surface changes and size management [4].

Electrical Conductivity: Materials made of graphene, such as GQDs, have a high electrical conductivity. By adding GQDs to textiles, the fabric can become more conductive, opening up new possibilities for uses like heating elements, sensors, and wearable electronics. GQD performance for particular textile applications can be optimized by varying their size and doping to control their conductivity [5].

Mechanical Properties: Although the incorporation of GQDs into textiles shouldn't materially impair the fabric's flexibility and mechanical integrity, their mechanical qualities may not directly affect textile applications. For GQDs to be successfully used, they mustn't negatively impact the mechanical qualities of textiles [6].

N. Manjubaashini et al., *Graphene Quantum Dots*, Materials Horizons: From Nature to Nanomaterials, https://doi.org/10.1007/978-981-97-5722-0_15

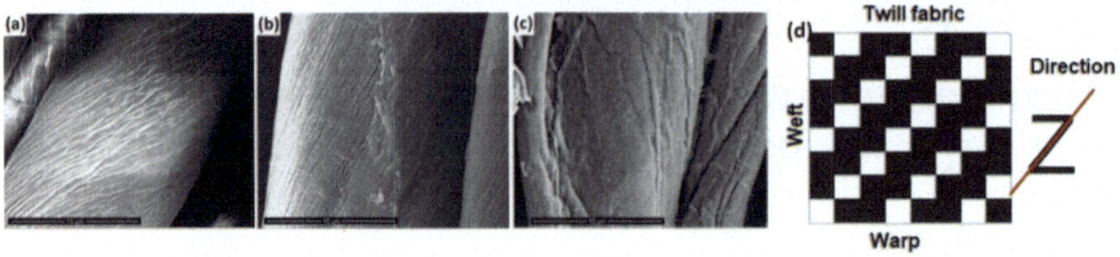

Fig. 15.1 FE-SEM images; **a** pristine CO, **b** PnGC, **c** PTnGC; and **d** structure and direction of cotton fabric [9]

Environmental Compatibility: For sustainable textile applications, considerations about the effects of graphene quantum dots (GQDs) and the techniques used in their synthesis are essential. To reduce damage to the environment, green synthesis techniques and non-toxic precursor materials must be given priority [7].

Scalability and Economically: The commercial feasibility of graphene quantum dots (GQDs) in textile applications hinges on the availability of scalable synthesis pathways and economical production methods. For GQDs to be widely used in the textile sector, effective synthesis methods that can generate GQDs in large numbers at a fair price are essential [8].

Titanium dioxide (TiO_2) and nitrogen-doped (N-GQD) nanoparticles (NPs) exhibit a wide range of UV absorption and are therefore attractive options for functional finishing on textile materials, providing anti-UV properties (Fig. 15.1a-d). Using polyvinyl alcohol as a binding agent, the high-pressure and high-temperature exhaustion technique was used to apply the resulting nanocoatings to cotton fabric by an industrial discontinuous method. Both compounds are good UV absorbers, and their UV protection factor (UPF) largely depends on the concentrations employed in the nanocoated cotton fabrics, according to the UV protection assessment [9].

15.2 Flexible Fabrications Using GQDs

With simple solvent extraction process tweaking, naturally occurring GQD-containing materials can yield surfaces with the necessary wettability qualities. Asphaltene-derived GQDs might thus be applied, in bulk, toward a wide range of tunable surface coatings due to their natural abundance, low toxicity, and intense fluorescence (Fig. 15.2a-b). Aerogels infused with type-OM GQDs promptly sank instead of floating on water, and fabrics spray-coated with these GQDs quickly absorbed moisture [10].

Utilizing an overlayer of nitrogen-doped (N-GQDs) over carbon fabric coated in reduced graphene oxide (CF-rGO), an extremely flexible and conductive electrode is created. To maximize the amount of edge surface that may participate in the catalytic activity, the visible light active photocatalyst N-GQDs self-assembled in a highly porous leaflet form [11]. Because of its binder-free, stable, flexible, and

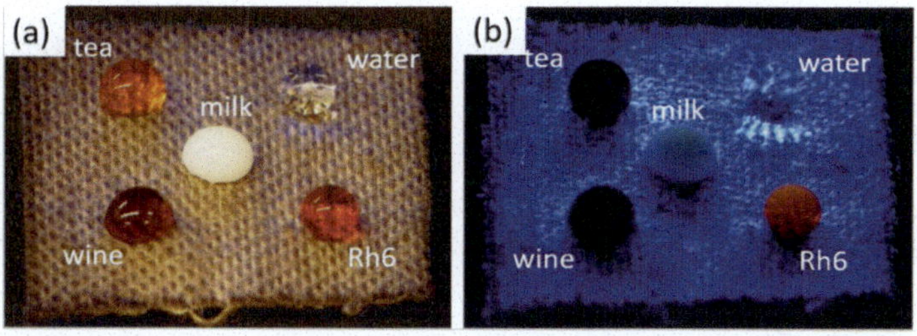

Fig. 15.2 Photographs of various droplets (tea, water, milk, wine, and a solution of rhodamine 6G in water) on cotton fabrics treated with type-AC GQDs; **a** normal light, and **b** under UV lamp at 365 nm [10]

highly conducting nature, the photo-electrocatalyst electrode is a good fit for flexible catalytic devices such as wearable supercapacitors and flexible solar cells (Fig. 15.3a-b).

When OH-functionalized graphene nanosheets (OH-GQDs) were sandwiched between graphene nanosheets, co-coated CNWT (GG/CNWT) with rGO showed improved electrochemical performance. Using a traditional three-electrode test, the electrochemical performances of G/CNWT, GG/CNWT, and PGG/CNWT-based electrodes were examined and studied. After 6000 cycles at a current density of 5 mA cm^{-2}, the Ca of PGG/CNWT reached 195 m F cm^{-2} at 0.1 mA cm^{-2} and had good electrochemical stabilities (Fig. 15.4B a-d). Their research has shown that a novel flexible supporting substrate for the fabrication of the rGO-functionalized textile electrode may be the ultra-hydrophilic CNWT [12]. Furthermore, the PGG/

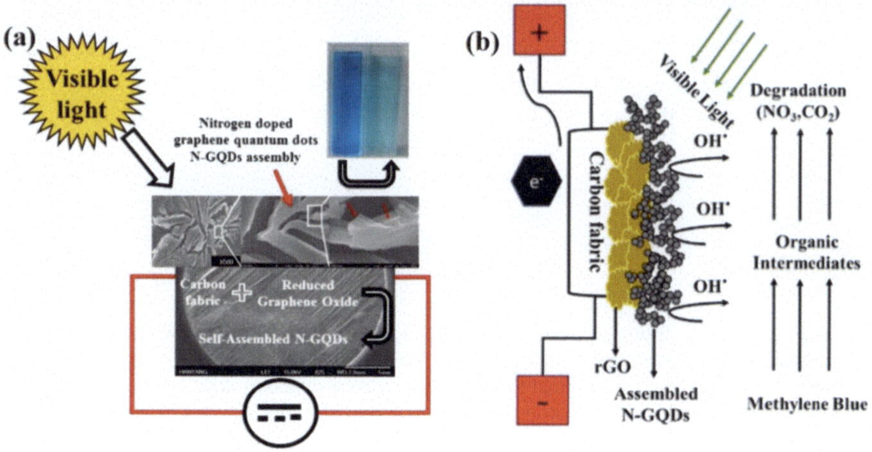

Fig. 15.3 **a** Nitrogen-doped GQDs were self-assembled on reduced graphene oxide-coated carbon fabric for water treatment, and **b** electro-photocatalytic activity of the proposed electrode. Reprinted with permission from @ Appl Surf Sci 480, 1035–1046 (2019)

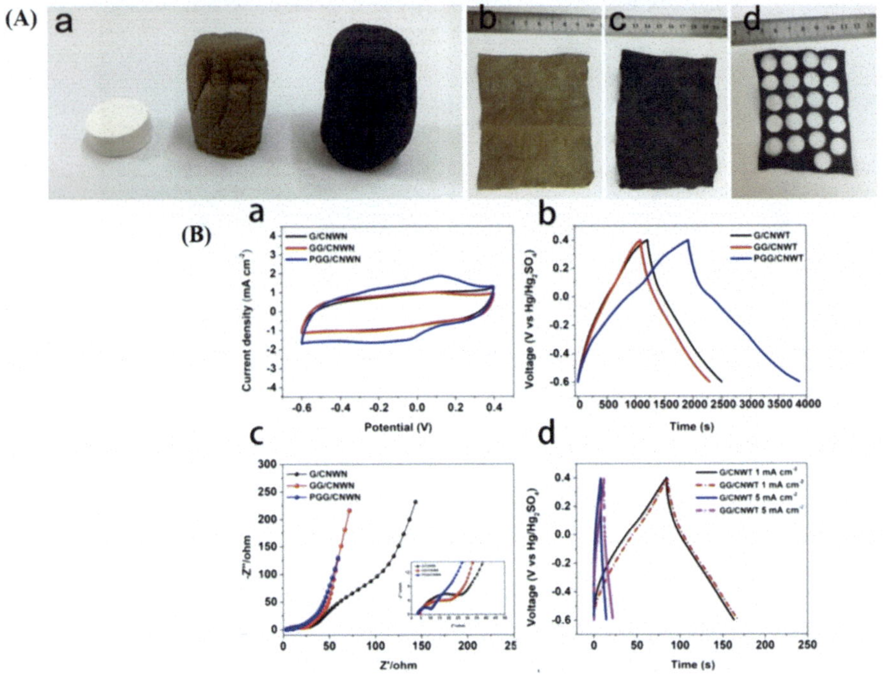

Fig. 15.4 **A** (a) Digital photos of CNWT, GO/CNWT, G/CNWT (right), (b-c) Digital photos of dried flexible GO/CNWT and G/CNWT, and (d) Making large through-holes on GG/CNWT; and **B** Electrochemical properties of G/CNWT, GG/CNWT and PGG/CNWT; (a) CV curves at 10 mV s^{-1}, (b) GCD curves at 0.1 mA cm^{-2}, (c) Nyquist plots, and (d) Ca comparison of G/CNWT and GG/CNWT. Reprinted with permission from @ Synthetic Metals 270, 116,571 (2020)

CNWT electrode is used in flexible and wearable electronics, among other applications, thanks to the current straightforward scale-up technique of fabrication (Fig. 15.4A a-d).

References

1. Zehua J, Peter O, Sidong L, Liehui G (2015) Graphene, graphene quantum dots and their applications in optoelectronics. Curr Opinion in Colloid & Interf Sci 20:439–453
2. Peyman GB, Kolsum MS, Ahmad PM (2023) Graphene quantum dots: synthesis, characterization, and application in wastewater treatment: a review. Mater Adv 4:4272–4293
3. Zhu X, Li Y, Cao P, Li P, Xing X, Yu Y, Guo R, Yang H (2023) Recent advances of graphene quantum dots in chemiresistive gas sensors. Nanomaterials 13:2880
4. Intae S, Seung-Rak S, Jongil A, Jin-Wook C, Soyern K, Woo Young L, Jun Hyup L (2021) Photoluminescent surface-functionalized graphene quantum dots for spontaneous interfacial homeotropic orientation of liquid crystals. J Mol Liq 332:115901

5. Mahmuda AS, Minzhi C, Nazakat AK, Md Taslimur R, Bidhan B (2020) Graphene coated textile based highly flexible and washable sports bra for human health monitoring. Mater Des 193:108792
6. Ashfaqul HK, Touhid UH, Masudur Rahman ANM, Shamima Akter S, Shah A (2022) Fabrication, properties, and performance of graphene-based textile fabrics for supercapacitor applications: a review. J Energy Storage 56:105988
7. Gulati S, Baul A, Amar A, Wadhwa R, Kumar S, Varma RS (2023) Eco-friendly and sustainable pathways to photoluminescent carbon quantum dots (CQDs). Nanomaterials 13:554
8. Li Y, Chopra N (2015) Progress in large-scale production of graphene. Part 1: chemical methods. J Miner, Metals & Mater Soc 67:34–43
9. Felipe BHS, Cabral RLB, Ladchumananandasivam R, Zille A, Kim S, Fechine PBA, Nascimento JHO (2022) Nanocoating on cotton fabric with nitrogen-doped graphene quantum dots/ titanium dioxide/PVA: an erythemal UV protection and photoluminescent finishing. J Mater Res Technol 18:2435–2450
10. Razieh FH, Maryam A, Hassan H, Qiuyun L, Xuehua Z, Jonathan GCV, Alkiviathes (2023) Asphaltene-derived graphene quantum dots for controllable coatings on glass, fabrics, and aerogels. ACS Omega 8:43610−43616.
11. Rabia R, Mumtaz A, Iftikhar AS, Alvira AA, Maiyalagan T, Aima SA, Min Jae K, Sung HJ (2019) Self-assembled nitrogen-doped graphene quantum dots (N-GQDs) over graphene sheets for superb electro-photocatalytic activity. Appl Surf Sci 480:1035–1046
12. Feng S, Nantao H, Yanjie S, Hong L, Bin L, Cheng Z, Gang L, Zhi Y, Yafei Z (2020) PANI/ Graphene quantum dots/graphene co-coated compressed non-woven towel for wearable energy storage. Synth Met 270:116571

Chapter 16
Healthcare and Biological Application of Graphene Quantum Dots

16.1 Introduction

GQDs are carbon compounds composed of larger-than-100-nm-thick graphene layers. These are extremely tiny materials with good optical, electrical, chemical, and physical characteristics. They also shine. Additionally, using them in living things is safe. Green quantum dots (GQDs) are gaining a lot of attention as a novel class of eco-friendly materials [1]. GQDs have a wide range of applications in the medical domain. Biosensing, imaging intracellularly, drug delivery, infection detection, tissue construction, and light-induced thermal therapy are a few of them [2]. This audit provides information on the application of GQD blends, their potential for medical harm, their interactions with cells, and their molecular effects on the body [3].

In the disciplines of chemistry, material science, and biology, the application of graphene quantum dots to the therapy of disease is a novel and cutting-edge strategy. Hybrid quantum dots have demonstrated encouraging results in preclinical tests when used to treat cancer, which calls for more in-depth investigation into their therapeutic potential [4]. To make sure hybrid quantum dots do not build up in healthy tissues and cause harm, it is crucial to address safety concerns regarding their usage in cancer treatment [5]. Numerous medical domains, such as intracellular imaging, drug transport, chemical sensing, disease treatment, and tissue engineering, employ biocompatible quantum dots extensively [6]. Specialized applications such as the detection of chemicals inside the body have led to the development of modified graphene quantum dots, which have also been investigated for their potential applications in imaging and therapy [7]. Target-modified graphene quantum dots are used in drug delivery and cancer treatment to reduce the effect on normal cells. Different kinds of modified graphene quantum dots mostly concentrate in the liver, heart, lungs, spleen, and kidneys, according to pharmacokinetic studies [8].

N. Manjubaashini et al., *Graphene Quantum Dots*, Materials Horizons: From Nature to Nanomaterials, https://doi.org/10.1007/978-981-97-5722-0_16

16.2 Detection of Cancer Cells

GQDs specifically harm the DNA of cancer cells to inhibit the growth of tumors. The nucleus-targeting TAT peptides (TAT-NGs) were added to the amine-functionalized GQDs, and they were then grafted with folic acid (FA) modified PEG via disulfide linkage, which targets cancer cells (FAPEG-TNGs) (Fig. 16.1). The resultant FAPEG-TNGs demonstrated effective targeting of cancer cells, nucleus uptake, and biocompatibility. Through $\pi-\pi$ and electrostatic interactions, they adsorb on DNA, causing DNA damage, upregulating proteins linked to cell death, and eventually suppressing the development of cancer cells [9].

The molecular processes by which GQDs in an inflammatory milieu cause human periodontal ligament stem cells (PDLSCs) to proliferate and undergo osteogenic differentiation (Fig. 16.2). By stimulating the Wnt/β-catenin signaling pathway in the inflammatory milieu, GQDs may enhance PDLSCs' capacity for osteogenic differentiation [10]. When exposed to ultraviolet light, GQDs produced a strong green luminescence, with a fluorescence quantum yield of up to 24.6%. The GQDs had a weight yield of around 11.1 ± 0.9%. GQDs had mostly entered the cytoplasmic area of the HeLa cells after penetrating them. The apoptotic assays for Annexin-V, LDH, ROS, and CKK-8 were used to assess the cytotoxicity of GQDs. Low-dose GQD internalization (12.5 and 25 μg/mL) did not result in appreciable cytotoxicity, as seen by the high cell viability, low level of LDH release, low level of internal cellular ROS, and little degree of cell apoptosis (Fig. 16.3a–c). The in vivo experiment also showed that zebrafish embryos and larvae were not significantly affected

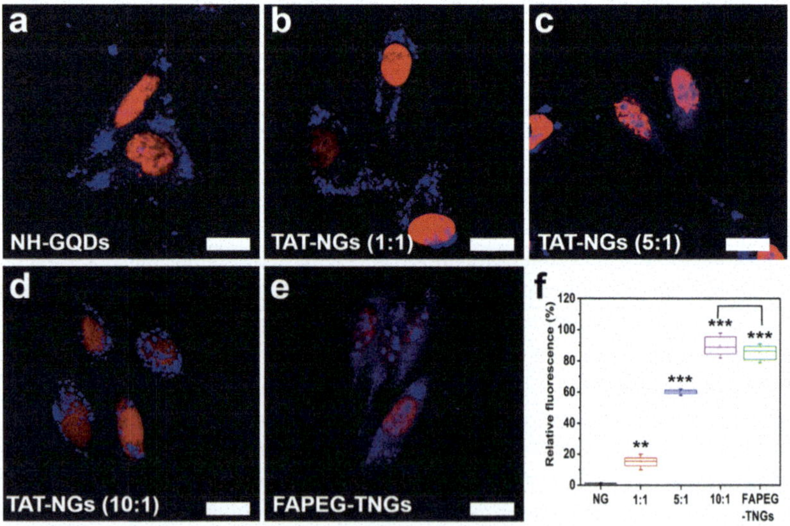

Fig. 16.1 Imaging of; **a** NH-GQDs, **b** TAT-NGs (1:1), **c** TAT-NGs (5:1), **d** TAT-NGs (10:1), and **e** FAPEG-TNGs **f** Fluorescence intensity of TAT-NGs and FAPEG-TNGs relative to that of NH-GQDs [9]

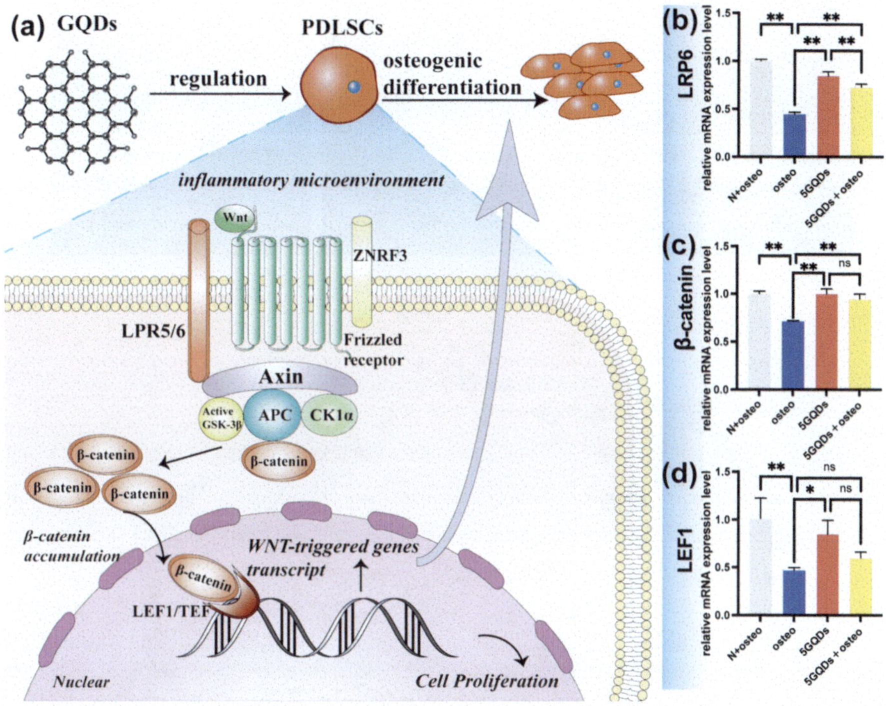

Fig. 16.2 GQDs on the proliferation of PDLSCs under the media mimicking proinflammatory environment [10]

by modest doses of GQDs [11]. Therefore, there was a lot of promise for biological and medicinal uses for as-prepared GQDs with strong PL and good biocompatibility.

Metformin is delivered to specified tissues using Graphene Oxide Quantum Dots (GOQD-HA) nanocomposite that is CD44 targeted and functionalized with Hyaluronic Acid. In both diet-induced obese mice and palmitic acid-induced RAW264.7 cells, metformin-loaded GOQD-HA (GOQD-HA-Met) effectively downregulated the production of proinflammatory cytokines and restored antioxidant status at lower dosages than free metformin (Fig. 16.3d). In addition to serving as a drug delivery vehicle, the GOQD-HA nanocarrier also acted as a therapeutic agent, which increased the effectiveness of metformin. Because HA and GOQD both have anti-inflammatory and antioxidant qualities, respectively, GOQD-HA has therapeutic benefits [12]. In meta-inflammation, where the nanocarrier itself functions as a therapeutic agent, the study elucidates the role of GOQD-HA as a targeted drug delivery alternative for metformin.

The antitumor drug doxorubicin (DOX) was loaded into a novel drug delivery system based on arginine-glycine-aspartic acid (RGD)-conjugated GQDs for targeted cancer fluorescence imaging as well as tracking and monitoring drug delivery without the need for external dyes. GQDs unlinked with DOX were not harmful to U251

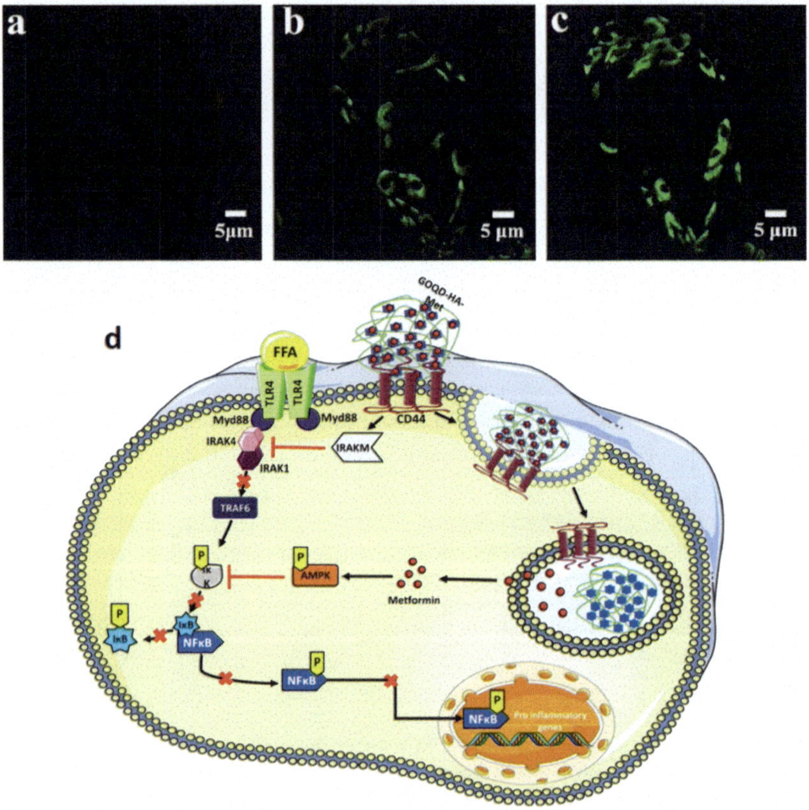

Fig. 16.3 Confocal fluorescence images of histopathological slices in heart region at excitation of 405 laser: **a** control, **b** larvae treated with 100 μg/mL, **c** larvae treated with 200 μg/mL [11], and **d** Proposed anti-inflammatory mechanism of action of GOQD-HA-Met on free fatty acid-induced activated macrophages [12]

human glioma cells when used as a nanocarrier (Fig. 16.4a–e). Within a wide range of DOX doses, DOX-GQDs-RGD conjugates showed significant cytotoxicity to U251 glioma cells in comparison to free DOX. The drug efficacy of DOX enhanced after using GQDs as drug carriers without requiring an increase in DOX dosage. The results of DOX-GQDs-RGD conjugates' cellular uptake show that, after 16 h of incubation, some GQDs as well as DOX entered the cell nucleus [13]. When paired with effective nuclear delivery, this improvement significantly increased DOX's cytotoxicity. The pathogen sensor is portable and inexpensive, with a paper-based analytical device (μPAD) that can conduct immunoassays and measure analyte concentration through integration with an automated color-detecting system that evaluates the μPAD's color intensity. The portable color detection system at the center of the proposed sensor is capable of reading the red, green, and blue color of the paper that is emitted by fluorescent nanomaterials, such as gold nanoclusters (AuNCs) and GQDs, which are

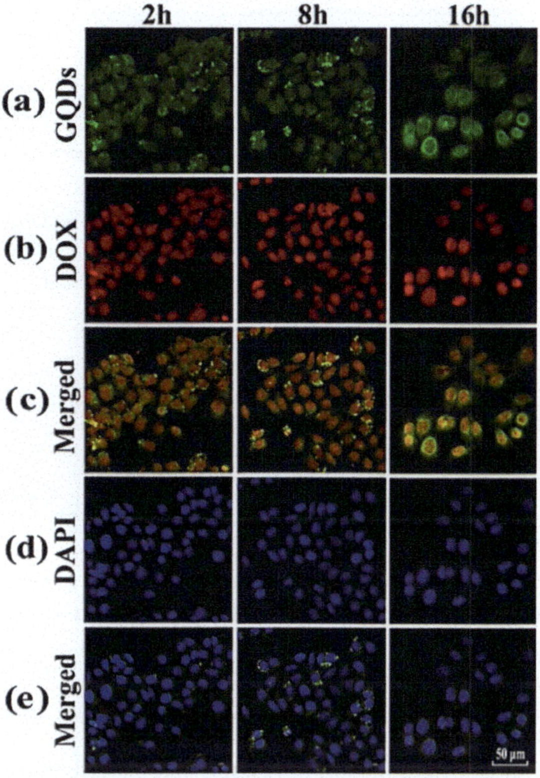

Fig. 16.4 CLSM images of U251 Glioma cells treated with DOX-GQDs-RGD: **a** GQD signals collected in the range of 525 ± 50 nm, **b** DOX signals collected in the range of 595 ± 50 nm, **c** merged images of GQDs and DOX, **d** cell nuclei stained with DAPI and signals collected from 450 ± 50 nm, **e** merged images of GQDs, DOX and DAPI. Reprinted with permission from @ Sens Actuators B 256, 616–623 (2018)

conjugated with antibodies to indicate the immunoassay results and transform the presence of a pathogen into a colorful fluorescence signal.

The paper-based detection method reduces the detection limit to as low as a sub nanogram/mL by using GQDs and AuNCs with high quantum yield and relatively high fluorescence intensity as the sensing signal. Moreover, GQDs and AuNCs have little background interference from cellulose and can exhibit distinct fluorescence when exposed to the same light source (UV light), making it possible to detect many analytes at once with a single UV light. Moreover, a little 10-min reaction time is required, which facilitates prompt and highly sensitive diagnosis [14]. Because of its increased sensitivity, multiplexity, and low equipment requirements, the suggested handheld pathogen sensor may identify pathogens with greater speed and accuracy (Fig. 16.5). This makes it appropriate for usage in environments with limited resources and inadequate medical infrastructure.

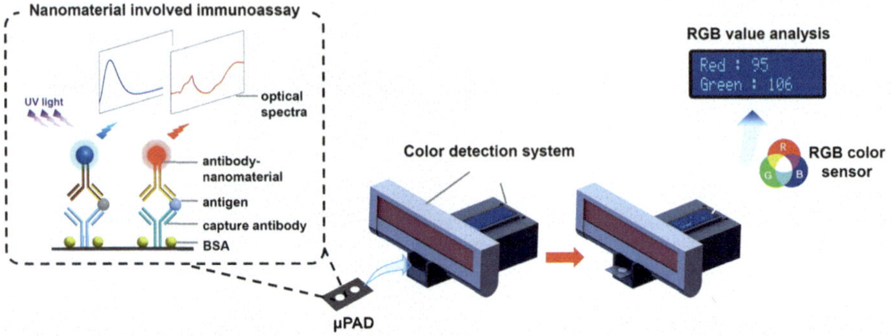

Fig. 16.5 Proposed pathogen sensor consisted of a color detection system and a μPAD, which integrates with nanomaterials, decreasing the detection limit and enabling multi-target detections. Reprinted with permission from @ Sens Actuators B 363, 131824 (2022)

16.3 Bioimaging and Antibacterial Efficacy

GQDs have been investigated as cellular bioimaging agents and dual-functional targeted medication carriers [15]. GQDs-scFvB10 were conjugated to a single chain variable fragment of antibody (scFv) that was tailored to bind to the epidermal growth factor receptor (EGFR) with high affinity (B10) through amide covalent bonds. Western blot analysis and confocal imaging of MDA-MB-231, EGFR-overexpressing breast cancer cells, were used to show the targeting ability of scFvB10 following conjugation to GQDs and the possible use of GQDs-scFvB10 in targeted bioimaging (Fig. 16.6).

With a high cisplatin (CDDP) loading capacity of 50% and a pH-dependent release with a slower release rate at neutral conditions, the targeted GQDs' surface may be able to lessen CDDP's widely reported systemic toxicity. On MDA-MB-231 cells, the targeted CDDP-loaded nanocarriers ((CDDP)GQDs-scFvB10) showed noticeably greater toxicity than the non-targeted ones, indicating their effective absorption through EGFR. On the other hand, cells expressing saturated EGFR exhibited reduced absorption and lethal impact of (CDDP)GQDs-scFvB10, indicating the nanocarriers' specificity for EGFR-overexpressing cells [16]. Through its interactions with EGFRs, the scFvB10-functionalized GQD offers a viable platform for targeted cellular imaging and CDDP administration.

As a light-triggered antibacterial agent, Mila M studied the antibacterial activity of S, N-doped GQDs. Using L-cysteine amino acid as a source of heteroatoms, gamma irradiation was utilized as a technique to modify GQDs in a single step. Human serum albumin can bind to L-cysteine-modified dots, according to fluorescence quenching experiments. Eight bacterial strains were used to assess the antibacterial activity of GQDs in combination with blue light (470 nm) irradiation for one and 6 h (Fig. 16.7a, b). The most effective sample was GQD-cys-25, which showed minimum inhibitory concentrations (MIC) against *S. aureus*, *E. faecalis*, and *E. coli*

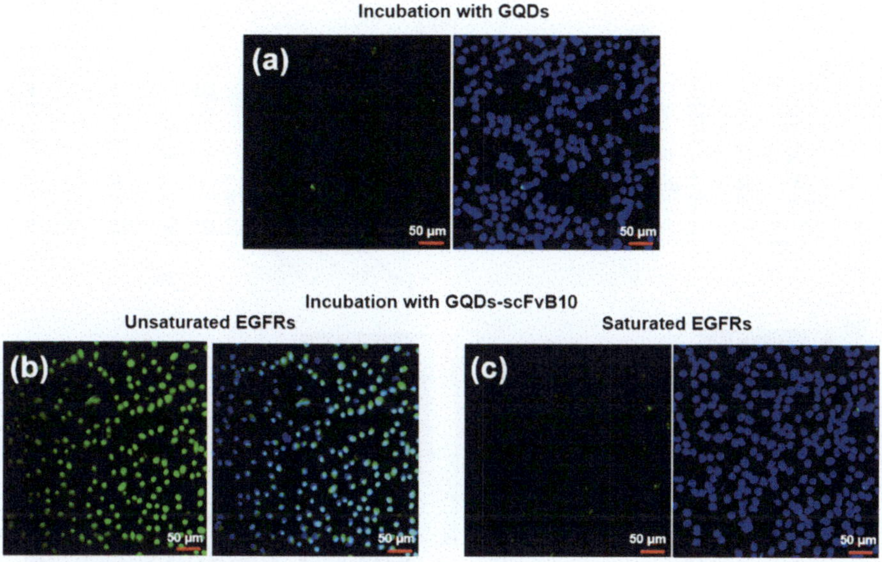

Fig. 16.6 Confocal fluorescence images of MDA-MB-231 cells after incubation; **a** 20 μg ml^{-1} GQDs dispersion for 40 min, **b** 20 μg ml^{-1} GQDs-scFvB10 dispersion for 40 min, **c** 20 μg ml^{-1} GQDs-scFvB10 and 30 μg ml^{-1} free scFvB10 for 40 min. Reprinted with permission from @ Mater Sci Eng C 94, 247–257 (2019)

as low as 125 μg/mL after just one hour of exposure to blue light [17]. A human monoclonal antibody (mAb) called pembrolizumab (GQDs–pembrolizumab) was conjugated with graphene quantum dots (QDs) by Raziyeh GA to create an imaging probe for breast cancer that targets the immune checkpoint programmed death receptor (PD-1) and its programmed death ligand-1 (PD-L1). In a BALB/c mouse model with 4T1 tumors, biodistribution, and SPECT imaging were used to evaluate the pharmacokinetics and targeting effectiveness of the 99mTc-GQDs-pembrolizumab. The good in vitro stability and high radiochemical purity (RCP > 95%) show how very promising GQDs–pembrolizumab complex formation with Technetium-99 m. The results of imaging and biodistribution experiments showed that the presence of the PD-1 receptor was responsible for the high activity of 99mTc-GQDs–pembrolizumab at the tumor site (8.4%ID/g) (Fig. 16.7c, d). Furthermore, the liver and spleen, which are also lymphoid organs, showed enhanced radiotracer uptake. The appropriate characteristics and actions of 99mTc-GQDs-pembrolizumab indicate that it could be a promising new probe for the creation of immune checkpoint monoclonal antibodies based on radiopharmaceuticals [18]. It may also help with immunotherapy, which is a treatment option for many different types of cancer.

For the first time, the antibacterial efficacy of zinc oxide (ZnO) nanoparticles combined with GQDs made from brewery wasted grain (BSG) against methicillin-resistant Staphylococcus aureus (MRSA) isolated from clinical wound specimens is examined. ZnO/GQD composites showed superior antibacterial action against

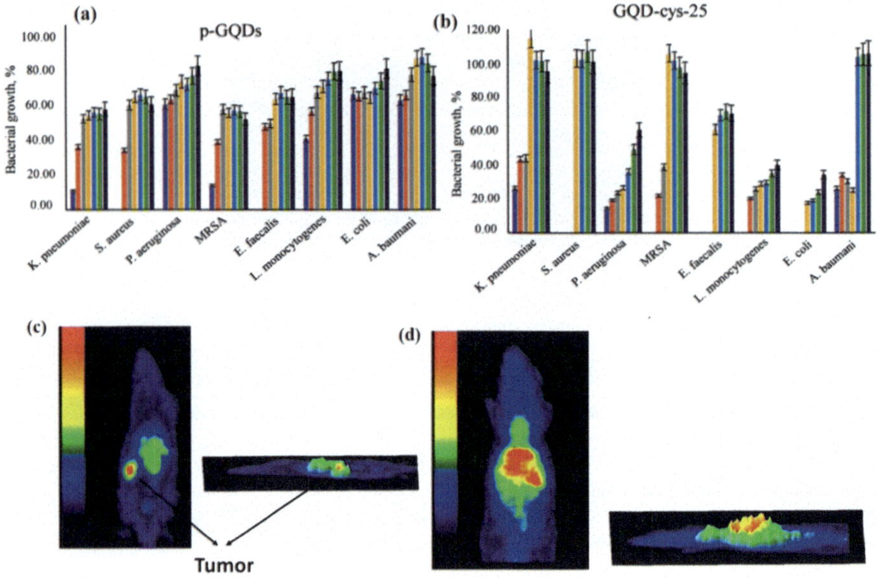

Fig. 16.7 Concentration-dependent bacterial growth in percentages: **a** p-GQDs, **b** GQD-cys-25. Reprinted with permission from @ J Photochem Photobiol 250, 112818 (2024); SPECT imaging performed one hour after the injection of 7.4 MBq of 99mTc-GQDs–pembrolizumab into the tail vein of BALB/c mice bearing 4 T1 tumors (**c**) and normal mice (**d**). Reprinted with permission from @ Arab J Chem 17, 105518 (2024)

Staphylococcus aureus and Escherichia coli as compared to individual GQDs and ZnO [19]. Significant efficacy was also shown by ZnO/GQDs in disinfecting MRSA (ATCC and clinical isolates from wound specimens). It has been demonstrated through cytotoxicity tests using 3T3 mouse embryo fibroblasts that ZnO/GQDs are not harmful to mammalian cells (Fig. 16.8).

For repeated PDT and long-term tumor imaging, acidity-activated nano transformers (GQD NT) are being developed as PS carriers. GQD NT actively targets tumor tissues under the direction of the Arg-Gly-Asp peptide. As a result, the tumor acidity becomes more soluble and expands, perhaps leading to prolonged tumor retention. GQD NT then fragments into tiny bits to enhance tumor penetration (Fig. 16.9). When exposed to laser light, GQD NT produces a mild kind of hyperthermia, which increases the permeability of cell membranes and facilitates the uptake of PSs. The most remarkable thing is that the as-prepared GQD NT performs efficient repetitive PDT in addition to "turning on" fluorescence and magnetic resonance signals. Notably, compared to other reported works, the overall PS dose is lowered to 3.5 μmol kg^{-1}, which is 10–30 times lower [20]. Overall, the approach overcomes the overdose challenge in recurrent PDT by using a smart vehicle to improve PS accumulation, retention, and release in tumors through planned deformation.

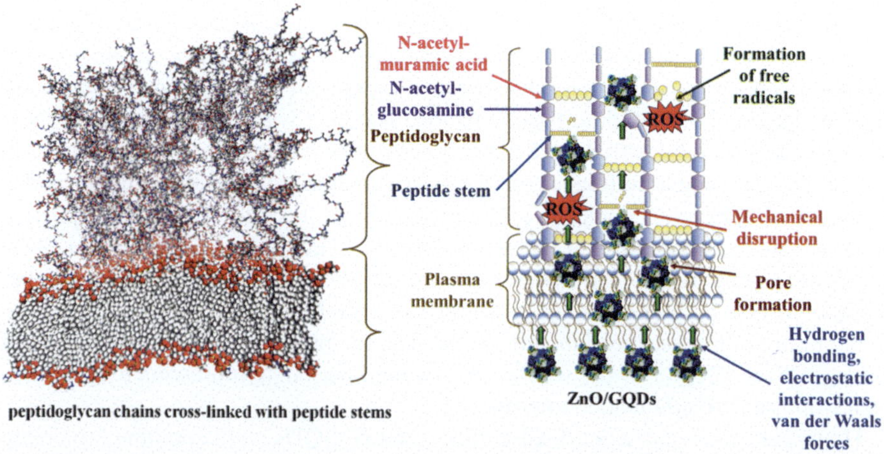

Fig. 16.8 Schematic representation of *MRSA* outer membrane structure with its key peptidoglycan chains cross-linked with peptide stems compared with all-atom structure representation and the bacterial disruption mechanism using ZnO/GQDs. Reprinted with permission from @ J Environ Chem Eng 12, 112263 (2024)

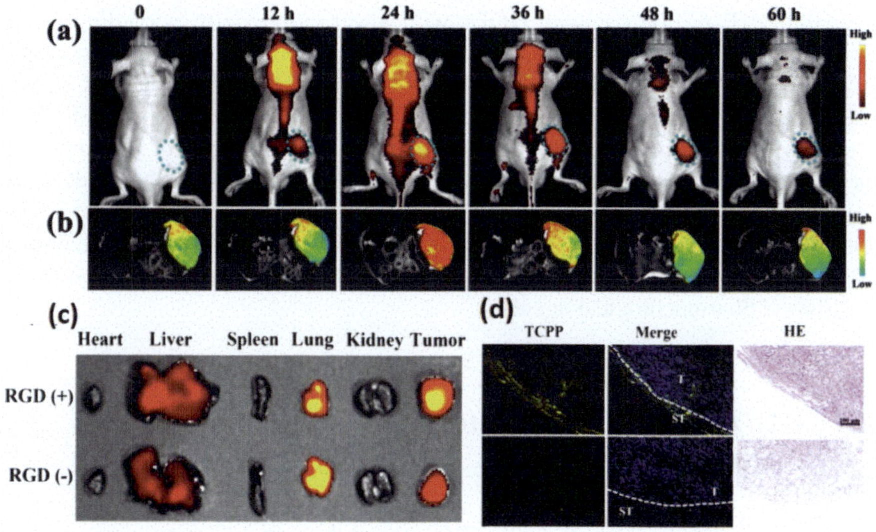

Fig. 16.9 **a** In vivo FLI of A549 tumor-bearing mice after intravenous injection of TCPP@GQD NT, **b** In vivo MR images of A549 tumor-bearing mice after intravenous injection of Mn-TCPP@GQD NT, and **c** Ex vivo FLI **d** Ex vivo FL images of tumor tissues. Reprinted with permission from @ Adv Mater 35, 2211337 (2023)

Nitrogen-functionalized (N-GQD) integrated gelatin methacryloyl (GelMA) hydrogel microspheres that can be monitored over an extended period while maintaining or improving the other advantageous aspects of 3D cell encapsulation are produced using a microfluidic-based method. The goal was to create monodisperse N-GQD-loaded GelMA hydrogel microspheres that encapsulated skeletal muscle cells (C2C12) using a multilayer droplet-based microfluidic system. The microfluidic device's flow rates could be adjusted to control the size of the microspheres. Encapsulated skeletal muscle cells showed excellent cell viability from day 1 (82.9 $\pm$ 6.50%) to day 10 (92.1 $\pm$ 3.90%). In comparison to the GelMA microgels, the N-GQD-loaded microgels encasing the cells showed increased metabolic activity. Using immunofluorescence labeling on the day, the presence of sarcomeric α-actin was confirmed [21]. Over the whole experimental period, a fluorescence signal was seen from the N-GQD-loaded microgels.

Hydrogel kits, fluorescent flexible films, and N-GQDs were combined to create a complete sensing system. Therefore, it is anticipated that the produced N-GQDs will be a useful analytical tool for tracking the amounts of ATP and Fe^{3+} in biological matrices [22]. Vanadium coordination compounds [VO(p-dmada)] were densely packed on one side of the GQDsheets in a new GQD-VO (p-dmada) complex, potentially by the $\pi-\pi$ stacking process. GQD-VO(p-dmada) demonstrated membrane permeability (Papp) on par with GQDs with lower cytotoxicity, according to the in vitro experiments. In vivo tests conducted on mice with type 2 diabetes revealed that, after three weeks of therapy, GQD-VO(p-dmada) showed a delayed glucose-lowering profile but more significant effects on insulin augmentation and β-cell protection than VO(p-dmada) alone. Furthermore, it was found for the first time that GQD alone could significantly lower the blood lipid levels in the db/db mice [23]. The overall pharmacokinetic performance and hypoglycemic effects of GQD-VO(p-dmada) were improved, and the use of GQD as a nanoplatform for drug administration may open up a wide range of possibilities for the continued development of metal-based pharmacological agents.

16.4 GQD-Based HIV Inhibitors and Gene Delivery Platforms

The RTI-conjugated molecule GQD-CHI499 exhibits an EC_{50} value in cells of 0.066 μg/mL and an IC50 of 0.09 μg/mL, making it a promising option for HIV treatment. A time of addition (TOA) approach was also used to explore the target of action in the HIV replicative cycle of the drug-conjugated samples GQD-CHI499 and GQD-CDF119 (Fig. 16.10a). The results showed that both conjugated samples had a mechanism of action akin to those of NNRTI medicines [24]. GQDs prevent α-syn from fibrillizing and interact directly with mature fibrils to cause their disintegration. Furthermore, GQDs can mitigate mitochondrial dysfunctions, recover synaptic loss and neuronal death, lessen Lewy body and neurite production, and stop neuron-to-neuron transmission of α-syn disease caused by α-syn produced fibrils (Fig. 16.10b–d). GQDs have been shown to cross the blood–brain barrier in vivo and guard against behavioral abnormalities, Lewy body/Lewy neurite pathology, and dopamine neuron loss brought on by α-syn preformed fibrils [25].

They create nanocarriers derived from GQD and add polyethylenimine (PEI) to provide the system with more functional groups for customization and improved biocompatibility. This technology not only delivers genes, such as green fluorescent protein (GFP) but also targets colon cancer cells with antibodies and binds to the epidermal growth factor receptor (EGFR) covalently to improve cell uptake of nanocarriers and cell membrane penetrability (Fig. 16.11a). The anticancer medication doxorubicin (Dox) is noncovalently encapsulated to provide pH-induced drug release at tumor sites and provides room for additional functional gene alteration to achieve multistrategy cancer therapy [26]. In comparison to free Dox therapies, this nanoparticle acts as a multifunctional gene delivery system that improves cytotoxicity and extends the duration of inhibitory capacity in colon cancer cells. Furthermore,

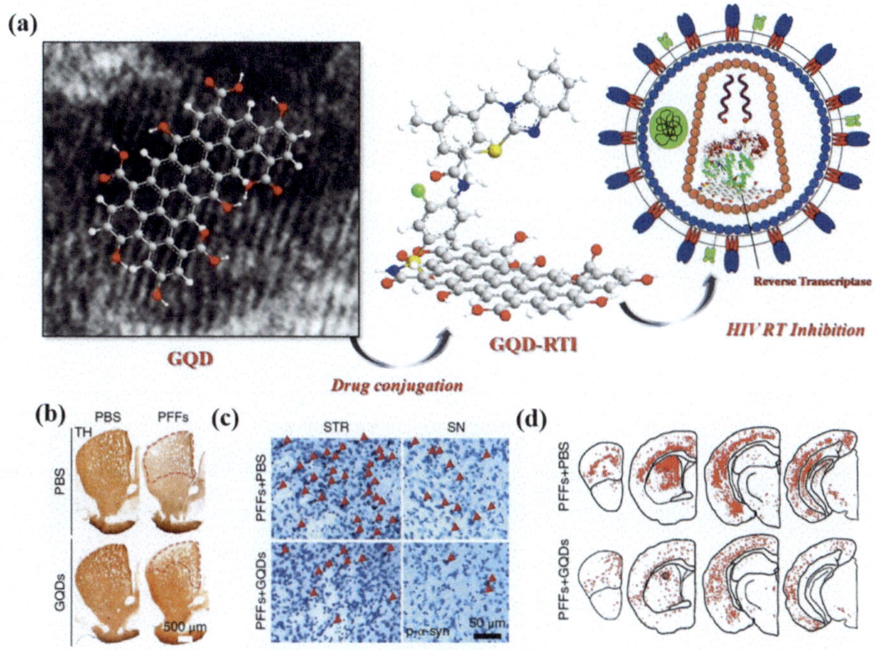

Fig. 16.10 **a** GQD-based systems as HIV Inhibitors. Reprinted with permission from @ Bioconjug Chem 29, 3084–3093 (2018); **b** Representative TH immunohistochemistry images in the striatum of α-syn PFF-injected hemisphere; **c** Representative p-α-syn immunostaining images in the striatum (STR) and substantia nigra (SN) of α-syn PFF-injected hemisphere, and **d** Distribution of LB/LN-like pathology in the CNS of α-syn PFF-injected hemisphere. Reprinted with permission from @ Nat Nanotechnol 13: 812–818 (2018)

in an investigation using xenograft mice, GQD composites show comparable tumor suppression effectiveness with noticeably less toxicity as compared to the free Dox treatment group (Fig. 16.11b). In primary neurons, C-GQDs show very little toxicity, and their effectiveness as a therapy for Parkinson's disease was almost identical to that of CF-GQDs. These findings point to an effective and environmentally responsible way to synthesize new nanomedicine, as naturally produced medications may be more patient-friendly [27].

To investigate the viability of in vivo imaging with GQDs, fluorescence imaging of GQDs loaded in polyacrylamide gel and subcutaneously implanted close to the thoracic region of a mouse that had been put to sleep was performed (Fig. 16.12a–c). From the implanted location, deep red fluorescence (610 nm emission filter) was seen with a comparatively low autofluorescence background [28]. With its porous silica shell and emitting and photothermally active GQDs, carbanosilica is a promising nano theranostics agent that shows 68.75% tumor shrinkage as opposed to 34.48% without NIR light exposure (Fig. 16.12d–e). The heat generated modifies the tumor's permeability, which promotes the build-up of nanotheranostics inside the tumor

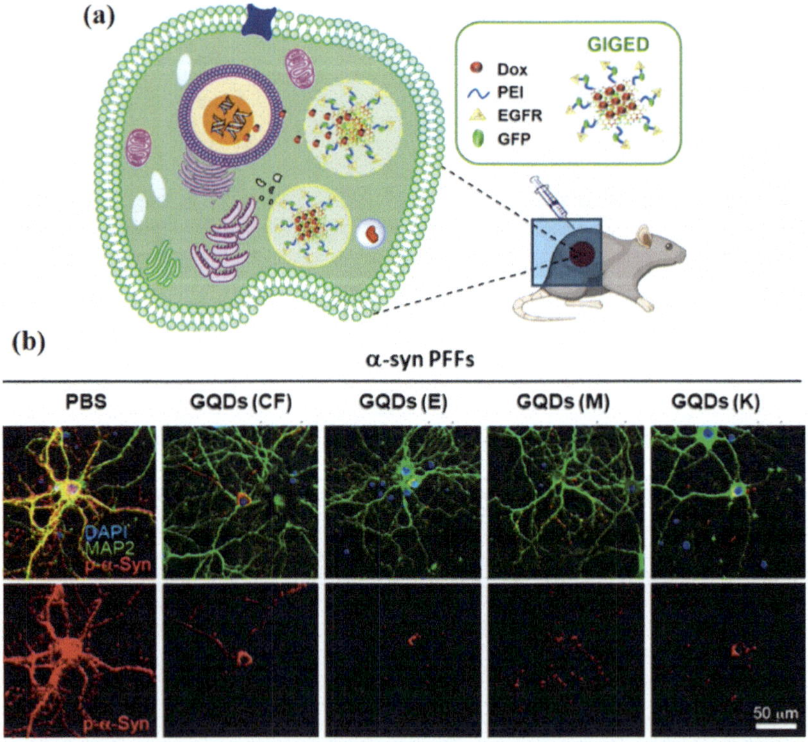

Fig. 16.11 **a** GQD for gene delivery platform for colon cancer in vivo studies. Reprinted with permission from @ ACS Appl Bio Mater 2020, 5948–5956 (2020); **b** Representative p-α-syn immunostained neurons [27]

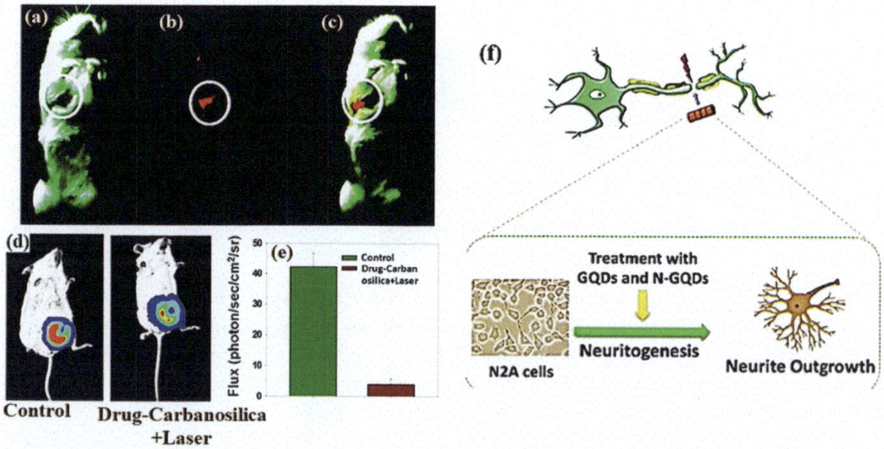

Fig. 16.12 In vivo imaging; **a** Mouse under white light after implanting GQD loaded in acrylamide gel, **b** Fluorescence image acquired with long pass filter, **c** Blended image confirms the source of fluorescence from the implanted region, **d** Control and drug loaded with carbanosilica under laser light for qualitative analysis, **e** Quantitative analysis of drug-loaded carbanosilica under NIR light irradiation [29], **f** Schematic of GQD and N-GQD acting as Neurotrophic Agents. Reprinted with permission from @ ACS Appl Bio Mater 6, 2237–2247 (2023)

environment [29]. The efficacy of GQDs and N-GQDs as neurotrophic agents was assessed in vitro using the N2A cell line. Standard tests like the SRB assay and neurite outgrowth assay were carried out (Fig. 16.12f). Neurite outgrowth assay results were confirmed by confocal imaging studies, quantitative real-time PCR (qPCR) studies, and immunohistochemistry results [30].

Utilizing GQD-decorated luminous porous silicon (GQDs@PSi), which was then loaded with peptide and implanted in chitosan (CS) film, the stimuli-responsive smart dressing for theranostics of diabetic wounds was created. Along with CS's breakdown in the acidic and oxidative milieu, the smart dressing demonstrated stimuli-responsive medication release in diabetic wounds with slightly acidic and strongly oxidative circumstances (Fig. 16.13a–d). Results from both in vitro and in vivo experiments showed that the smart dressing greatly mended diabetic wounds and improved cell migration and proliferation [31]. By measuring the rate at which the dressing fluoresces, it is also possible to determine in real-time whether diabetic wounds are getting worse or are healing.

16.5 Wound Healing Applications

It has been proposed to combine a low concentration of H_2O_2 with GQDs, the "safe" carbon nanomaterials, to create an antibacterial system. The results of all the in vitro studies show that this intrinsic activity significantly increases the antibacterial activity

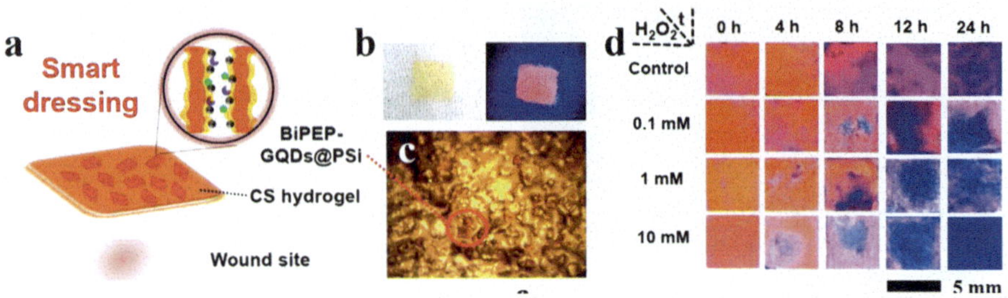

Fig. 16.13 **a** Structure of smart dressing based on BiPEP-GQDs@PSi/CS matrix, **b** Photographs of wound dressing with or without the UV excitation (377 nm), **c** Optical microscopic image of the dressing comprising BiPEP-GQDs@PSi and CS matrix, and **d** Fluorescence photos of the smart dressing on simulated skin with different concentration of H_2O_2. Reprinted with permission from @ Acta Biomaterialia 131, 544–554 (2021)

of H_2O_2, and the system that was built had a wide range of antibacterial activity against both Gram-positive (Staphylococcus aureus) and Gram-negative (Escherichia coli) bacteria (Fig. 16.14a, b). More crucially, GQD-Band-Aids are made and exhibit outstanding antibacterial properties with the help of H_2O_2 at low doses in vivo to evaluate the antibacterial performance of the developed system in actual wound disinfection (Fig. 16.14c–e) [32].

Alina RV created a range of nitrogen-containing quantum dots (N-GQDs) that are lightly metal-doped (iron oxide, silver, thulium, neodymium, cerium oxide, cerium chloride, and molybdenum disulfide). These NGDs show high contrast characteristics in the ultrasound brightness mode and can be used for visible and/or near-infrared fluorescence imaging. For the majority of metal-doped N-GQDs in the vascular phantom, agarose gel, and animal tissue, a successful tenfold ultrasonic

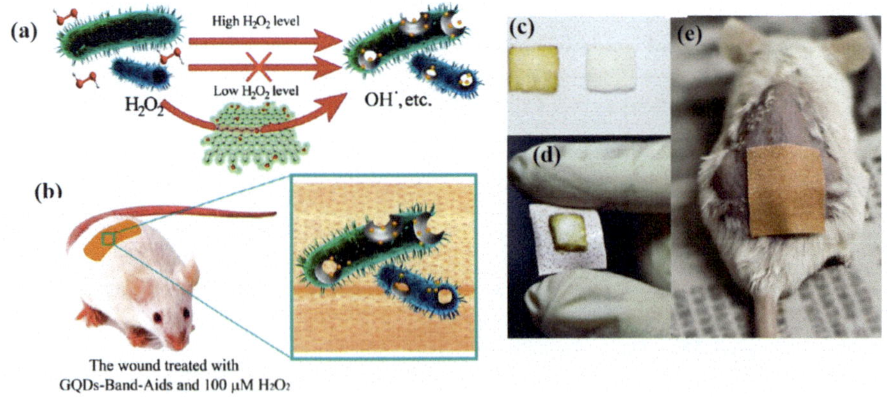

Fig. 16.14 **a** GQDs and low level of H_2O_2 for the antibacterial system, **b** GQD-Band-Aids used in wound disinfection in vivo, **c** The cotton fabric absorbed with (left) and without (right) GQDs, **d** The obtained GQD-Band-Aid, and **e** The mouse treated with GQD-Band-Aid. Reprinted with permission from @ ACS Nano 8, 6202–6210 (2014)

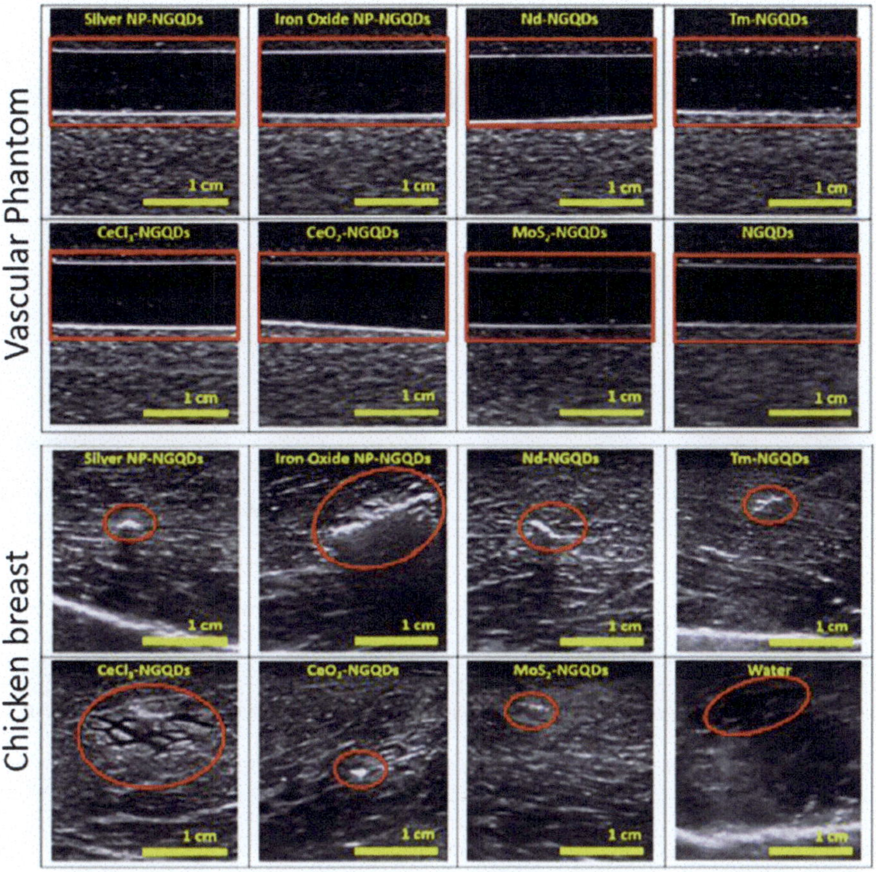

Fig. 16.15 Ultrasound images of metal-doped N-GQDs in the vascular phantom and chicken breast after injection. Reprinted with permission from @ ACS Biomater Sci Eng 8, 4965–4975 (2022)

signal amplification is observed at 0.5–1.6 mg/mL [33]. Metal-doped N-GQDs are a promising agent for many theragnostic applications because of their ability to combine high-precision fluorescence tracking with non-invasive ultrasound imaging (Fig. 16.15).

Pei-Chun Wong group incorporates a nitrogen-doped GQD (N-GQD) hydrogel produced from bioresources into the Mg ZK60 alloy to create a GQD hydrogel–magnesium (Mg) composite scaffold enhanced by plasma nanotechnology for functional bone defect repair (Fig. 16.16a–c) [34]. Results both in vitro and in vivo showed that, when compared to AgNPs alone, GQDs@Ag hybrids with improved biosafety had a noticeably greater effect in killing MRSA and speeding up MRSA-infected wound healing. All of these findings point to a high degree of therapeutic promise for the biocompatible nanosystem of GQDs@Ag in the treatment of MRSA infection [35].

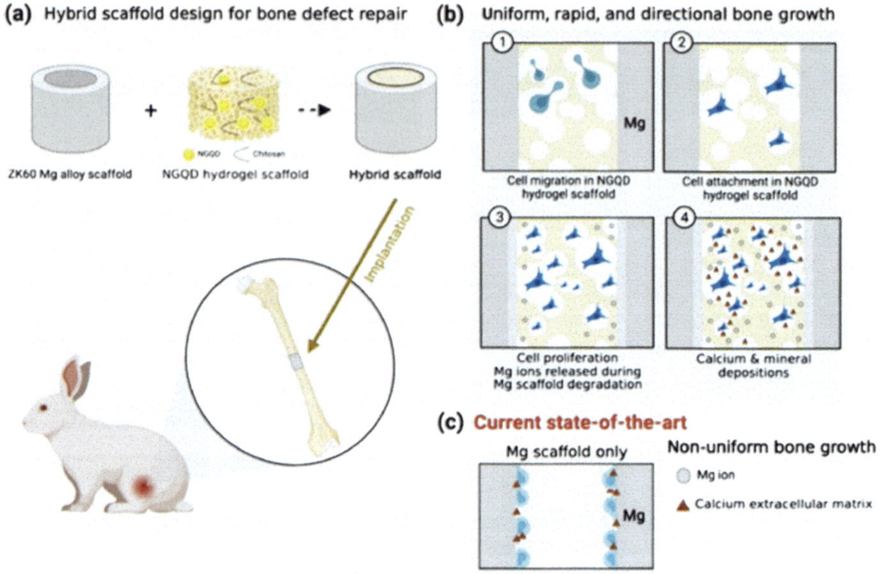

Fig. 16.16 **a** Design of the highly functional hybrid scaffold employing the synergy between N-GQD hydrogels and the Mg-based alloy for bone defect repair in vivo, **b** Integrated N-GQD hydrogels providing a suitable template for cell migration; (1) cell attachment (2) cell proliferation induced by the released Mg^{2+} from the metallic scaffold, (3) calcium and mineral depositions, (4) Promote a uniform, rapid, and directional bone growth across the hydrogel, **c** Nonuniform bone growth in the pristine Mg-based scaffold due to the absence of the ECM and pore structures. Reprinted with permission from @ ACS Appl Mater Interfaces 15, 44607–44620 (2023)

References

1. Yichun X, Bin W, Yu Y, Xuejing C, Yan X, Liang HG (2019) Cytotoxicity and autophagy induction by graphene quantum dots with different functional groups. J Environ Sci 77:198–209
2. Ziwei H, Thomas G, Junfei T, Wei S, Garnier G (2018) Effect of bovine serum albumin treatment on the aging and activity of antibodies in paper diagnostics. Front Chem 6:161
3. Mansuriya BD, Altintas Z (2020) Applications of graphene quantum dots in biomedical sensors. Sensors 20:1072
4. Mohammad HK, Majid A, Abbas R, Sadanand P (2024) Graphene quantum dots: background, synthesis methods, and applications as nanocarrier in drug delivery and cancer treatment: an updated review. Inorg Chem Commun 161:112032
5. Fei C, Weiyin G, Xiaopei Q, Hong Z, Lianhua L, Pu L, Weiling F, Yang L (2017) Graphene quantum dots in biomedical applications: recent advances and future challenges. Front Lab Med 1:192–199
6. Abdellatif AAH, Younis MA, Alsharidah M, Al Rugaie O, Tawfeek HM (2022) Biomedical applications of quantum dots: overview, challenges, and clinical potential. Int J Nanomed 17:1951–1970
7. Shital S, Aniket P, Ravindra G (2022) Graphene quantum dots: a pharmaceutical review. Asian J Pharma Res 12:341–348
8. Li K, Liu W, Ni Y, Li D, Lin D, Su Z, Wei G (2017) Technical synthesis and biomedical applications of graphene quantum dots. J Mater Chem B 5:4811–4826

9. Qi L, Pan T, Ou L et al (2021) Biocompatible nucleus-targeted graphene quantum dots for selective killing of cancer cells via DNA damage. Commun Biol 4:214
10. Gao W, Liang Y, Wu D et al (2023) Graphene quantum dots enhance the osteogenic differentiation of PDLSCs in the inflammatory microenvironment. BMC Oral Health 23:331
11. Jiang D, Chen Y, Li N, Li W, Wang Z, Zhu J et al (2015) Synthesis of luminescent graphene quantum dots with high quantum yield and their toxicity study. PLoS One 10:e0144906
12. Sarkar K, Bank S, Chatterjee A et al (2023) Hyaluronic acid-graphene oxide quantum dots nanoconjugate as dual-purpose drug delivery and therapeutic agent in meta-inflammation. J Nanobiotechnol 21:246
13. Dong J, Wang K, Sun L, Sun B, Yang M, Chen H, Dong L (2018) Application of graphene quantum dots for simultaneous fluorescence imaging and tumor-targeted drug delivery. Sens Actuators B 256:616–623
14. Hao Y, Jia-Hui L, Zhi-Shun D, Wei-Ting C, Yau Kei C, Yi Chun Y, Huan-Tsung C, Chien-Fu C (2022) Detection of pathogens using graphene quantum dots and gold nanoclusters on paper-based analytical devices. Sens Actuators B 363:131824
15. Fatemeh N, Rouholah ZD, Morteza H, Mohammad RG (2018) Fluorescence turn-on sensing of thiamine based on Arginine—functionalized graphene quantum dots (Arg-GQDs): central composite design for process optimization. Sens Actuators B 255:2078–2085
16. Fatemeh N, Yun Rui K, Peng C, Jaleh V, Abbas AK, Sierin L (2019) Targeting graphene quantum dots to epidermal growth factor receptor for delivery of cisplatin and cellular imaging. Mater Sci Eng C 94:247–257
17. Mila M, Gabriele C, Aurelio B, Michelangelo S, Olivera M, Tatjana V, Biljana TM, Svetlana J (2024) Blue-light-driven photoactivity of L-cysteine-modified graphene quantum dots and their antibacterial effects. J Photochem Photobiol 250:112818
18. Raziyeh GA, Ghazaleh S, Seyed ESE, Morteza PH, Mostafa S, Mohammad S, Ali NK, Atena N, Mehdi SA (2024) Development of novel radiolabeled antibody-conjugated graphene quantum dots for targeted in vivo breast cancer imaging and biodistribution studies. Arab J Chem 17:105518
19. Lekshmi GS, Agnieszka K, Sathyavathi S, Steven JH, Antonina Z, Piotr P, Joanna P, Chakraborty S, Wouter H, Joseph Nathanael A, Bartłomiej J, Lukasz K, Lukasz K, Vignesh K (2024) Engineering of brewery waste-derived graphene quantum dots with ZnO nanoparticles for treating multi-drug resistant bacterial infections. J Environ Chem Eng 12:112263
20. Yuqi Y, Baolong W, Xu Z, Hongchuang L, Sen Y, Yifan Z, Yunhuang Y, Maili L, Chaohui Y, Peng H, Xin Z (2023) Activatable graphene quantum-dot-based nanotransformers for long-period tumor imaging and repeated photodynamic therapy. Adv Mater 35:2211337
21. Aida ZT, Carlos CG, Abu Bakar S, Johana BM, Masoud M, Grissel Trujillo-de S, Mario MA, Ashit Kumar P, Eduardo MG, Lawrence K, Marc JM, Sergio OM, Mallar R (2023) Nitrogen-functionalized graphene quantum dot incorporated GelMA microgels as fluorescent 3Dtissue constructs. Nanoscale 15:16277
22. Hongyuan Z, Jieqiong W, Shanshan W, Chenzhao W, Xiangyu Y, Xuewei S, Chunzhu J, Guoying S (2023) Nitrogen-doped graphene quantum dot-based portable fluorescent sensors for the sensitive detection of Fe^{3+} and ATP with logic gate operation. J Mater Chem B 11:6082
23. Jiaojiao D, Bo F, Yaqiong D, Ming Z, Xiaoda Y (2020) Vanadium coordination compounds loaded on graphene quantum dots (GQDs) exhibit improved pharmaceutical properties and enhanced antidiabetic effects. Nanoscale 12:9219–9230
24. Daniela I, Alessandro P, Stefania F, Laura DL, Annamaria M, Roberto R, Maria RB, Christophe P (2018) Graphene quantum dots based systems as HIV inhibitors. Bioconjugate Chem 29:3084–3093
25. Donghoon K, Je MY, Heehong H, Junghee L, Su HL, Seung PY, Myung JP, MinJun L, Seulah C, Sang HK, Saebom L, Seung-Hwan K, Sangjune K, Yong JP, Misaki K, Young-Ho L, Seokmin S, Seung RP, Sung JL, Seulki L, Byung HH, Han SK (2018) Graphene quantum dots prevent α-synucleinopathy in Parkinson's disease. Nat Nanotechnol 13:812–818
26. Pei-Ying L, Guang-Yu L, Jia-Huei Z, Jen-Hsien H, Er-Chieh C, Kuen Chan L (2020) GFP plasmid and chemoreagent conjugated with graphene quantum dots as a novel gene delivery platform for colon cancer inhibition in vitro and in vivo. ACS Appl Bio Mater 2020:5948–5956

27. Dong JK, Je MY, Yeonjoon S, Donghoon K, Insung K, Joonhee M, Mina P, Juhee K, Kyung SK, Byung HH (2021) Graphene quantum dots from carbonized coffee bean wastes for biomedical applications. Nanomaterials 11:1423
28. Ashwin KN, Swathi LB, Tuhin SS, Ramachandra Rao MS, Ganapathy K (2017) Oxygenated graphene quantum dots (GQDs) synthesized using laser ablation for long-term realtime tracking and imaging. RSC Adv 7:53822
29. Rajendra P, Nishant KJ, Amit SY, Manali J, Nalukurthi Naga VR, Mahadeo G, Gopal CK, Joao C, Rohit S (2021) Ultrahigh penetration and retention of graphene quantum dot mesoporous silica nanohybrids for image guided tumor regression. ACS Appl Bio Mater 4:1693–1703
30. Akshaya R, Mydhili R, Kalyani S, Pratishtha W, Arvind K, Sumana C, Sutapa G (2023) Biological evaluation of graphene quantum dots and nitrogen-doped graphene quantum dots as neurotrophic agents. ACS Appl Bio Mater 6:2237–2247
31. Yaoxuan C, Wei D, Yao J, Fangjie W, Fengna X, Jianmin W (2021) Graphene quantum dot-decorated luminescent porous silicon dressing for theranostics of diabetic wounds. Acta Biomater 131:544–554
32. Hanjun S, Nan G, Kai D, Jinsong R, Xiaogang Q (2014) Graphene quantum dots-band-aids used for wound disinfection. ACS Nano 8:6202–6210
33. Alina RV, Olga SZ, Bong HL, Olivia F, Steven N, Roberto GR, Giridhar RA, Anton VN (2022) Dual-mode fluorescence/ultrasound imaging with biocompatible metal-doped graphene quantum dots. ACS Biomater Sci Eng 8:4965–4975
34. Pei-Chun W, Darwin K, Jia-Lin W, Wei-Ru W, Kuan-Hao C, Chieh-Ying C, Ying-Chun C, Loganathan V, Chi-Ching K, Kostya KO, Wei-Hung C (2023) Plasma-enabled graphene quantum dot hydrogel—magnesium composites as bioactive scaffolds for in vivo bone defect repair. ACS Appl Mater Interfaces 15:44607–44620
35. Xianghua Z, Chunyi T, Tiansheng L, Li L, Xu L, Yuejun Y, Rushi L, Bin L (2020) Silver nanoparticles coated by green graphene quantum dots for accelerating the healing of MRSA-infected wounds. Biomater Sci 8:6670–6682

Chapter 17
Application of Graphene Quantum Dots in the Environmental Field

17.1 Introduction

Because of their excellent optoelectronic properties, great biocompatibility, and affordable production costs, GQDs are a strong contender to replace the widely used metal chalcogenides-based quantum dots. Furthermore, due to the bonding below and above the atomic plane, graphene exhibits extraordinary thermal and electrical conductivity when compared to normal semiconductor quantum dots [1]. Along with other physical characteristics like high surface area and surface grafting assistance through the conjugated network, GQDs have a wide range of excellent properties, including tunable photoluminescence (PL), remarkable spin property, exceptional UV-blocking capability, high photostability [2]. GQDs have a wide range of potential uses in several industries, such as energy-related, medicinal, sensing, light-emitting diodes, and catalysis [3].

17.2 Photocatalytic Application

GQDs with a layered graphene-like structure have an ultra-high surface area that makes them highly sensitive to environmental changes. They are also a viable (cell viability ~86.7%) eco-friendly choice for dye removal applications [4]. The N-GQD/CN that was synthesized for this study had a noticeably higher photocatalytic activity. The findings show that for RhB, the degradation rate is $k = 0.0303\ min^{-1}$ at 0.02 wt% N-GQD/CN. Even so, it demonstrated remarkable photocatalytic effects with BPA and HCl-TC. As co-catalysts, the N-GQDs g-CN improve light absorption and increase the number of active sites available for photocatalytic processes. Highly efficient photogenerated carrier separation is facilitated by N-GQD/CN photocatalysis. N-GQD/CN also showed outstanding reusability and stability. The most active substances in the reaction process in this system are $^{\bullet}O_2$ and $^1O_2$, however h^+ and

N. Manjubaashini et al., *Graphene Quantum Dots*, Materials Horizons: From Nature to Nanomaterials, https://doi.org/10.1007/978-981-97-5722-0_17

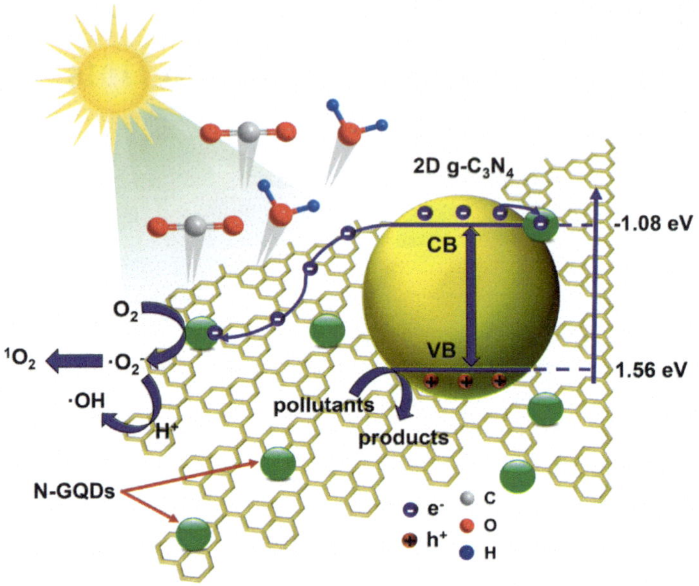

Fig. 17.1 Photocatalytic mechanism for N-GQD/CN system. Reprinted with permission from @ Appl Surf Sci 649, 159171 (2024)

•OH also have a big impact (Fig. 17.1). Consequently, broader light absorption and effective electron and hole separation are made possible by the production of composites utilizing non-metallic catalysts [5]. The creation of 0D/2D photocatalysts and effective pollutant photocatalytic degradation have enormous potential for use in the environmental water purification industry.

The goal of using quantum dots as a photocatalyst is to increase hydrogen generation by harnessing a significant percentage of the solar energy spectrum. The quantum dots under investigation are all suitable for photocatalytic hydrogen evolution based on the alignment of their frontier molecular orbitals (FMOs) concerning the water's redox potential. Visible light absorption is demonstrated by the energy gap between FMOs, which changes from 3.01 to 2.32 eV depending on the quantum dot size (Fig. 17.2). Except hydrogen adsorption, the Gibbs free energy estimate indicates that the production of intermediate species at $pH = 7$ is energetically unfavorable in the absence of light [6].

Using an easy impregnation technique, blue luminous GQD are integrated into NiTi-based layered double hydroxide nanosheets (NiTi-LDH). GQD/NiTi-LDH exhibits superior performance compared to ordinary TiO_2 P25, with outstanding NO photo-removal activity (61%) and little NO_2 release (<0.6%). Remarkably, independent of the duration of the prior irradiation, GQD/NiTi-LDH maintained a NO removal efficacy higher than 20% for up to 75 min following the conclusion of the light irradiation phase. This De-NO_x post-photocatalytic method is currently the most

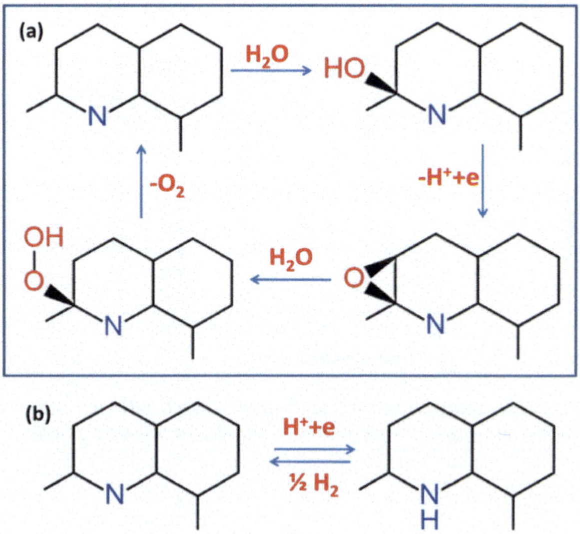

Fig. 17.2 **a** Oxidation, and **b** reduction of water, which leads to overall water splitting of N-GQD (28). Reprinted with permission from @ Int J Hydrog Energy 53, 1076–1082 (2024)

successful that has been documented. The electron migration from GQD to NiTi-LDH increases the electron storage capacity of NiTi-LDH under light excitation, which explains the dark/light mechanism. When the light is turned off, electrons are liberated and contribute to the production of superoxide radicals. Quantum yields for the GQDs ranged from 0.3 to 4.9%. The best photocatalytic performance was demonstrated by GQDs generated for 14 h at 150 °C. There are two stages in the first-order mechanism that caused the photodegradation of the MO. According to Luana et al. [7], step 1 is ascribed to the MO's adsorption by the GQDs, while stage 2 is ascribed to the MO's degradation by e-BC and/or O_2 radicals. Green synthesis was used to create the GQDs with photocatalytic activity, which work together to promote a sustainable and circular dynamic in industrial processes (Fig. 17.3).

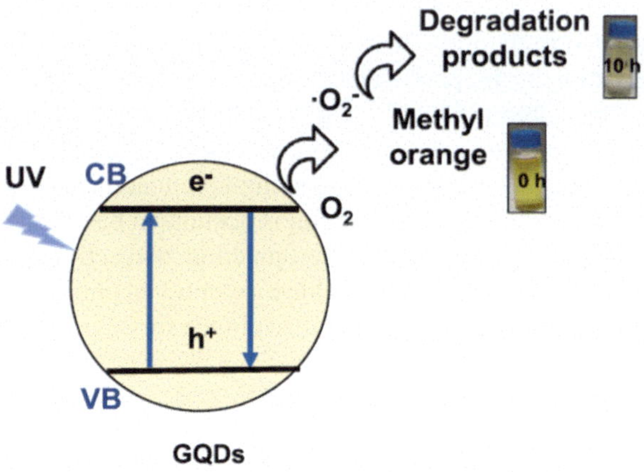

Fig. 17.3 Mechanism of methyl orange degradation by GQDs under UV light irradiation. Reprinted with permission from @ Colloids Surf A 651, 129442 (2022)

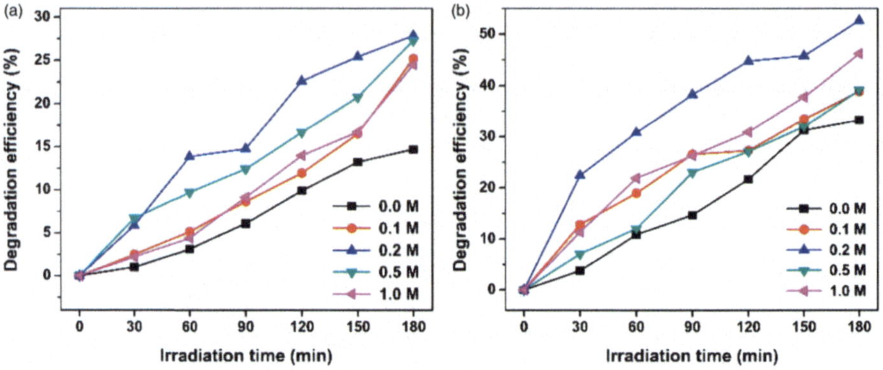

Fig. 17.4 Degradation efficiency **a** MO, and **b** MB after the photocatalytic process. Reprinted with permission from @ J Reinf Plast Compos 39, 81–94 (2020)

The improved carrier lifetime for enhanced photocatalytic degradation is thought to be significantly impacted by the high crystallinity of graphene quantum dots-ZnO nanocomposites. To improve the functional properties of photocatalytic material, the enhanced crystallinity lengthens the carrier lifespan on the material surface. Under UV light, a photocatalytic test involving methyl orange and methylene blue was carried out. For both organic dyes, the degradation rate constant reaches its maximum value at the proper preparation conditions of GQDs-ZnO nanocomposites. To decompose commercial glyphosate herbicide pollutants for an agricultural wastewater treatment experiment, the graphene quantum dots-ZnO nanocomposites were then applied. When compared to pure ZnO, the graphene quantum dots-ZnO nanocomposites exhibit a 23% increase in the degradation rate constant, indicating an improvement in the photocatalytic process [8]. Because of its low cost and environmentally acceptable nature, graphene quantum dots-ZnO nanocomposites can be employed as a photocatalytic material for decomposing organic colors and commercial pesticide pollutants (Fig. 17.4).

These ZnO-GQD heterojunctions' enhanced photocatalytic activity may be related to their effective charge carrier separation, which lowers the rate of recombination at the photocatalyst interface. The ZGQD2 heterojunction's high specific surface area (353.447 m^2/g^{-1}) not only enhances light absorption from the UV to the visible range but also has a substantial adsorption capacity for contaminants over the catalyst surface, leading to high photoactivity (Fig. 17.5). A tenable mechanism for the photocatalytic reaction is postulated based on the observed results, band gap alignment at ZnO-GQD heterojunction, and active species trapping experiments [9]. ZnO-GQD heterojunctions have exceptional photostability and recyclability, making them an attractive photocatalyst choice for environmental remediation applications.

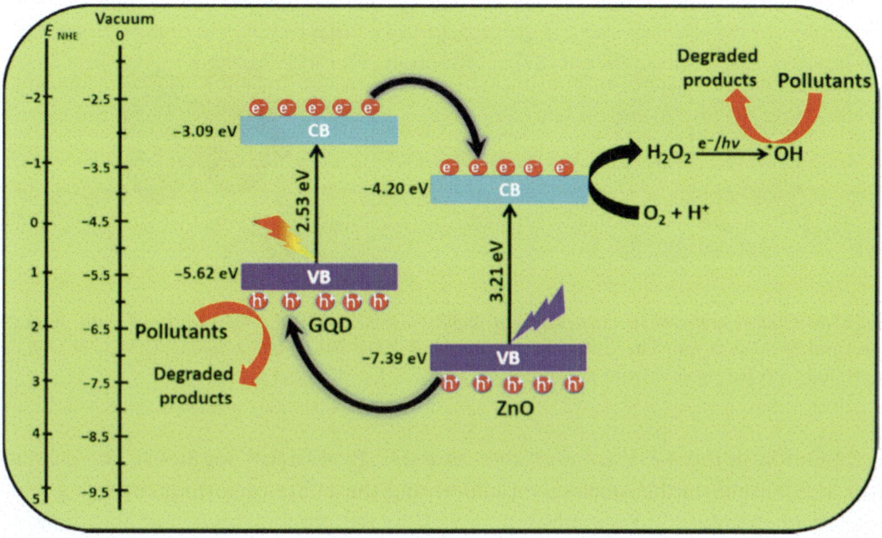

Fig. 17.5 Proposed charge transfer mechanism in GQD/ZnO leading to dye degradation. Reprinted with permission from @ Appl Surf Sci 447: 802–815 (2018)

17.3 Heterojunction Photocatalytic Oxidation and Reduction System

Using the photodegradation of bisphenol A as a model system, the photoactive performance of the proposed ternary heterojunctions in aqueous solution is examined. $BiOCl/BiVO_4$/N-GQD ternary heterojunctions also exhibit a significantly increased rate of photodegradation (Fig. 17.6a). Through electrochemical studies, photoluminescence, and electron spin resonance, the suggested charge separation and transfer process of $BiOCl/BiVO_4$/N-GQD ternary heterojunctions for improved photoactivity was inferred. The findings show that N-GQDs and $BiOCl/BiVO_4$ binary heterojunctions created a Z-scheme charge process, which resulted in effective charge carrier separation and potent photocatalytic performance [10].

In the simultaneous photocatalytic oxidation and reduction system, the 5%GQDs/TCN-0.4 sample exhibits the greatest rate of organic pollutant degradation and photocatalytic hydrogen production under simulated solar irradiation. Additionally, over the 5%GQDs/TCN-0.4 catalyst, the photocatalytic H_2 evolution rates in the solutions of 4-NP, CIP, and DEP are all greater than those in the pure water system. Furthermore, compared to the solutions of CIP and DEP, the H_2 evolution rate in the 4-NP solution is lower. As a result, 4-NP photodegrades at a faster rate than CIP and DEP (Fig. 17.6b). According to investigations using liquid chromatography-mass spectrometry and density functional theory, some photogenerated electrons were employed in the 4-NP photodegradation process, but not in the CIP or DEP photodegradation processes. Additionally, it results in a lower photocatalytic rate of

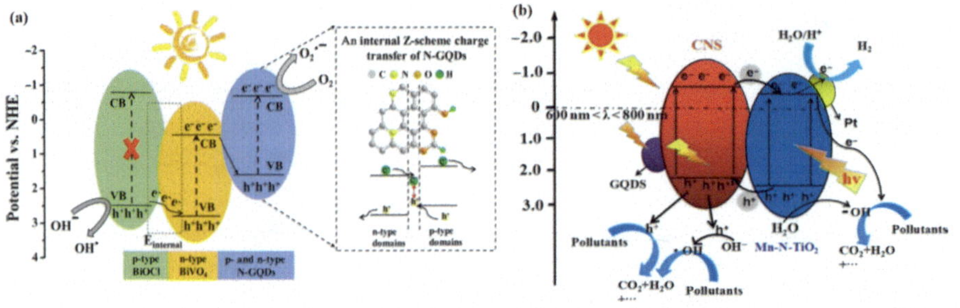

Fig. 17.6 Charge transfer mechanism of **a** GQDs/Mn-N-TiO_2-g-C_3N_4. Reprinted with permission from @ ACS Appl Mater Interfaces 9, 38832–38841 (2017); and **b** BiOCl/$BiVO_4$/N-GQDs. Reprinted with permission from @ Appl Catal B 227, 312–321 (2018)

H_2 evolution in the 4-NP solution than in the CIP and DEP solutions. The current work sheds light on the mechanism underlying the effect of various organic pollutants on photocatalytic H_2 evolution as well as the photocatalytic enhancement of the GQDs/TCN-0.4 catalyst for the first time [11].

Rhodamine B (RhB) was successfully degraded by photocatalysis using amine-modified GQD/CdS (GQDs/CdS) nanobelts when exposed to visible light. After 150 min, the degradation efficiency of RhB sample over 0.05 g of catalysts was 97.40%, which is much higher than the efficiency of pure CdS. Using DMPO as a spin trap, electron paramagnetic resonance (EPR) spectroscopy directly demonstrated that COH and CO_2^- were the reactive oxidative species (Fig. 17.7a). A potential mechanism of RhB photocatalytic degradation by GQDs/CdS in visible light was hypothesized, which is in line with the experimental data. This research could lead to the development of eco-friendly photocatalysts for the breakdown of organic dyes and water purification [12]. Electrostatic interactions were used to create mesoporous carbon nitride (mpg-C_3N_4) composites modified with hydroxyl-GQDs (Fig. 17.7b). In addition to being evenly distributed across the mpg-C_3N_4 surface, OH-GQDs also functioned as electron transfer routes, thereby boosting photocatalytic activity. $O_2^{\bullet-}$ is the primary active species for the photodegradation of contaminants, according to the ESR investigation and several trapping experiments [13].

The steady-state and time-resolved photoluminescence tests demonstrate that the TiO_2-GQD heterojunction demonstrates increased photocatalytic degradation (97%) of MB because of the simple interfacial charge separation. It's interesting to note that the TiO_2-GQD heterojunction's photoluminescence intensity partially faded, suggesting that electrons were transferred from GQDs to TiO_2. TiO_2-GQD hybrid has a first-order degradation rate constant that is 5.2 times higher than TiO_2's (Fig. 17.8a). The results of the free radical scavenger test indicated that the %OH radical was more important in the degradation of MB than the $^{\bullet-}O_2$ radical [14]. The increased photocatalytic performance was attributed to the combined impacts of the produced nanojunction and the electron collection effect and up-conversion photoluminescence capabilities of the N-GQDs. An effective molecular oxygen activation process was accomplished by creating a bulk-to-surface conduit between N-GQDs

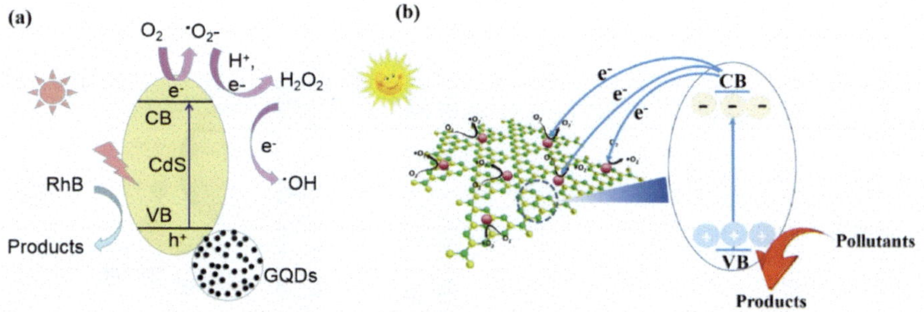

Fig. 17.7 **a** Photocatalytic mechanism of RhB degradation by GQDs/CdS under visible light irradiation [12]; and **b** Schematic model for the photocatalytic mechanism of GQDs/mpg-C_3N_4 composites. Reprinted with permission from @ Appl Catal B 207, 429–437 (2017)

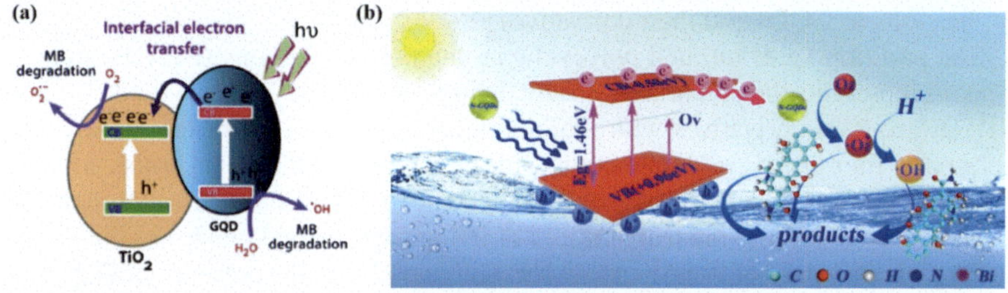

Fig. 17.8 **a** Relative MB concentration as a function of irradiation time in the absence and presence of different catalyst-scavengers. Reprinted with permission from @ Appl Catal B 224: 960–972 (2018); and **b** Schematic illustration of the proposed photocatalytic mechanism over the GBO-0.4 composite toward TC degradation under full-spectrum light irradiation. Reprinted with permission from @ Appl Catalysis B 277, 119218 (2020)

and BiO_{2-x} (Fig. 17.8b). The N-GQDs/BiO_{2-x} photocatalyst has significant application possibilities, as confirmed by its high photostability and mineralization ability toward tetracycline degradation [15].

17.4 Heterogeneous Photocatalytic System

For heterogeneous photocatalysis and photoelectrocatalysis under ambient conditions, recent developments of GQDs-semiconductor composited systems include non-selective pollutant degradation, selective organic transformation, hydrogen or oxygen production (Fig. 17.9a), CO_2 reduction, and photoelectrochemical (PEC) water splitting [16, 17]. The current review article is expected to offer new insights for investigating a wide range of GQDs-semiconductor composited systems for advanced

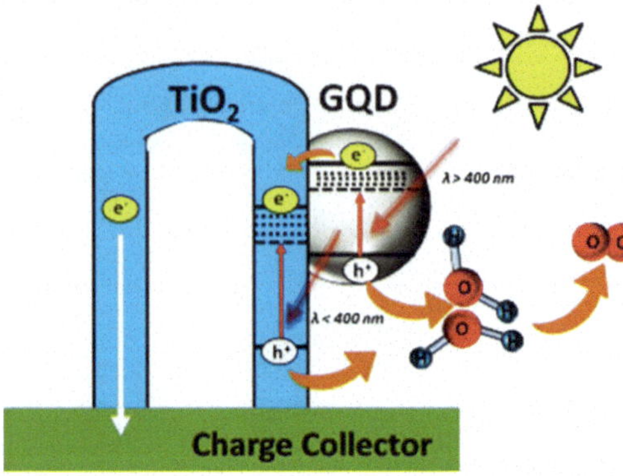

Fig. 17.9 Mechanism of photocharge carrier transfer at GQD/TiO_2 interface. The transfer process produces oxygen gas molecules by oxidation using photo holes from valence band of GQD and TiO_2. Reprinted with permission from @ Electrochim Acta 187, 249–255 (2016)

solar energy conversion, while also providing enriched information on the promising applications of GQDs and their derivatives [18].

The partitioning of nitrated polycyclic aromatic hydrocarbons (NPAHs) in crop systems represents the potential exposure to human health through the food channel. NPAHs are PAH derivatives with more harmful impacts on ecosystems. Based on fiber-optic fluorimetry in conjunction with GQDs as a fluorescent probe, a novel technique for in situ detection of 9-nitro anthracene (9-NAnt) and 3-nitrofluoranthene (3-NFla) adsorbed onto the leaf surfaces of living soybean and maize seedlings was developed. Qe values of 9-NAnt and 3-NFla increased with the presence of GQDs by 22.8 and 16.2%, respectively, compared to 28.7 and 20.3% for soybeans and maize. This was primarily due to the expansion of the surface area for adsorbing NPHAs caused by GQDs and the stronger electrostatic interaction between the $–NO_2$ of NPAH molecules and the functional groups (e.g., –COOH, –OH) of GQDs outer surfaces (Fig. 17.10a, b).

Furthermore, the steric effects that led to the easier accessibility of 3-NFla's $–NO_2$ to the outer surface of GQDs may account for the various enhancement degrees in the order of 3-NFla > 9-NAnt. In conclusion, crop security may be threatened by the GQDs' higher retention of NPAHs on crop leaf surfaces [19]. Materials for the breakdown of methyl orange (MO) under visible light irradiation ($\lambda > 400$ nm): N, S co-doped graphene quantum dots (N, S-GQDs)-reduced graphene oxide (rGO) TiO_2 nanotubes (TiO_2NT) (Fig. 17.10c) [20].

A straightforward defect engineering method that involves embedding GQDs during the self-assembly process of MIL-101(Cr) to modify the framework, cause mesoporosity, and create huge pore volumes. In comparison to the conventional MIL-101 (V_{meso}: 0.17 and V_{tot}: 1.22 cm^3/g), MIL-101@GQD-3 (V_{meso}: 0.68 and V_{tot}: 1.87 cm^3/g) demonstrated 300.0 and 53.3% greater meso and total pore volume. This led to 1.7- and 2.8-times greater benzene and toluene loading at 1 bar and 25 °C. Furthermore, we discovered that MIL-101@GQD-3 maintained its superiority with excellent cycle capacity and energy-efficient regeneration throughout a broad range of VOC concentrations and operating temperatures (25–55 °C). Based on blue luminous GQDs, a safe, sensitive, and selective nanoprobe was created for the detection

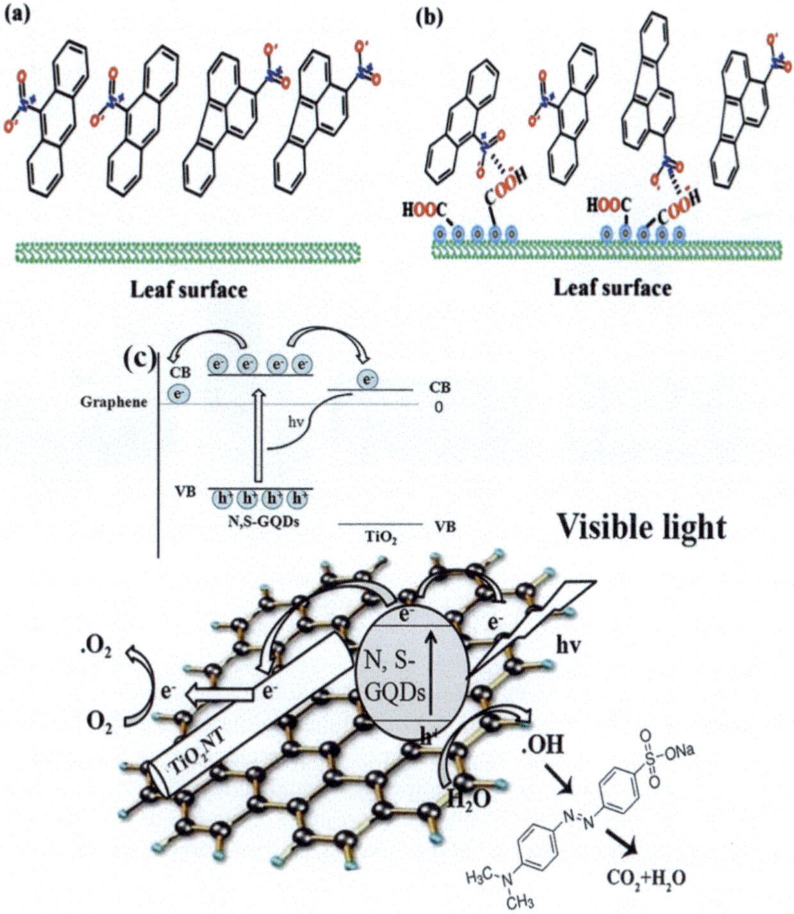

Fig. 17.10 **a, b** The interactions between adsorbent surface and NPAH molecules. Reprinted with permission from @ Ecotoxicol Environ Saf 162, 10–16 (2018); and **c** Photocatalytic activity of N, S co-doped GQD-graphene-TiO_2 nanotubes composite. Reprinted with permission from @ J Alloy Compd 691, 369–377 (2017)

of the endocrine disruptors, o- and p-nitrophenols (ONP and PNP, respectively). Excellent linearity is demonstrated by the fluorescent nanoprobe throughout broad dynamic ranges of 0.30–60.0 and 0.20–40.0 μg/mL, with quantification limits of 0.23 and 0.11 μg/mL and detection limits of 0.07 and 0.03 μg/mL for ONP and PNP, respectively. Using real samples from tap, river, and rainwaters, the sensing system was completely validated and used successfully to detect ONP and PNP with outstanding selectivity and recovery (98.6–101.5%) (Fig. 17.11). For the first time, a free piece of software was used in the construction of a paper-based analytical instrument that uses smartphone image processing to help with the in-field study of environmental waters quickly [21].

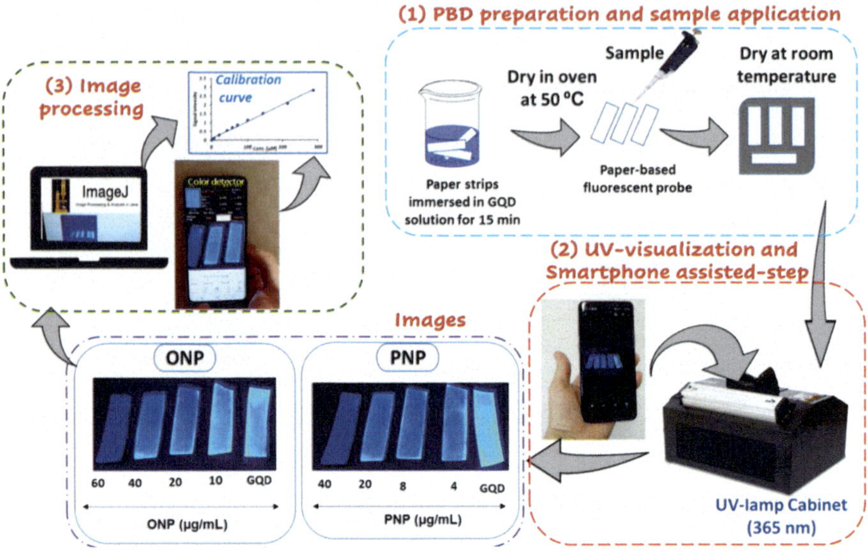

Fig. 17.11 Illustration of the steps for image processing-assisted paper-based sensing of ONP and PNP in water samples. Reprinted with permission from @ Microchem J 158, 105241 (2020)

The produced green nanomaterials, GQDs, were investigated for their catalytic function using a cationic dye as a model molecule. Mahmoud et al. [22] conducted a study on the effects of various experimental parameters on the degrading efficiency of GQD (Fig. 17.12a). These parameters included the pH of the dye solution, contacting time, dosage of GQDs, and initial concentration of NF dye. GQDs with excitation-independent emissions, but the larger size distribution of saccharides, such as fructose, chitosan, lignin, cellulose, and starch, enables the synthesis of GQDs with excitation-dependent emissions. Furthermore, the one-step synthesis of N, S-co-doped GQDs is made easier by the presence of heteroatoms like N and/or S in the chemical structures of lignin and chitosan as well as the highly reactive species produced by the plasma (Fig. 17.12b). These GQDs provide selective detection of hazardous environmental contaminants with a low limit of detection of 7.4 nM [23].

BiOCl nanosheets with oxygen vacancies were loaded with graphene quantum dots (GQDs) to create an effective GQDs/BiOCl-VO heterojunction photocatalyst. With visible light irradiation, the as-prepared GQDs/BiOCl-VO sample demonstrated enhanced efficiency for RhB degradation, owing to the combined benefits of oxygen vacancy, GQDs, and dye sensitization. In 20 min, a 2 wt% GQDs/BiOCl-VO composite successfully broke down 98% of RhB (Fig. 17.13a). Through ESR analysis and radical trapping tests, the primary active species were determined to be hole (h^+) and superoxide radical ($^{\bullet}O_2^-$) [24]. GQDs were used to investigate the photocatalytic degradation of Celestine Blue (CB) under visible light irradiation. A thorough investigation was conducted into the effects of pH, contacting time, GQD dosage, and initial dye concentration on the CB degradation kinetics (Fig. 17.13b).

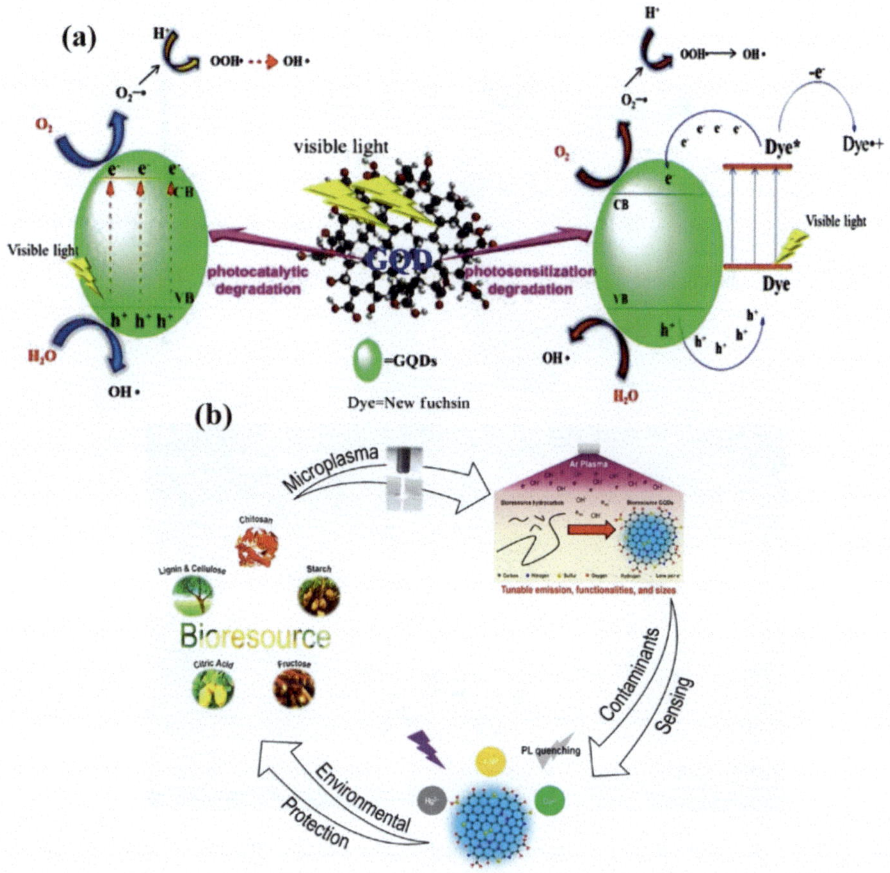

Fig. 17.12 **a** Schematic representation of the proposed mechanism for degradation of NF under visible light irradiation. Reprinted with permission from @ J Mol Catal A 409: 102–109 (2015); and **b** Microplasma-enabled sustainable synthesis of bioresource-derived heteroatom-doped GQDs as nanoprobes for diverse environmental contaminants. Reprinted with permission from @ ACS Appl Mater Interfaces 2022, 52289–52300 (2022)

The breakdown products that evolved during the photocatalytic degradation were identified using mass spectrometry studies [25].

A titanium dioxide (TiO_2) nanocubic-like photocatalyst that is sensitive to sunlight and N-doped (N-GQDs) has been created to improve light harvesting, particularly in the visible and near-infrared regions. Its exceptional quality allows it to optimize absorption over the whole sunlight spectrum. On the other hand, large oxygen vacancies brought about by the rise in N-GQD weight % inhibit the production of active radicals (Fig. 17.14a). As a result, the process of eliminating BPA took longer than it did for N-GQDs with a lower weight percentage [26]. GQDs in the TiO_2 lattice functioned as an effective electron transporter and photosensitizer, which is

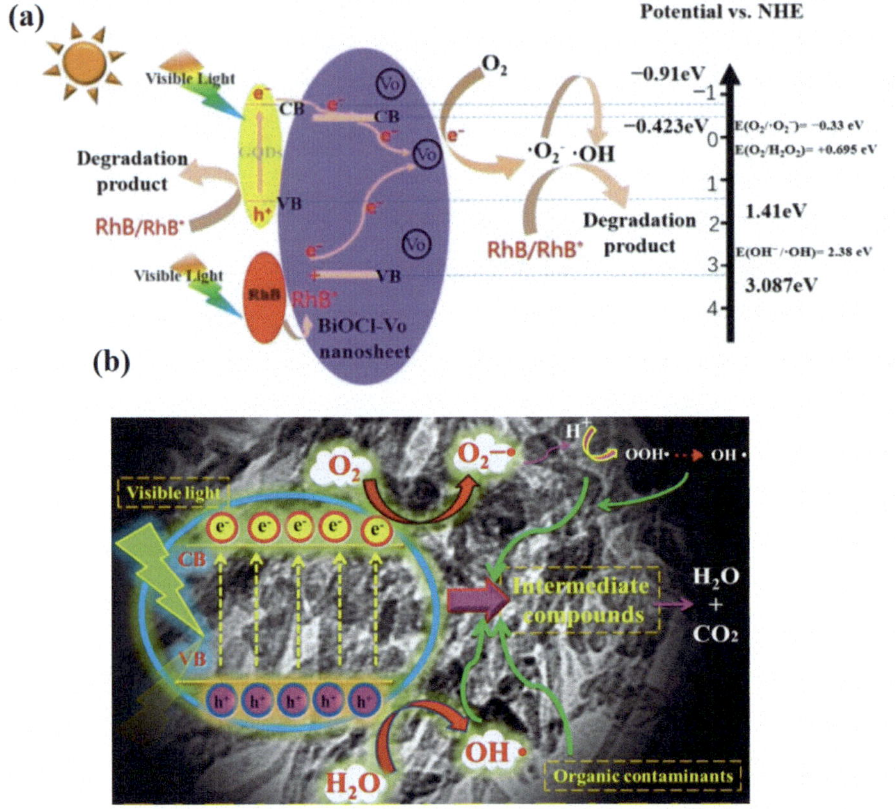

Fig. 17.13 **a** Degradation mechanism of RhB catalyzed by 2 wt% GQDs/BiOCl-VO under visible light irradiation (420–800 nm). Reprinted with permission from @ Molecules 29: 1362 (2024); and **b** Mechanism of CB degradation under visible light irradiation of the photocatalytic pathways. Reprinted with permission from @ J Mater Sci Mater Electron 28: 5135–5143 (2017)

why the produced nanocomposite against RB5 underwent excellent photocatalytic decolorization (Fig. 17.14b) [27].

Doping the GQDs improved their adsorption capabilities by favoring a multilayer adsorption process over monolayer adsorption (KF & Kd). At approximately 1377 mg g^{-1}, hexadecylamine sulfur and nitrogen co-doped GQDs (C16-SNGQDs) exhibited the largest adsorption capacity (qm). Phenanthrene interacts endothermically with pure GQDs, whereas it adsorbs exothermically spontaneously on modified GQDs. Phenanthrene adsorption is largely influenced by the presence of phosphate and sulfate anions, and it is optimal at room temperature and an acidic pH (Fig. 17.15). The physicochemical and adsorption performance of C16-SNGQDs was improved by co-doping with nitrogen and sulfur groups and by changing their hydrophilic to hydrophobic form. As a result, these materials may be used to remove hydrophobic organic contaminants from water [28].

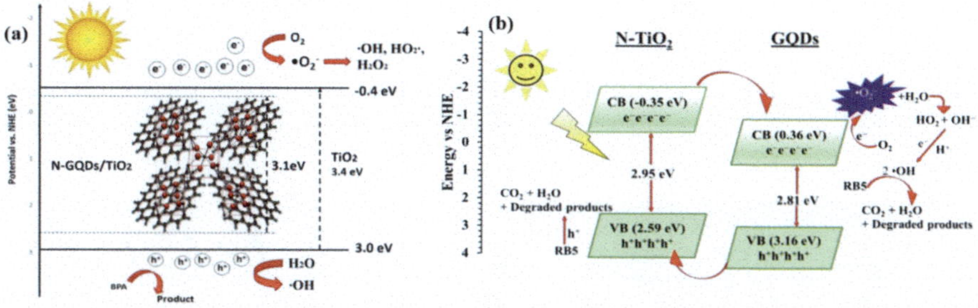

Fig. 17.14 **a** Photocatalytic mechanism of 0.5 N-GQDs/TiO_2 under sunlight irradiation. Reprinted with permission from @ Environ Sci Pollut Res 26, 3455–3464 (2019); **b** Effect of different trapping agents on the photocatalytic decolorization of RB5 using GQD–20N–TiO_2–300. Reprinted with permission from @ Environ Sci Pollut Res 30, 124992–125005 (2023)

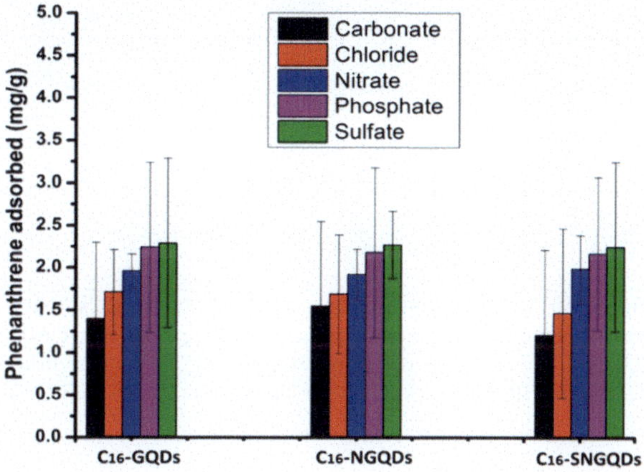

Fig. 17.15 Effect of various anions on the adsorption of phenanthrene by hexadecyl-GQDs [28]

17.5 Z-Scheme Photocatalytic System

A GQD's narrow bandgap increases its light absorption range and fraction. The intramolecular Z-scheme structure helps to couple electron–hole pairs that have been photoexcited from recombining on the same molecule (Fig. 17.16a–d). The functionalized GQDs are used for CO_2 reduction and water splitting under visible light by utilizing these two properties. When compared to alternative photocatalysts, GQDs have several benefits. They are made using inexpensive, readily available, and environmentally benign carbon materials. Additionally, GQDs' molecular size allows for good dispersion in aqueous solution and fully exposed active sites [29].

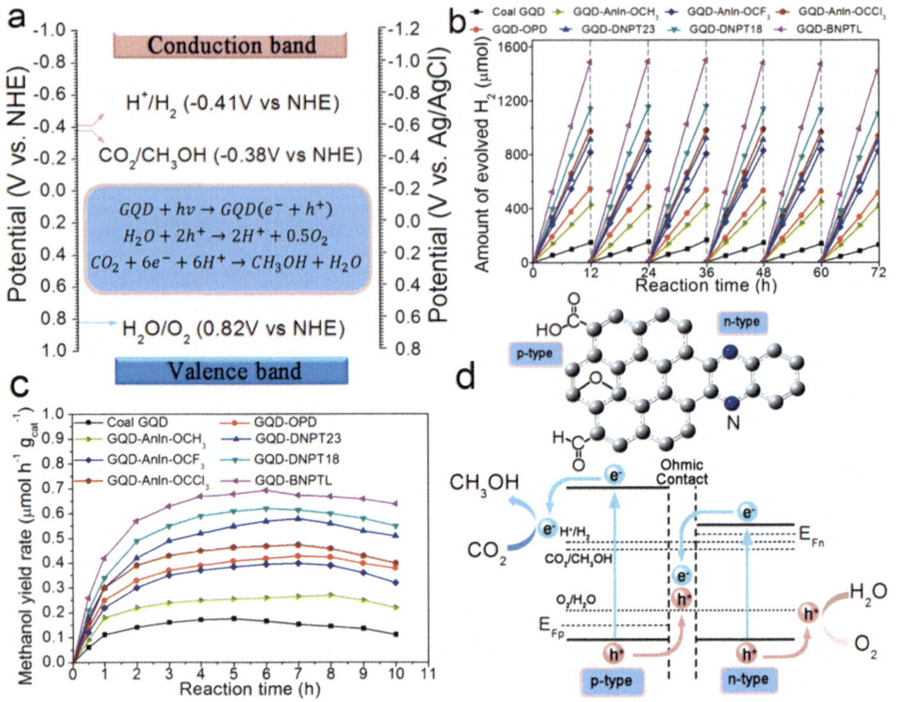

Fig. 17.16 **a** Illustration of standard redox potentials, CBM and VBM of GQD, and CO_2 reduction process; Time courses of H_2 evolution of **b** CO_2 reduction, **c** All GQD types under visible light (420–800 nm), and **d** Schematic illustration of GQD Z-scheme-based photocatalysis of CO_2 reduction. Reprinted with permission from @ ACS Nano 12, 3523–3532 (2018)

References

1. Anuja B, Sowbaranigha C, Folarin E (2021) TiO_2-graphene quantum dots nanocomposites for photocatalysis in energy and biomedical applications. Catalysts 11:319
2. Kaur A, Pandey K, Kaur R, Vashishat N, Kaur M (2022) Nanocomposites of carbon quantum dots and graphene quantum dots: environmental applications as sensors. Chemosensors 10:367
3. Kurniawan D, Chiang WH (2020) Chapter 7—Synthesis and applications of graphene quantum dots in ambipolar materials and devices. In: Zhou Y, Han S (eds) Ambipolar materials and devices. The Royal Society of Chemistry, pp 131–173
4. Niladri S, Gyanaranjan S, Sarat KS (2020) Graphene quantum dot decorated magnetic graphene oxide filled polyvinyl alcohol hybrid hydrogel for removal of dye pollutants. J Mol Liq 302:112591
5. Huimin H, Bin W, Qi T, Shihao J, Jinyuan L, Huaming L, Chongtai W, Hui X, Yingjie H (2024) Non-metallic nitrogen-doped graphene quantum dots coupled with g-C_3N_4 achieve efficient photocatalytic performance. Appl Surf Sci 649:159171
6. Bikash M, Prasanjit S (2024) Nitrogen-doped graphene quantum dots for extensive production of hydrogen: a hybrid density functional study. Int J Hydrog Energy 53:1076–1082
7. Luana VD, de Camilla Rocha OF, da Jean Castro C, Mayra AN, de Andre Fernando O, Renata PL (2022) Green synthesis optimization of graphene quantum dots by Doehlert design for dye photodegradation application. Colloids Surf A 651:129442

8. Suchada P, Pichitchai P, Sawitree W, Supphadate S, Sutthipoj W (2020) Modified graphene quantum dots-zinc oxide nanocomposites for photocatalytic degradation of organic dyes and commercial herbicide. J Reinf Plast Compos 39:81–94
9. Kumar S, Dhiman A, Sudhagar P, Krishnan V (2018) ZnO-graphene quantum dots heterojunctions for natural sunlight-driven photocatalytic environmental remediation. Appl Surf Sci 447:802–815
10. Zhu M, Liu Q, Chen W, Yin Y, Ge L, Li H, Wang K (2017) Boosting the visible-light photoactivity of BiOCl/$BiVO_4$/N-GQD ternary heterojunctions based on internal Z-scheme charge transfer of N-GQDs: simultaneous band gap narrowing and carrier lifetime prolonging. ACS Appl Mater Interfaces 9:38832–38841
11. Nie Y, Yu F, Wang L, Xing Q, Liu X, Pei Y, Zou J, Dai W, Li Y, Sui SL (2018) Photocatalytic degradation of organic pollutants coupled with simultaneous photocatalytic H_2 evolution over graphene quantum dots/Mn-N-TiO_2/g-C_3N_4 composite catalysts: performance and mechanism. Appl Catal B 227:312–321
12. Xiangfeng L, Yu C, Diwen Z, Menglin C, Weixia L, Huazhang G (2024) Aminated graphene quantum dots/CdS nanobelts for enhanced photocatalytic degradation of RhB dye under visible light. RSC Adv 14:255
13. Jinyuan L, Hui X, Yuanguo X, Yanhua S, Jiabiao L, Yan Z, Liang W, Liying H, Haiyan J, Huaming L (2017) Graphene quantum dots modified mesoporous graphite carbon nitride with significant enhancement of photocatalytic activity. Appl Catal B 207:429–437
14. Gone R, Jitendra K, Giri PK (2018) Interfacial charge transfer in oxygen-deficient TiO_2-graphene quantum dot hybrid and its influence on the enhanced visible light photocatalysis. Appl Catal B 224:960–972
15. Chen F, Liu LL, Zhang YJ, Wu JH, Huang GX, Yang Q, Chen JJ, Yu HQ (2020) Enhanced full solar spectrum photocatalysis by nitrogen-doped graphene quantum dots decorated BiO_{2-x} nanosheets: ultrafast charge transfer and molecular oxygen activation. Appl Catal B 277:119218
16. Zhiping Z, Shufeng C, Timothy Thatt YT, Fang-Xing X (2018) Graphene quantum dots (GQDs) and its derivatives for multifarious photocatalysis and photoelectrocatalysis. Catalysis 315:171–183
17. Zeng Z, Xiao FX, Gui X, Wang R, Liu B, Yang Tan TT (2016) Layer-by-layer assembly of nitrogen-doped graphene quantum dots monolayer decorated one-dimensional semiconductor nano architectures for solar-driven water splitting. J Mater Chem A 4:16383–16393
18. Sudhagar P, Herraiz-Cardona I, Park H, Song T, Noh SH, Gimenez S, Sero IM, Fabregat-Santiago F, Bisquert J, Terashima C, Paik U, Kang YS, Fujishima A, Han TH (2016) Exploring graphene quantum dots/TiO_2 interface in photoelectrochemical reactions: solar to fuel conversion. Electrochim Acta 187:249–255
19. Haifeng S, Yanli N, Ruijie F, Ruiyao M (2018) Novel method for in situ investigation into graphene quantum dots effects on the adsorption of nitrated polycyclic aromatic hydrocarbons by crop leaf surfaces. Ecotoxicol Environ Saf 162:10–16
20. Hongwei T, Kai S, Xiaoying H, Liang Q, Weitao Z (2017) N, S co-doped graphene quantum dots-graphene-TiO_2 nanotubes composite with enhanced photocatalytic activity. J Alloy Compd 691:369–377
21. El-Shaheny R, Yoshida S, Fuchigami T (2020) Graphene quantum dots as a nanoprobe for analysis of *o*- and *p*-nitrophenols in environmental water adopting conventional fluorometry and smartphone image processing-assisted paper-based analytical device: In-depth study of sensing mechanisms. Microchem J 158:105241
22. Mahmoud R, Maryamosadat M, Hamid RR (2015) Graphene quantum dots as novel and green nano-materials for the visible-light-driven photocatalytic degradation of cationic dye. J Mol Catal A 409:102–109
23. Darwin K, Neha S, Michael RR, Yu-Yuan C, Yan-Teng C, Guan-Xian W, Yen-Yu Y, Pei-Chun Y, Kostya Ken O, Wei-Hung C (2022) Plasma nanoengineering of bioresource-derived graphene quantum dots as ultrasensitive environmental nanoprobes. ACS Appl Mater Interfaces 2022:52289–52300

24. Zisheng S, Wei C, Yin H, Fen Z, Lingling W, Dan Z, Xuanye C, Sugang M (2024) Boosting visible-light photocatalytic activity of BiOCl nanosheets via synergetic effect of oxygen vacancy engineering and graphene quantum dots-sensitization. Molecules 29:1362
25. Mahmoud R, Maryamosadat M, Ali D, Hamid RR (2017) Application of graphene quantum dots as green homogenous nanophotocatalyst in the visible-light-driven photolytic process. J Mater Sci Mater Electron 28:5135–5143
26. Ping FL, Kah HL, Lan CS, Azrina AA, Pichiah S (2019) Amalgamation of N-graphene quantum dots with nanocubic like TiO_2: an insight study of sunlight sensitive photocatalysis. Environ Sci Pollut Res 26:3455–3464
27. Muhammad SK, Nadia R, Saeed R, Liu C, Ahson JS, Muhammad A, Iftikhar Z, Muhammad A, Farhan H, Muhammad B (2023) Improved photocatalytic decolorization of reactive black 5 dye through synthesis of graphene quantum dots–nitrogen-doped TiO_2. Environ Sci Pollut Res 30:124992–125005
28. Gugu K, Adedapo OA, Forbes PBC (2022) Hexadecylamine functionalised graphene quantum dots as suitable nano-adsorbents for phenanthrene removal from aqueous solution. RSC Adv 12:23922
29. Yibo Y, Jie C, Nan L, Jingqi T, Kaixin L, Jizhou J, Jiyang L, Qinghua T, Peng C (2018) Systematic bandgap engineering of graphene quantum dots and applications for photocatalytic water splitting and CO_2 reduction. ACS Nano 12:3523–3532

Chapter 18
Biocompatibility and Toxicity of Graphene Quantum Dots

18.1 Introduction

GQDs have attracted a lot of interest lately because of their special qualities and possible uses in many industries, including biomedicine. For their safe use in biomedical applications, it is essential to comprehend their toxicity and biocompatibility [1].

18.1.1 *Biocompatibility*

Numerous investigations have revealed that GQDs have outstanding biocompatibility, particularly when produced using the right procedures and surface functionalization strategies. Functionalized GQDs are good prospects for a variety of biomedical applications, including drug administration, biosensing, and bioimaging, due to reports of their great biocompatibility and minimal cytotoxicity in vitro. Surface alterations are essential for increasing biocompatibility since they lessen the possibility of cytotoxicity and increase stability in biological settings [2].

18.1.2 *Toxicity*

There are still worries about GQD toxicity despite its encouraging biocompatibility, especially in light of its long-term impacts and interactions with biological systems. Size, surface functionalization, concentration, length of exposure, and mode of delivery are among the variables affecting GQD toxicity. Research conducted in vitro has indicated that extended exposure to high concentrations of GQDs can lead to cytotoxicity through the production of reactive oxygen species (ROS), disruption of cell

N. Manjubaashini et al., *Graphene Quantum Dots*, Materials Horizons: From Nature to Nanomaterials, https://doi.org/10.1007/978-981-97-5722-0_18

membranes, and damage to DNA. Regarding the systemic toxicity of GQDs, in vivo investigations have produced contradictory findings; some have shown little negative effects, while others have suggested that organ buildup and inflammatory reactions may occur. Long-term toxicity profiles of GQDs are significantly influenced by their biodistribution and clearance mechanisms in vivo [3].

18.2 Mitigation Strategies

Enhancing GQDs' biocompatibility and lowering their potential toxicity can be achieved through surface functionalization with biocompatible moieties like polymers, peptides, or biomolecules. To guarantee the synthesis of GQDs with consistent characteristics and low impurities, which can increase toxicity, adequate characterization and quality control are crucial. To fully assess the biocompatibility and long-term toxicity of GQDs, more investigation is required, particularly in complex biological systems and pertinent animal models [4].

GQDs are innovative biomaterials that have a wide range of uses in the biological sciences. GQDs showed minimal cytotoxicity to living cells and were biocompatible. To find out whether the GQDs can be employed in people, however, animal research would be more helpful [5]. Because GQDs are so thin just one carbon atom thick and a few nanometers in size they can be easily absorbed by the body and subsequently eliminated. GQDs can also be applied topically or orally as medication carriers. As ultrasmall nanoparticles, GQDs are in contrast to GO or rGO. Consequently, it is difficult to see injected GQDs in mouse organs when using an optical microscope [6]. GQDs administered intravenously or intraperitoneally to mice or rats showed little to no toxicity [7]. Nevertheless, GQDs may cause harm to tissues and cells when exposed to laser radiation. From ultraviolet (UV) to red or near-infrared (NIR) wavelengths, GQDs have a wide absorption spectrum [8]. ROS would be produced by GQDs under appropriate laser light. According to Markovic et al. [9], ROS such $^{\bullet}OH$ and $^{1}O_2$ are harmful to biomolecules like DNA and can cause necrosis or apoptosis in cells. GQDs' phototoxic quality is employed in localized photodynamic and photothermal cancer treatments. Light irradiation would only harm tumors or other tissues that need medical attention when high amounts of GQDs are present. Systemic toxicity may not be caused by GQD phototoxicity [10].

HEK-293 cells were used in an MTT colorimetric experiment with a standard MTT reagent (5-diphenyl tetrazolium bromide) to evaluate the cytotoxicity of the produced GQDs [11]. Except for Ho-GQDs, which produce over 80% cell viability at over 1 mg/mL, all GQDs in this assay provide over 80% cell viability at over 0.5 mg/mL [12]. Certain GQDs, such as those doped with Ce and N, support more than 80% of cell viability at up to 3 mg/mL. This confirms that the GQDs created in this work can be employed for in vitro imaging and radical scavenging and are highly biocompatible when compared to several other nanomaterials, such as carbon nanotubes and graphene oxide [13].

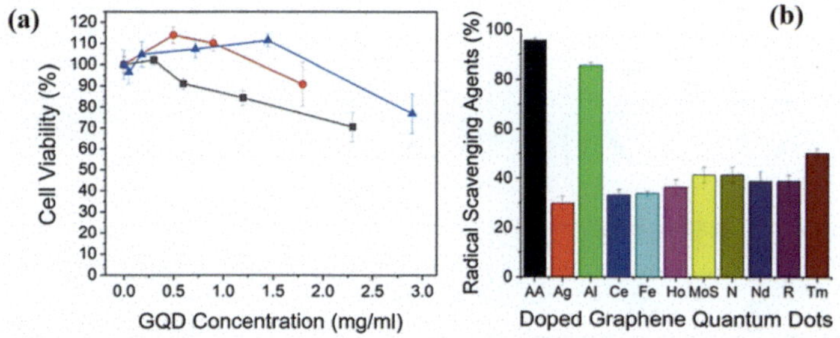

Fig. 18.1 **a** MTT assay- HEK-293 cell viability achieved with various concentrations, and **b** DPPH radical scavenging assay [17]

GQDs have been found to possess antioxidant qualities, which render them suitable for numerous significant uses such as the management of diabetes, rheumatoid arthritis, cancer, and Alzheimer's disease. These nanomaterials can be used at significant therapeutic doses because of their high biocompatibilities, which are typically greater than 1 mg/mL (Fig. 18.1) [14]. To make sure the antioxidants reach their intended site, the intrinsic fluorescence of the GQDs within the cells and biological tissues can be used as a tracking mechanism. Lastly, the big GQD platform provides concurrent medication delivery capabilities that, when combined with its antioxidant qualities, can help treat the ailments indicated above [15]. Metal and even nitrogen doping can make these nanomaterials more effective at fending off the impacts of free radicals, which are frequently encountered in biological systems, as all doped GQDs perform at least 30% better than the undoped cores [16]. To fully realize the promise of GQDs for radical scavenging, more research is necessary to determine the ideal doping levels and underlying mechanisms [17].

When GO is administered to mice, the results include granulomatous lesions, inflammatory cell infiltration, pulmonary toxicity, and pulmonary edema fibrosis; these effects are contingent upon the dose of GO [18]. Mice can be given varying amounts of GO to assess their toxicity in vivo (Fig. 18.2). According to the findings, GO causes lung granulomas in mice and its toxicity varies with dosage [19]. GO causes specific cardiovascular issues,zebrafish embryos were used in the study. GO induces cardiotoxicity, an accelerated heartbeat, and various other cardiovascular abnormalities at concentrations between 0.4 and 1 mg/mL [20]. The in vivo toxicity tests primarily address issues such as the testing substance's biodistribution across the organs of a whole living human, its potential excretory system elimination, potential organ damage, etc. [21]. A standard test involves injecting mice or other living organisms with a specific dosage of GQDs and then watching them for a set amount of time, such as 24 h (Fig. 18.3). GQD excretion, organic buildup, and biodistribution are assessed during this time [22, 23].

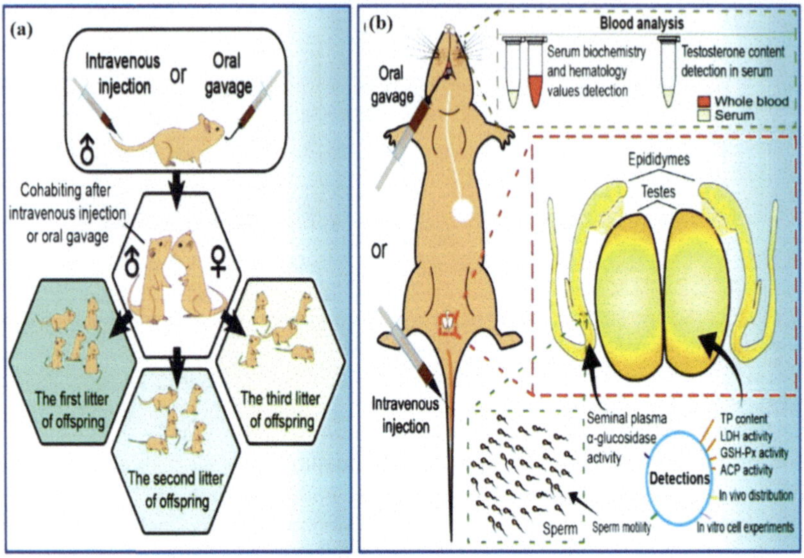

Fig. 18.2 **a** Supervision of GQD and evaluation of mouse sexual behaviors and offspring health, and **b** Toxicity studies of GQD in male mouse germ cells. Reprinted with permission from @ Woodhead Publishing (2020)

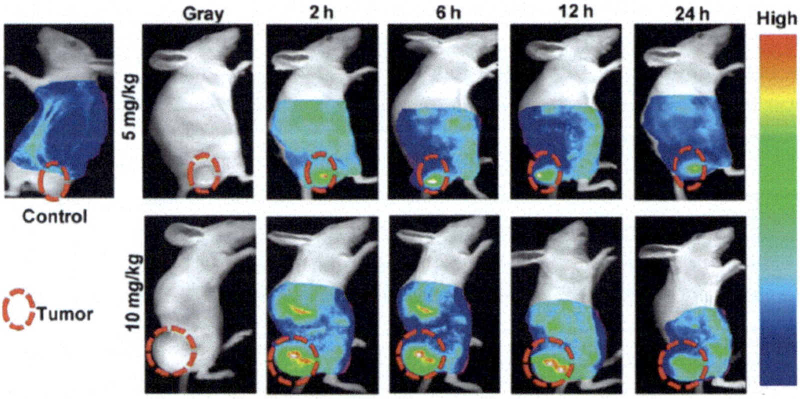

Fig. 18.3 In vivo toxicity tests using the Mice model [23]

18.2.1 Application of GQDs as an Inhibitor

The application of GQDs as a strong inhibitor to prevent human islet amyloid polypeptide (IAPP), a marker of type 2 diabetes, from aggregating and becoming toxic in vivo. Through hydrogen bonding, hydrophobic, and electrostatic interactions, GQDs made contact with IAPP (Fig. 18.4). This caused the peptide fibrillization to

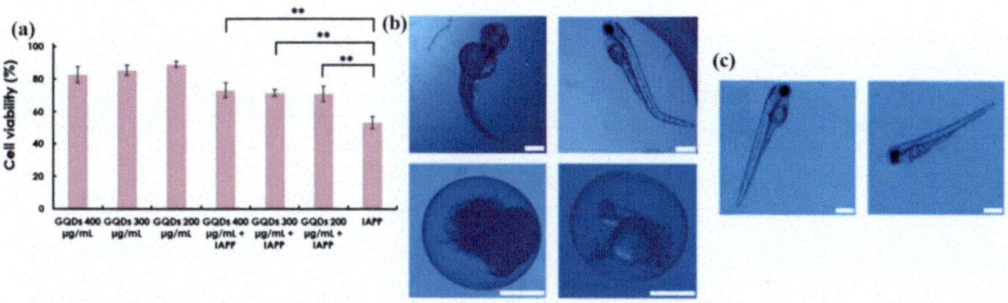

Fig. 18.4 **a** βTC 6 pancreatic beta cell viability determined by Alamar Blue after treatment of GQDs and IAPP as well as GQDs pre-incubated, and **b**, **c** Bright-field imaging of abnormality upon IAPP microinjection into the yolk of embryos and mitigation of toxicity by pre-incubating IAPP with GQDs. Reprinted with permission from @ Nanoscale 10, 19995 (2018)

go off course, removing the harmful intermediates [24]. A thioflavin T kinetic assay, fluorescence quenching, circular dichroism spectroscopy, a cell viability assay, and in silico discrete molecular dynamics simulations were used to investigate these interactions. The results showed that the embryonic zebrafish significantly recovered from the damage caused by IAPP in vivo, as evidenced by improved hatching as well as reduced production of reactive oxygen species, abnormality, and organism mortality [25].

The complete investigation of GQDs' toxicity and biosafety is still pending, and there are still questions regarding the way these nanomaterials interact with biological substrates because this also depends on the unique physicochemical properties of various formulations. The pharmacokinetic and pharmacodynamic factors that are impacted by the dosage, exposure duration, and route of administration will also have an impact on in vivo toxicity. Following a 48-h incubation period with either human bronchial epithelial cell line, 16HBE14o^{-}, or murine NIH3T3 fibroblasts, GQDs were tested for in vitro cell viability at concentrations ranging from 1.2 to 100 μg/ml; both cell types continued to proliferate in the presence of GQDs. No obvious toxicity was observed (Fig. 18.5). The potential application of GQDs in photodynamic therapy (PDT) of cancer is contingent upon their capacity to produce elevated concentrations of ROS. Here, we show that, in comparison to traditional photosensitizers, the GQDs significantly increase the formation of singlet oxygen and other ROS [26].

Highly biocompatible (HGQD) particles were biodistributed and used for fluorescence imaging in mice. The produced HGQDs were injected into KM mice, and the fluorescence imaging system was used to view ex vivo images of isolated organs (Fig. 18.6). According to the findings, the HGQDs were concentrated in several organs and dispersed throughout the body via systemic circulation [27]. The buildup of HGQDs in the liver, kidney, and brain is probably the reason why the PL intensity of these organs was noticeably higher than that of PBS. The buildup of HGQDs in the liver, kidney, and brain is demonstrated by the quantitative study of biodistribution [28]. These findings also suggested that HGQDs could cross the blood–brain barrier, opening doors for the diagnosis and treatment of disorders linked to the brain. Ex vivo

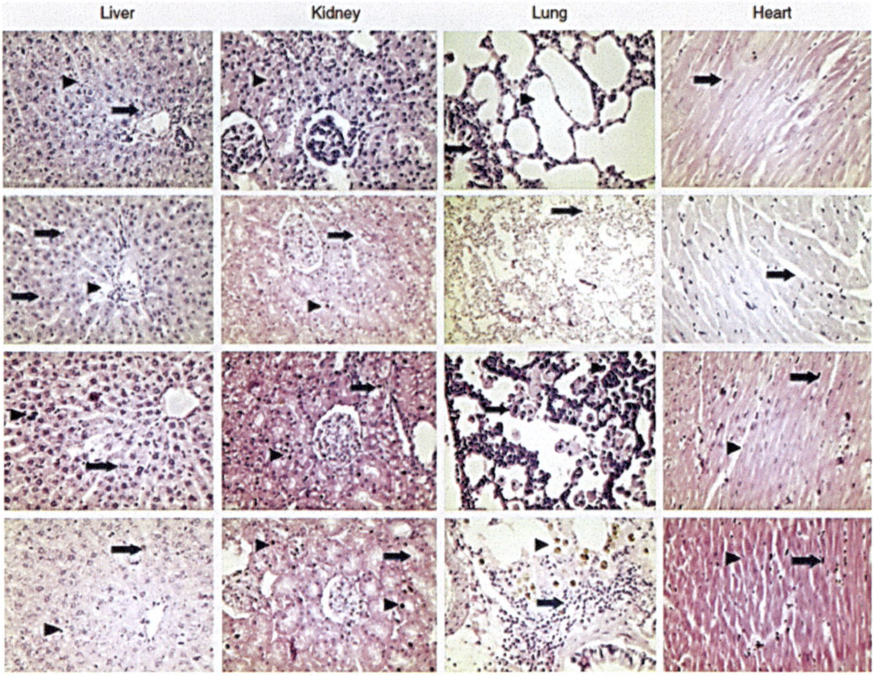

Fig. 18.5 Histological evaluation of the vital organs of rats at 4 weeks after intravenous injection of GQDs. Reprinted with permission from @ Nanomedicine 13, 1923–1937

fluorescence imaging of separated organs revealed that 24 h after intravenous injection, HGQDs accumulated in the liver, kidney, and brain [29]. Twenty days following the administration of the produced HGQDs, no signs of inflammation were seen in the kidney, lung, spleen, liver, or heart. These results are very promising because they demonstrate the safety and significant selectivity of GQDs for use in biomedical applications [30].

The impact of GQDs on $CD34^+$ hematopoietic stem cells' transcriptome following a 36-h incubation period. Compared to $CD34^+$ hematopoietic stem cells cultured without GQDs, only one of the 20,800 documented gene expressions the selenoprotein W, 1 is altered by GQDs (Fig. 18.7). The expression of 1171 genes is only marginally impacted, according to a meta-analysis that only considers the most pronounced alterations caused by the cell culture [31]. We have discovered eight corresponding mildly impacted signaling pathways, including but not limited to apoptosis induction. These findings imply that, after 36 h of incubation, GQDs with sizes in the range of a few nanometers barely affect the transcriptome level of $CD34^+$ cells, indicating their great applicability for in vivo investigations, like fluorescence labeling or delivery protocols, without having a significant impact on the functional state of the cells. We used confocal fluorescence microscopy to visualize the uptake

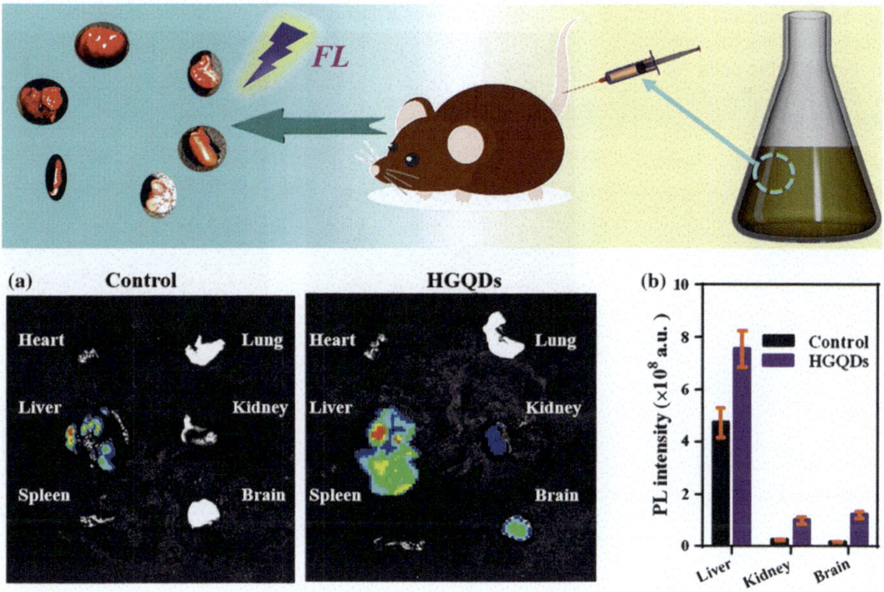

Fig. 18.6 **a** Ex vivo images of isolated organs of mice at 24 h after injection of HGQDs and PBS solution (control), and **b** PL intensities of the HGQDs from isolated organs (2 mg kg^{-1}). Reprinted with permission from @ J Mater Sci 55, 1198–1215 (2020)

of GQDs into the cells on living $CD34^{+}$ cells that had been labeled with Hoechst 33342 to represent the nuclei [32].

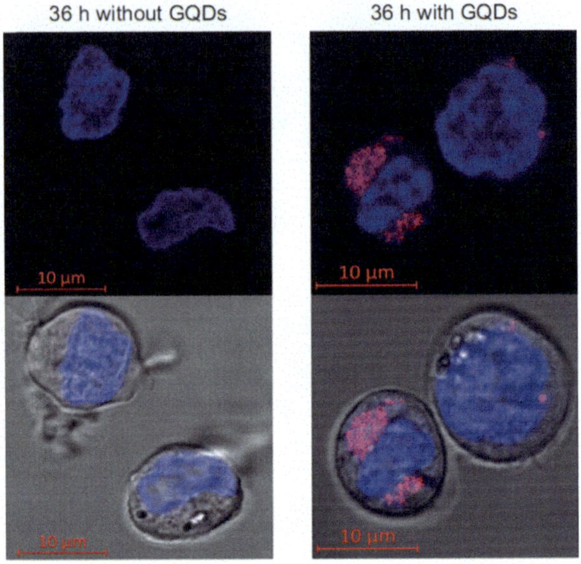

Fig. 18.7 Confocal fluorescence microscopy images of living cells cultured for 36 h without GQDs [31]

References

1. Liang L, Peng X, Sun F, Kong Z, Shen JW (2020) A review on the cytotoxicity of graphene quantum dots: from experiment to simulation. Nanoscale Adv 3:904–917
2. Shuvam S, Somyajit P, Aritra P, Subhajit M, Kunal S, Das M (2023) A review on synthesis, toxicity profile and biomedical applications of graphene quantum dots (GQDs). Inorg Chim Acta 557:121677
3. Sangiliyandi G, Jin-Hoi K (2016) Synthesis, toxicity, biocompatibility, and biomedical applications of graphene and graphene-related materials. Int J Nanomed 11:1927–1945
4. Miruna-Silvia S, Beatrice GS, Bogdan T, Anca D (2021) Graphene quantum dots: physico-chemical characterization and in vitro biological effects. Mater Proc 4:27
5. Lv O, Tao Y, Qin Y, Chen C, Pan Y, Deng L, Liu L, Kong Y (2016) Highly fluorescent and morphology-controllable graphene quantum dots-chitosan hybrid xerogels for in vivo imaging and pH-sensitive drug carrier. Mater Sci Eng C 67:478–486
6. Sui X, Luo C, Wang C, Zhang F, Zhang J, Guo S (2016) Graphene quantum dots enhance anti-cancer activity of cisplatin via increasing its cellular and nuclear uptake. Nanomed Nanotechnol 7:1997–2006
7. Maoquan C (2017) Toxicity of graphene quantum dots. In: The world scientific Encyclopedia of nanomedicine and bioengineering, vol II, pp 127–137
8. Wo F, Xu R, Shao Y, Zhang Z, Chu M, Shi D, Liu S (2016) Multimodal cancer therapeutics via magneto-mechanical force and near-infrared laser-triggered doxorubicin-loaded graphene quantum dot-coated hollow magnetic nanospheres. Theranostics 6:485–500
9. Markovic ZM, Ristic BZ, Arsikin KM, Klisic DG, Harhaji-Trajkovic LM, Todorovic-Markovic BM, Kepic DP, Kravic-Stevovic TK, Jovanovic SP, Milenkovic MM, Milivojevic DD, Bumbasirevic VZ, Dramicanin MD, Trajkovic VS (2012) Graphene quantum dots as autophagy-inducing photodynamic agents. Biomaterials 33:7084–7092
10. Bozich JS, Lohse SE, Torelli MD, Murphy CJ, Hamers RJ, Klaper RD (2014) Surface chemistry, charge and ligand type impact the toxicity of gold nanoparticles to Daphnia magna. Environ Sci Nano 1:260–270

11. Aoki K, Saito N (2020) Biocompatibility and carcinogenicity of carbon nanotubes as biomaterials. Nanomaterials 10:264
12. Gonzalez-Rodriguez R, Campbell E, Naumov A (2019) Multifunctional graphene oxide/iron oxide nanoparticles for magnetic targeted drug delivery dual magnetic resonance/fluorescence imaging and cancer sensing. PLoS One 14:e0217072
13. Lee B, Stokes GA, Valimukhametova A, Nguyen S, Gonzalez-Rodriguez R, Bhaloo A, Coffer J, Naumov AV (2023) Automated approach to in vitro image-guided photothermal therapy with top-down and bottom-up-synthesized graphene quantum dots. Nanomaterials 13:805
14. Arslan J, Jamshed H, Qureshi H (2020) Early detection and prevention of Alzheimer's disease: role of oxidative markers and natural antioxidants. Front Aging Neurosci 12:231
15. Varesi A, Campagnoli LIM, Carrara A, Pola I, Floris E, Ricevuti G, Chirumbolo S, Pascale A (2023) Non-enzymatic antioxidants against Alzheimer's disease: prevention, diagnosis and therapy. Antioxidants 12:180
16. Bekhet OH, Eid ME (2021) The interplay between reactive oxygen species and antioxidants in cancer progression and therapy: a narrative review. Transl Cancer Res 10:4196–4206
17. Adam B, Steven N, Bong HL, Alina V, Roberto GR, Olivia S, Abby D, Anton VN (2023) Doped graphene quantum dots as biocompatible radical scavenging agents. Antioxidants 12:1536
18. Wang K, Ruan J, Song H, Zhang J, Wo Y, Guo S et al (2010) Biocompatibility of graphene oxide. Nanoscale Res Lett 6:1–8
19. Zhang S, Yang K, Feng L, Liu Z (2011) In vitro and in vivo behaviors of dextran functionalized graphene. Carbon 49:4040–4049
20. Henna TK, Pramod K (2020) Chapter 16, Biocompatibility of graphene quantum dots and related materials. In: Handbook of biomaterials biocompatibility. Woodhead Publishing Series in Biomaterials, pp 353–367
21. Wu X, Tian F, Wang WX, Chen J, Wu M, Zhao JX (2013) Fabrication of highly fluorescent graphene quantum dots using L-glutamic acid for *in vitro/in vivo* imaging and sensing. J Mater Chem C 1:4676–4684
22. Wang ZG, Zhou R, Jiang D, Song JE, Xu Q, Si J, Chen YP, Zhou X, Gan L, Li JZ, Zhang H, Liu B (2015) Toxicity of graphene quantum dots in zebrafish embryo. Biomed Environ Sci 28:341–351
23. Shujun W, Ivan SC, Qin L (2016) The toxicity of graphene quantum dots. RSC Adv 6: 89867–89878
24. Miaoyi W, Yunxiang S, Xueying C, Guotao P, Ibrahim J, Aleksandr K, Thomas PD, Sijie L, Jingquan L, Feng D, Pu Chun K (2018) Graphene quantum dots against human IAPP aggregation and toxicity in vivo. Nanoscale 10:19995
25. Jian L, Xizhi Z, Jing J, Yanjie W, Haiye J, Junhua Z, Xinmin N, Bin L (2019) Systematic assessment of the toxicity and potential mechanism of graphene derivatives in vitro and in vivo. Toxicol Sci 167:269–281
26. Tanveer AT, Chris JS, Daniel CJF, Liangxu L, van der Anienke V, Sophie L, Muhammad A, Farhat J, Muhammad A, Paul GW, Shaowei Z (2018) Biocompatibility and toxicity of graphene quantum dots for potential application in photodynamic therapy. Nanomedicine 13:1923–1937
27. Nurunnabi M, Khatun Z, Huh KM, Park SY, Lee DY, Cho KJ, Lee YK (2013) In vivo biodistribution and toxicology of carboxylated graphene quantum dots. ACS Nano 7:6858–6867
28. Chaoren Y, Xiaoling H, Ping G, Tongtong H, Peng C, Dewei W, Xinli Z, Jian W, Chaoli W (2020) Highly biocompatible graphene quantum dots: green synthesis, toxicity comparison and fluorescence imaging. J Mater Sci 55:1198–1215
29. Jack CR Jr, Bennett DA, Blennow K, Carrillo MC, Dunn B, Haeberlein SB, Elliott C, Masliah E, Ryan L, Liu E, Silverberg N (2018) NIA-AA research framework: toward a biological definition of Alzheimer's disease. Alzheimer's Dement 14:535–562
30. Weinshenker D (2018) Long Road to ruin: noradrenergic dysfunction in neurodegenerative disease. Trends Neurosci 41:211–223
31. Stefan F, Lisa Z, Ron-PC ML, Bastian M, Christoph J, Thomas H, Rainer H (2019) The low toxicity of graphene quantum dots is reflected by marginal gene expression changes of primary human hematopoietic stem cells. Sci Rep 9:12028

32. Yuqi L, Zonglin G, Hailiang C, Yaoxing H (2019) Potential toxicity of graphene (oxide) quantum dots via directly covering the active site of anterior gradient homolog 2 protein. Sci Rep 14:7091

Chapter 19
Future Research on Graphene Quantum Dots

19.1 Introduction

A promising platform for electrically tunable qubits in a two-dimensional material is bilayer graphene. The capacity to store quantum information in the valley degree of freedom two-fold orbital degeneracy resulting from the hexagonal crystal structure's symmetry is of special significance. Since established spin- and orbital-mixing mechanisms are unlikely to be at play for valleys, using them could be beneficial and lead to more resilient qubits [1]. Valley states have a Berry curvature that makes it possible to electrically alter their energies, indicating potential paths toward coherent qubit manipulation. The relaxation duration of valley states, which ultimately determines the coherence properties of these qubits and, thus, their applicability as practical qubits, is still unknown. In gate-defined bilayer graphene quantum dot (GQD) devices, we quantify the characteristic relaxation periods of these spin and valley states here [2]. It is possible to discriminate between different valley states with a fidelity of more than 99%. In comparison to spin states, the relaxation time between valley triplets and singlets is more than an order of magnitude longer, lasting over 500 ms. This demonstrates that bilayer graphene may host electrically regulated, long-lived valley qubits, paving the way for future investigations on valley-qubit coherence [3].

As a result of the large-scale production of GQDs, which may find use in foldable, small-factor light-emitting diodes in the future as well as several other fields, particularly in the development of 5G and 6G networks in conjunction with AI, these fluorescent materials could be a viable option for next-generation LED technology, which could bring about new advancements in device optimization and applications based on carbon dots as well as useful applications in day-to-day life shortly [4]. If there is a constant supply of fuel and oxygen ions (usually from the air) to sustain the intricate chemical process, fuel cells can generate power continually. Such energy technology is sustainable and ecologically friendly. Recent studies have shown that functionalized GQDs have also been employed as fuel cell electrodes. The internal

N. Manjubaashini et al., *Graphene Quantum Dots*, Materials Horizons: From Nature to Nanomaterials, https://doi.org/10.1007/978-981-97-5722-0_19

charge transfer has enhanced the stability and catalytic activity of the functionalized GQDs. Subsequently, there have been other attempts to employ functionalized GQDs as materials for fuel cell cathodes [5].

19.2 Futuristic Perspectives of GQDs-Based Materials

1. Large-scale GQD synthesis still has to be accomplished to be used in commercial applications. Scalable purification should be taken into account in addition to scalable production.

2. For a given application, a thorough characterization of GQDs is necessary, along with an extremely exact understanding of their fluorescence and other characteristics.

3. Surface passivation and heteroatom doping may increase the quantum yield of GQDs.

4. For biological applications, GQD functionalization and dispersion stability must be enhanced.

5. For bandgap tunability, which is necessary for solar cells and energy-related applications, controlled size and shaped GQDs must be produced [6].

In terms of automation, mobility, real-time, and on-site detection, the sensor sector is undergoing a revolution thanks to the integration of smartphones and artificial intelligence. Without the assistance of trained staff, smartphones enable reliable point-of-care (POC) analyte detection (Fig. 19.1). Optical biosensing using cell phones has been documented in many environmental, health, and security domains. With its advantages over traditional detection methods, electrochemiluminescence (ECL) is a potent technique that finds extensive use in commercial sensors of industrial, biomedical, and environmental areas. Its advantages include ultra-high sensitivity, good specificity, low background noise, and high reproducibility. The widespread use of smartphones, artificial intelligence, and the Internet of Medical Things (IoMT) in conjunction with sensing substrates can help bring QDs-based ECL sensors closer to the market by enabling the real-time, point-of-care, reliable, and affordable detection of analytes of food, industry, environment, and clinical origin. An overview of electrochemiluminescent sensors' potential for the future. Consequently, there is a great deal of potential for commercial use in the rapidly developing field of electrochemiluminescent QD-based sensors [7].

Because viral vectors may transfer genetic material to target cells, they are used in gene therapy and delivery. GQDs assist in removing the immunogenicity-related challenges that viral vectors have encountered, which have prevented their clinical application. They are great candidates to be utilized as a traceable transfection vector to transport genes to both in vitro and in vivo environments because of their tiny size, capacity to traverse biological barriers, high stability, and luminescent features. To advance GQDs as a non-viral gene vector, the proper conjugation of components required loading of nucleic acid (RNA, DNA) like signaling molecules and cationic polymers on the basal plane needs to be resolved. When compared to other drug delivery systems (DDS), the potential for using nano-drugs administered orally in

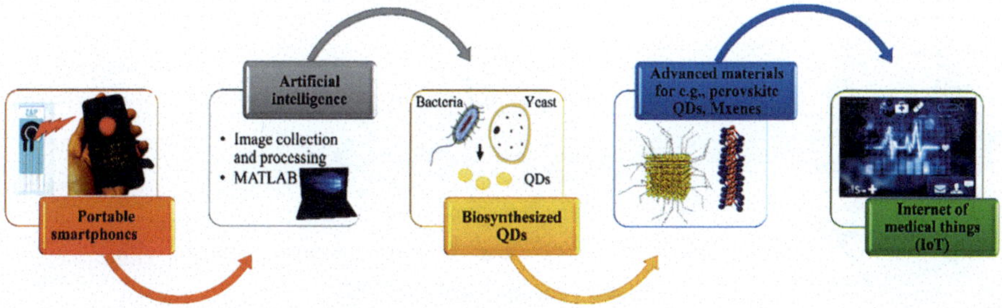

Fig. 19.1 Schematic diagram showing prospects of ECL sensors. Reprinted with permission from @ Chem Eng J 477, 146958 (2023)

very small amounts to prevent irritation from other administration methods has also drawn attention and is perceived as a developing advantage. Because of the protection offered by the functionalized part of the GQD-DDS, medications from acidic environments can be administered orally [8]. In a short time, GQDs will be able to dominate the pharmaceutical, biomedical, and other industries thanks to systematic research and study documentation.

It's interesting to note that GQDs are a promising next-generation carbon nanomaterial with a wide range of biomedical uses, including biosensing various serious health conditions. Furthermore, there has to be increased emphasis on the simultaneous identification of several prostate cancer (PC) biomarkers in a single sample utilizing GQDs. Lastly, a specific focus is necessary for the accuracy and viability of GQD-based biosensors for PC early diagnosis. Therefore, GQDs are an innovative and promising fluorescent carbon nanomaterial that provides a platform for the quick, accurate, and high-sensitivity detection of PC [9].

Significant questions are raised when GQDs are used in clinical settings as opposed to lab settings, mostly concerning safety and biodistribution. These worries call for a thorough investigation of several aspects, such as internalization, metabolization, and possible alterations of these nanomaterials inside of cells, especially when it comes to dealing with their toxicity. Moreover, it is critical to evaluate the possibility that GQDs' physical and chemical characteristics will change when they are incorporated into the biological environment, particularly inside cancer cells, since these modifications could have a significant impact on the efficacy and safety of the particles during clinical trials. Furthermore, it is crucial to design GQDs in a way that makes it easier for them to successfully cross the BBB in the particular instance of Glioblastoma multiforme (GBM) therapy. It is necessary to get past this obstacle to use them for GBM diagnosis and treatment. To put it simply, the effective integration of GQDs into clinical practice requires careful attention to safety concerns, thorough investigation of how they interact with biological systems, and the creation of customized strategies to guarantee safe and effective BBB penetration. In the context of GBM care, such endeavors are essential to maximizing the diagnostic and therapeutic potential of these nanomaterials [10].

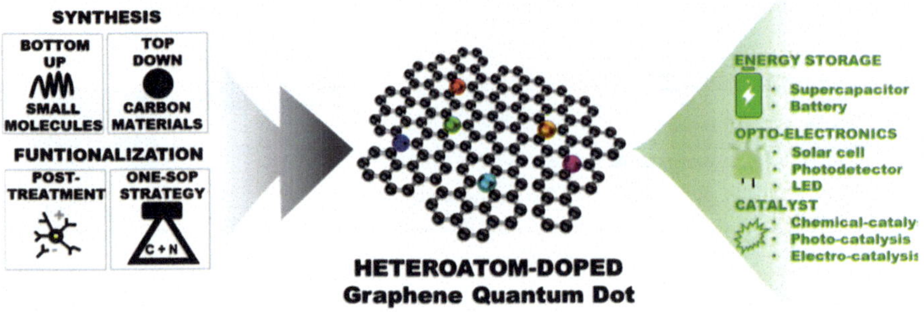

Fig. 19.2 Schematic of heteroatoms doped GQDs. Reprinted with permission from @ Small 19, 2304497 (2023)

Previous research on GQDs has mostly focused on the exploration of novel precursors for the synthesis of nano-sized carbon structures, rather than examining the critical variables affecting the structural, morphological, and electrical features of the Ds. Therefore, future research should focus on analyzing the structural and morphological differences between planar and spherical 0D-like CQDs in greater detail and examining the impact of crystallinity on their stability and properties to improve the fundamental properties and practical applicability of GQDs (Fig. 19.2). Theoretical computations may also be utilized to anticipate reaction paths, allowing for more effective and consistent control over the shape, size, morphology, and edge functional groups of different GQD materials. The field of GQDs would benefit from such developments [11].

GQDs have gained recognition recently as promising solutions to a wide range of issues in several fields. GQDs have been used in a variety of progressive applications due to their unique combination of noteworthy properties, which include excellent optoelectronic properties, readily available raw materials, very little cytotoxicity, excellent photostability, uniform dispersibility, and high solubility in several solvents. We have thoroughly reviewed the several uses of graphene quantum dots (GQDs) in industry, photocatalysis, textiles, healthcare and biomedicine, and the environment in this textbook.

GQDs, both undoped and various element doped, outperform other conservative materials when it comes to developing bioimaging techniques for the detection of biomolecules with varying scales and heavy metal ions with deserving selectivity and sensitivity which are essential for the remediation of human health. Furthermore, the structurally modified GQDs' altered surface area and capabilities enable simple interaction with biological components and hazardous metal ions. It is important to highlight that the latest findings reveal the use of GQDs in therapeutic applications, including illness detection and tailored drug delivery systems. Even though GQDs have a wide variety of applications, their practical uses in numerous domains are still pacified by their features such as shifting of emission, low quantum yield, and unknown processes. Therefore, to fully utilize GQDs in materials science, nanoscience, and nanobiotechnology, effective, methodical, and focused research is required. The development of GQDs is accelerating due to the connection of new possible surface functionalities with their optoelectronic and electrochemical capabilities, which may allow them to be used in additional applications.

References

1. Ashish S, Kalyan PS (2022) Valley degree of freedom in two-dimensional van der Waals materials. J Phys D 55:303003
2. Banszerus L, Hecker K, Moller S et al (2022) Spin relaxation in a single-electron graphene quantum dot. Nat Commun 13:3637
3. Rebekka G, Chuyao T, Jocelyn T, Max JR, Jonas DG, Lisa MG, Kenji W, Takashi T, Thomas I, Klaus E, Wei WH (2024) Long-lived valley states in bilayer graphene quantum dots. Nat Phys 20:428–434
4. Ashish G, Amrita J, Santosh Kumar T (2022) Review on fluorescent carbon/graphene quantum dots: promising material for energy storage and next-generation light-emitting diodes. Materials 15:7888
5. Narinder K, Reena D, Rawat BS, Rawat PS, Diptarka R, Devendra S (2024) A review on synthesis, challenges as well as future prospects of graphene quantum dot (GQD). Pramana—J Phys 98:42
6. Pradip K, Chetna D, Neeraj D, Shiv S, Raju K, Sarika V, Archana S, Manoj Kumar G, Surender K, Rajeev K, Avanish Kumar S (2022) Graphene quantum dots: a contemporary perspective on scope, opportunities, and sustainability. Renew Sust Energy Rev 157:111993
7. Jyotsana M, Neeraj D, Nitin Kumar S, Giovanna M, Ajeet K, Sandeep K (2023) Electrochemiluminescent quantum dots as emerging next generation sensing platforms. Chem Eng J 477:146958
8. Deepak KG, Anirudh S, Khushboo D, Sunita N (2022) Targeted drug delivery using graphene quantum dots: approaches, limitations and future perspectives. ECS Trans 107:16081–16098
9. Jidnyasa P, Sachin C, Rahul T, Zamir K, Matin S, Trupti P, Prashant VS, Sopan N (2022) Graphene quantum dot based ultrasensitive probe for biosensing of prostate cancer biomarkers: current updates and future challenges. Adv Nat Sci: Nanosci Nanotechnol 13:013001
10. Sopan N, Sachin C, Pravin P (2023) Design of carbon and graphene quantum dots based nanotheranostics applications for glioblastoma management: Recent advanced and future prospects. J Drug Delivery Sci Tech 89:105060
11. Min JI, Jin IK, Seok-Ki H, Byung JM, Sukang B (2023) From pristine to heteroatom-doped graphene quantum dots: an essential review and prospects for future research. Small 19:2304497

SPRINGER NATURE

GPSR Compliance

The European Union's (EU) General Product Safety Regulation (GPSR) is a set of rules that requires consumer products to be safe and our obligations to ensure this.

If you have any concerns about our products, you can contact us on ProductSafety@springernature.com

In case Publisher is established outside the EU, the EU authorized representative is:

Springer Nature Customer Service Center GmbH
Europaplatz 3
69115 Heidelberg, Germany

The manufacturer's authorised representative in the EU is Springer Nature Customer Service Centre GmbH, Europaplatz 3, 69115 Heidelberg, Germany. If you have any concerns regarding our products, please contact ProductSafety@springernature.com

Printed and bound by CPI Group (UK) Ltd, Croydon, CR0 4YY
01/10/2026
02225962-0010